Lecture Notes in Artificial Intelligence   8069

Subseries of Lecture Notes in Computer Science

For further volumes:
http://www.springer.com/series/1244

Ashutosh Natraj · Stephen Cameron
Chris Melhuish · Mark Witkowski (Eds.)

Towards Autonomous Robotic Systems

14th Annual Conference, TAROS 2013
Oxford, UK, August 28–30, 2013
Revised Selected Papers

 Springer

Editors
Ashutosh Natraj
Stephen Cameron
Department of Computer Science
University of Oxford
Oxford
UK

Mark Witkowski
Department of Electrical
and Electronic Engineering
Imperial College London
London
UK

Chris Melhuish
University of Bristol
and the West of England
Bristol
UK

ISSN 0302-9743
ISBN 978-3-662-43644-8
DOI 10.1007/978-3-662-43645-5
Springer Heidelberg New York Dordrecht London

ISSN 1611-3349 (electronic)
ISBN 978-3-662-43645-5 (eBook)

Library of Congress Control Number: 2014942252

LNCS Sublibrary: SL7 – Artificial Intelligence

Printed on acid-free paper

Springer is part of Springer Science+Business Media (www.springer.com)

Preface

This volume contains the papers for TAROS 2013, the 14th conference in the now annual Towards Autonomous Robotics and Systems conferences.

The TAROS series was started by Ulrich Nehmzow in Manchester in 1997 under the name TIMR (Towards Intelligent Mobile Robots). In 1999, Chris Melhuish and Ulrich formed the conference Steering Committee, which was joined by Mark Witkowski in 2003 when the conference adopted its current name. The Steering Committee has provided a continuity of vision and purpose to the conference over the years as it has moved around the UK, and under their stewardship TAROS has become a major conference on autonomous robotics, while also attracting an increasing international audience. Sadly, Ulrich died in 2010, but his contribution is commemorated in the form of the Ulrich Nehmzow Best Student Paper Award, sponsored this year by his family.

In the 14th edition of TAROS 2013, we received a total of 75 submissions. The submissions were offered under two categories, one: full-length papers, of which 58 submissions were received, and the other: extended abstracts, of which we received 17 submissions. All the submissions were reviewed by three members of our International Programme Committee. This year there were 20 oral paper presentations corresponding to an acceptance rate of 34 % plus two poster sessions that consisted of 28 poster presentations in the main part of the conference, with presenters from the UK, together with Belgium, Brazil, China, Ecuador, Germany, Ireland, Italy, Japan, The Netherlands, Spain, Taiwan, and the USA.

The Springer-Verlag Best Paper Award went to Nina Gaissert and her colleagues from FESTO for their paper "Inventing a Micro Aerial Vehicle inspired by the mechanics of dragonfly flight"; we were all slightly saddened to discover that she was unable to bring that particular vehicle with her to Oxford. The Best Student Paper Award went to Nima Keivan et al. for their paper titled "Realtime Simulation-in-the-loop Control for Agile Ground Vehicles".

We were also fortunate to hear keynote lectures from John Hallam of the Maersk Mc-Kinney Moller Institute in Odense and by Arnoud Visser of the University of Amsterdam, and a special lecture by Paul Newman of Oxford University sponsored by the IET. On the Friday a well-attended Industry Day took place with the theme of Horizon 2020.

As general chair I am deeply indebted to many others for making TAROS a success.

I would like to thank all those many members of the Program Committee for their (generally) prompt efforts in reviewing the papers; to Chris Melhuish and Mark Witkowski for their overall guidance; to Elizabeth Walsh and Andrea Pilot for running the show; to the local members of the Committee, Ashutosh Natraj the programme chair and Samuel Bucheli who made and maintained the web-site; and to all the attendees for making TAROS 2013 an interesting and vibrant event.

August 2013

Ashutosh Natraj
Stephen Cameron
Chris Melhuish
Mark Witkowski

Organization

Conference Chair

Stephen Cameron

Program Chair

Ashutosh Natraj

TAROS 2013 – Webmaster

Samuel Bucheli

Editorial Committee

Ashutosh Natraj
Stephen Cameron

Mark Witkowski
Chris Melhuish

Program Commitee

Ajith Abraham
Lyuba Alboul
Ronald C. Arkin
Fatima Benamar
Paul Bremner
Samuel Bucheli
Guido Bugmann
Stephen Cameron
Lola Canamero
Andrea Carbone
Anders Lyhne Christensen
Andrew Conn
Torbjorn Dahl
Zahra Daneshifar
Kerstin Dautenhahn
Mehrdad Davarifar
Geert De Cubber
Cédric Demonceaux

Sotirios Diamantas
Tony Dodd
Sanja Dogramadzi
Stéphane Doncieux
Marco Dorigo
Kerstin Eder
Mathew H. Evans
David Fofi
Ella Gale
Swen Gaudl
Ioannis Georgilas
Roderich Gross
Dongbing Gu
Abdi Ibrahim Hadi
Heiko Hamann
William Harwin
Phil Husbands
Ioannis Ieropoulos

Yaochu Jin
Maarja Kruusmaa
Haruhisa Kurokawa
Theocharis Kyriacou
Frederic Labrosse
Stasha Lauria
Mark Lee
Nathan Lepora
Dieu Sang Ly
Stéphane Magnenat
Walterio Mayol
Giorgio Metta
Ben Mitchinson
Lazaros Nalpantidis
Ashutosh Natraj
Paul Newman
Jekaterina Novikova

Calogero M. Oddo
Cagdas D. Onal
Ahlem Othmani
Martin Pearson
Marcos A. Rodrigues
Ferdinando Rodriguez Y. Baena
Erol Sahin
Thomas Schmickl
Matthew Studley
Jon Timmis
Elio Tuci
Bram Vanderborght
Pascal Vasseur
Hugo Vieira Neto
Marsette Vona
Sonia Waharte
Myra s. Wilson

TAROS Steering Committee

Chris Melhuish
Mark Witkowski

BARA-AFR Industry Day Chairs

Jacques Penders
Grant Collier

Organizing Committee

Stephen Cameron
Ashutosh Natraj
Samuel Bucheli

Elizabeth Walsh
Andrea Pilot

Sponsoring Institutions

Department of Computer Science University of Oxford, UK
Institution of Engineering and Technology (IET)
Springer Verlag

Contents

Artificial Intelligence

Dynamic Power Sharing for Self-Reconfigurable Modular Robots   3
  Chi-An Chen, Akiya Kamimura, Luenin Barrios, and Wei-Min Shen

Heuristically-Accelerated Reinforcement Learning: A Comparative
Analysis of Performance   15
  Murilo Fernandes Martins and Reinaldo A. C. Bianchi

TREEBOT: Tree Recovering Renewable Energy Robot   28
  Giovanni Gerardo Muscolo and Rezia Molfino

Coy-B, an Art Robot for Exploring the Ontology of Artificial Creatures   30
  Paul Granjon

Time Preference for Information in Multi-agent Exploration
with Limited Communication   34
  Victor Spirin, Stephen Cameron, and Julian de Hoog

Discrimination of Social Tactile Gestures Using Biomimetic Skin   46
  Hector Barron-Gonzalez and Tony Prescott

Bio-inspired and Aerial Robotics

A Personal Robotic Flying Machine with Vertical Takeoff Controlled
by the Human Body Movements   51
  Vittorio Cipolla, Aldo Frediani, Rezia Molfino,
  Giovanni Gerardo Muscolo, Fabrizio Oliviero, Domenec Puig,
  Carmine Tommaso Recchiuto, Emanuele Rizzo, Agusti Solanas,
  and Paul Stewart

Developing the Cerebellar Chip as a General Control Module
for Autonomous Systems   53
  Emma D. Wilson, Sean R. Anderson, Tareq Assaf, Jonathan M. Rossiter,
  Martin J. Pearson, and John Porrill

UAV Horizon Tracking Using Memristors and Cellular Automata
Visual Processing   64
  Ioannis Georgilas, Ella Gale, Andrew Adamatzky, and Chris Melhuish

Sensor Integrated Navigation for a Target Finding UAV   76
  Michael J. Park and Charles Coldwell

Inventing a Micro Aerial Vehicle Inspired by the Mechanics
of Dragonfly Flight... 90
   Nina Gaissert, Rainer Mugrauer, Günter Mugrauer, Agalya Jebens,
   Kristof Jebens, and Elias Maria Knubben

Computer Vision

Sensor Data Fusion Using Unscented Kalman Filter for VOR-Based Vision
Tracking System for Mobile Robots 103
   Muhammad Latif Anjum, Omar Ahmad, Basilio Bona,
   and Dong-il "Dan" Cho

Visual Homing of an Upper Torso Humanoid Robot Using a Depth Camera.... 114
   Alan Broun, Chris Beck, Tony Pipe, Majid Mirmehdi, and Chris Melhuish

Efficient Construction of SIFT Multi-scale Image Pyramids
for Embedded Robot Vision... 127
   Peter Andreas Entschev and Hugo Vieira Neto

An Evaluation of Image-Based Robot Orientation Estimation 135
   Juan Cao, Frédéric Labrosse, and Hannah Dee

A Heuristic-Based Approach for Flattening Wrinkled Clothes 148
   Li Sun, Gerarado Aragon-Camarasa, Paul Cockshott, Simon Rogers,
   and J. Paul Siebert

Vision-Based Cooperative Localization for Small Networked Robot Teams 161
   James Milligan, M. Ani Hsieh, and Luiz Chaimowicz

Active Vision Speed Estimation from Optical Flow 173
   Sotirios Ch. Diamantas and Prithviraj Dasgupta

Mechanical Weeding Using a Paddy Field Mobile Robot
for Paddy Quality Improvement 185
   Yasuhiro Yamada, Keisuke Iwakabe, Guanzuo Liu, and Toshiyoshi Uejima

Multi-modal People Detection from Aerial Video Footage.............. 190
   Helen Flynn and Stephen Cameron

Control

Design of a Modular Knee-Ankle-Foot-Orthosis Using Soft Actuator
for Gait Rehabilitation.. 195
   S.M. Mizanoor Rahman

Evaluation of Laser Range-Finder Mapping for Agricultural Spraying Vehicles ... 210
   Francisco-Angel Moreno, Grzegorz Cielniak, and Tom Duckett

Autonomous Coverage Expansion of Mobile Agents via Cooperative Control
and Cooperative Communication   222
    Said Al-Abri and Zhihua Qu

Estimation of Contact Forces in a Backdrivable Linkage
for Cognitive Robot Research   235
    Richard Thomas and William Harwin

A Simple Drive Load-Balancing Technique for Multi-wheeled
Planetary Rovers.   247
    James C. Finnis and Mark Neal

Control-Oriented Nonlinear Dynamic Modelling of Dielectric
Electro-Active Polymers   259
    Will Jacobs, Emma D. Wilson, Tareq Assaf, Jonathan M. Rossiter,
    Tony J. Dodd, John Porrill, and Sean R. Anderson

Emotionally Driven Robot Control Architecture for Human-Robot
Interaction   261
    Jekaterina Novikova, Swen Gaudl, and Joanna Bryson

Adaptive Control of Robot System of Half Passive Joints   264
    Chenguang Yang, Jing Li, Zhijun Li, Weisheng Chen, and Rongxin Cui

Realtime Simulation-in-the-Loop Control for Agile Ground Vehicles   276
    Nima Keivan and Gabe Sibley

Humanoid and Robotic Arm

Development of a Highly Dexterous Robotic Hand with Independent
Finger Movements for Amputee Training   291
    Ali H. Al-Timemy, Alexandre Brochard, Guido Bugmann,
    and Javier Escudero

A Robotic Suit Controlled by the Human Brain for People
Suffering from Quadriplegia.   294
    Alicia Casals, Pasquale Fedele, Tadeusz Marek, Rezia Molfino,
    Giovanni Gerardo Muscolo, and Carmine Tommaso Recchiuto

A Novel Acoustic Interface for Bionic Hand Control   296
    Richard Woodward, Marcus Gardner, Paolo Angeles, Sandra Shefelbine,
    and Ravi Vaidyanathan

An Approach to Navigation for the Humanoid Robot Nao in Domestic
Environments.   298
    Changyun Wei, Junchao Xu, Chang Wang, Pascal Wiggers,
    and Koen Hindriks

An Embodied-Simplexity Approach to Design Humanoid Robots
Bioinspired by Taekwondo Athletes   311
    Rezia Molfino, Giovanni Gerardo Muscolo, Domenec Puig,
    Carmine Tommaso Recchiuto, Agusti Solanas, and A. Mark Williams

Navigation

A Robotic Geospacial Surveyor   315
    Sam Wane

Motion Planning and Decision Making for Underwater Vehicles
Operating in Constrained Environments in the Littoral   328
    Erion Plaku and James McMahon

Using Range and Bearing Observation in Stereo-Based EKF SLAM.   340
    Yao-Chang Chen, Tsung-Han Lin, and Ta-Ming Shih

On New Algorithms for Path Planning and Control of Micro-rover Swarms....   353
    H.D. Ibrahim and C.M. Saaj

Expecting the Unexpected: Measure the Uncertainties for Mobile Robot
Path Planning in Dynamic Environment   363
    Yan Li, Brian Mac Namee, and John Kelleher

Swarm Robotics

Novel Method of Communication in Swarm Robotics Based
on the NFC Technology   377
    Ulf Witkowski and Reza Zandian

Gesturing at Subswarms: Towards Direct Human Control of Robot Swarms. ...   390
    Gaëtan Podevijn, Rehan O'Grady, Youssef S.G. Nashed, and Marco Dorigo

Profiling Underwater Swarm Robotic Shoaling Performance
Using Simulation   404
    Mark Read, Christoph Möslinger, Tobias Dipper, Daniela Kengyel,
    James Hilder, Ronald Thenius, Andy Tyrrell, Jon Timmis,
    and Thomas Schmickl

Path Planning for Swarms by Combining Probabilistic Roadmaps
and Potential Fields.   417
    Alex Wallar and Erion Plaku

Towards Exogenous Fault Detection in Swarm Robotic Systems   429
    Alan G. Millard, Jon Timmis, and Alan F.T. Winfield

Verification and Ethics

Ethical Choice in Unforeseen Circumstances   433
Louise Dennis, Michael Fisher, Marija Slavkovik, and Matt Webster

TheatreBot: A Software Architecture for a Theatrical Robot   446
Julián M. Angel Fernandez and Andrea Bonarini

ROBSNA: Social Robot for Interaction and Learning Therapies.   458
*Paulina Vélez, Katherine Gallegos, José Silva, Luis Tumalli,
and Cristian Vaca*

ARE: Augmented Reality Environment for Mobile Robots   470
Mario Gianni, Federico Ferri, and Fiora Pirri

Author Index   485

Artificial Intelligence

Dynamic Power Sharing for Self-Reconfigurable Modular Robots

Chi-An Chen$^{(\boxtimes)}$, Akiya Kamimura, Luenin Barrios, and Wei-Min Shen

Polymorphic Robotics Lab, Information Sciences Institute,
University of Southern California, Marina del Rey, CA 90292, USA
jotaro.chen@gmail.com

Abstract. Dynamic power sharing is used to extend the operation time of self-reconfigurable modular robots [1]. In this area of research, power and energy consumption has consistently been a critical factor in the determination of a robot's operation time. To this end, a method to dynamically share power would enable the robot to extend its operation time and allow it to function well beyond its individual energy life span. In this paper, a dynamic power sharing mechanism is proposed that provides such capabilities. It consists of five power sharing modes and is demonstrated on SuperBot, a self-reconfigurable modular robot developed at USC by the Polymorphic Robotics Laboratory at the Information Sciences Institute. The five modes include: (1) offering power (2) power bypass (3) receiving power (4) both charging the battery and receiving power and lastly (5) battery charging. The five modes that comprise the implementation will demonstrate how the self-reconfigurable modular robot SuperBot can share power dynamically. Finally, experiments were performed on SuperBot conclusively demonstrating that the operation time can be increased upwards of 30 % more compared with the original hardware lacking the power sharing capabilities.

Keywords: Robot · Dynamic · Power sharing

1 Introduction

Self-reconfigurable modular robots are autonomous machines created to deal with unpredictable, dynamic and unforeseen situations in environments and tasks. They are well suited to function in environments requiring high degrees of multitasking and adaptation such as robot arm, planetary exploration, building complex structures, fixing objects in space [3], etc. Depending on the task, they can form various shapes to maximize the work efficiency of each task. Figure 1 shows SuperBot in a rolling track configuration [2]. There are six modular robots connected together to form the shape and perform the rolling track task. In the initial design and implementation of SuperBot, each modular robot did not have a power sharing mechanism and relied solely on the energy contained within its battery pack. From Table 1, the first run (970 m) was terminated due to module

A. Natraj et al. (Eds.): TAROS 2013, LNAI 8069, pp. 3–14, 2014.
DOI: 10.1007/978-3-662-43645-5_1, © Springer-Verlag Berlin Heidelberg 2014

Fig. 1. SuperBot in a rolling track configuration [2].

Table 1. Average remaining battery voltage (every run was started from a fully charged battery 8.2 V) [2].

Module no.	1st run (970 m)	2nd run (1142 m)
Module 1	3.63	3.48
Module 2	7.41	5.19
Module 3	7.45	3.63
Module 4	7.43	7.23
Module 5	7.43	6.70
Module 6	7.44	7.63

1's low battery and the second run (1142 m) was terminated by module 1's or module 3's low batteries.

Based on the experimental results, it is evident that in order to extend the operation time, it is important to solve low power issues that cause failures throughout the entire system. For example, in the above experiment, module 1's battery malfunctioned resulting in a failure that threatened to fracture the entire rolling track configuration. Therefore, if SuperBot wants to extend its operation time, it would be conducive to have modules capable of offering their power to other modules so that work and operation can proceed normally.

In this paper, a prototype of dynamic power sharing circuits was designed and implemented in two SuperBot modules. The dynamic power sharing mechanism consists of 5 operation modes. The experimental results demonstrate the power sharing functionality and confirm the improved operation time on SuperBot modules.

2   Related Work

Power sharing is an essential problem in self-reconfigurable modular robots because the structures possible are large and diverse and require a solution that must function across all configurations. Due to the various kinds of self-reconfigurable modular robots, several kinds of power sharing structures and circuits have been developed. Figure 2(a) shows the Odin Modular Robot [4]

(a) Odin [4]   (b) Electrostatic Latches [5]   (c) Planar Catoms [6]

Fig. 2. The related work

developed by the University of Southern Denmark where the power is transferred by the power link across the bodies of links and joints. Figure 2(b) shows Electrostatic Latches developed by CMU [5] where the capacitors are used to transfer the power from a square wave generator. Figure 2(c) shows Planar Catoms [6] also developed by CMU where the transformer is used to transfer the power, providing power to the connected module. The robust power sharing capabilities which are able to handle the dynamic nature of self-reconfigurable modular robots especially in unpredictable situations are not shown. For example, if one of the batteries malfunctions or has no power. In such cases, the module might cease working properly and may also affect the other modules. Because self-reconfigurable modular robots are highly versatile and dynamic, the power sharing mechanism should be also designed in a dynamic fashion to handle all kinds of tasks.

3 System Design

A dynamic power sharing circuit is created based on SuperBot's modular structure. This section introduces the system design for the dynamic power sharing capabilities. As shown in Fig. 3(a) and (b), each module consists of one master board, one slave board, and six communication boards on the six faces. The modular robots can share power with each other by docking to one of the six communication boards.

3.1 Hardware Design

After several hardware versions [7], the power sharing circuit was created as shown in Fig. 3(b). Figure 3(c) shows switch 1's circuit. The switch in this circuit is mainly composed of the three components: STL8NH3LL Power FET, MIC5018 High-Side MOSFET Driver and 2N7002 MOSFET. STL8NH3LL Power FET is used for passing high electric current (Max. 8 A). MIC5018 High-Side MOSFET Driver is used for high output voltage (around 16 V). 2N7002 MOSFET is used for controlling MIC5018 High-Side MOSFET Driver to turn on or off. MIC5018 High-Side MOSFET Driver is turned on by default so that each module is ready to receive power any time and can be turned on if other modules are willing to help. Switch 0, switch 1, switch 3 are designed based on switch 1's structure, without the MIC5018 High-Side MOSFET Driver.

3.2 Software Design

The dynamic power sharing tasks have been developed based on the previous research [8]. The two classes have been modified: system tasks and behavior tasks.

1. System tasks [8]: system tasks consist of low-level programs which are used to control the hardware. For example, in power sharing mechanism, each

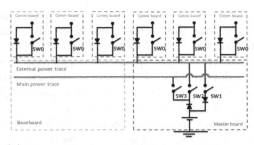

(a) SuperBot module [2].

(b) Power sharing simplified circuit in one module.

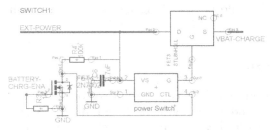

(c) Power sharing switch 1 circuit diagram.

Fig. 3. Hardware design.

module needs to negotiate their power needs before determining which power sharing mode they should adopt. Therefore, the power sharing signal needs to be embedded in the program, such as the "misc Task" which is used to communicate between the master and the slave boards.

2. Behavior tasks [8]: behavior tasks consist of high-level programs which are used to perform the robot's high-level behavior, such as rolling tracks, a caterpillar gait, a butterfly stroke, etc. For verifying SuperBot's dynamic power sharing capabilities, the query task has been created and used to determine which power sharing mode the modular robot should select. Behavior tasks can manipulate the functions from system tasks.

4   Dynamic Power Sharing

The SuperBot hardware shown in Fig. 3 that includes the dynamic power sharing circuit has 5 modes: (1) offering power (2) power bypass (3) receiving power (4) both charging the battery and receiving power and (5) battery charging.

4.1   Mode 1: Offering Power

The basic function for dynamic power sharing is that one module can offer power to other modules. When module 1 offers power to module 2; the blue arrow

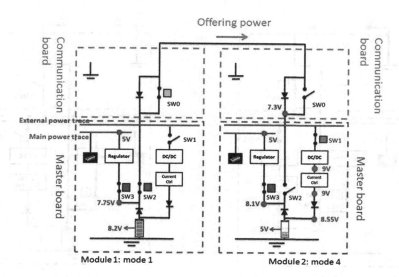

Fig. 4. Module 1, in mode 1, shares power to module 2, in mode 4 (Color figure online).

represents offering power to module 2's system and the green arrow represents charging module 2's battery. The yellow LED, red LED and blue LED stay on, which signify that switch 0 (SW0), switch 2 (SW2), and switch 3 (SW3) are closed respectively. Figure 4 shows the status of the offering power mode switches. The current from the battery goes through SW2, offering power to the external power path and also through SW0 to the docking board so that other modules can get power from module 1 by connecting to the docking board.

4.2   Mode 2: Power Bypass

If the module doesn't have enough power or doesn't want to offer its power to the other module(s), it enters "power bypass mode" and lets the modules which are not directly connected to the receiving power modules pass their power. Figure 5(a) shows the power bypass mode switches status. Because SW1 and SW2 are open, the battery only offers power to its own system. SW0 is closed, so the module can pass power through the external power path and offer power to the docking board.

4.3   Mode 3: Receiving Power

If the module with insufficient battery doesn't want to charge its own battery, for example, the battery malfunctions, it can enter "only receiving power mode". Figure 5(b) shows how the source comes from the docking board, through SW2 and SW3, to offer the receiving module's system power.

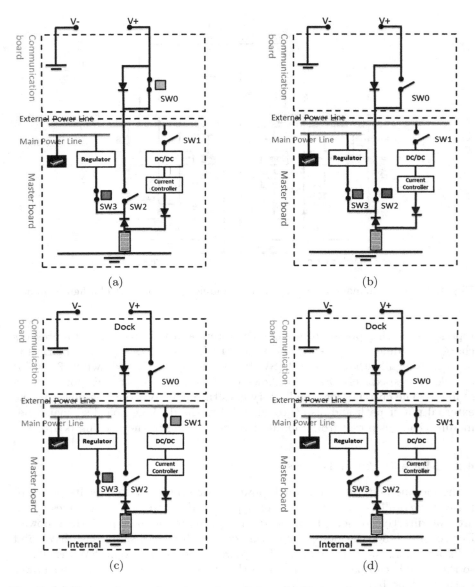

Fig. 5. (a) Status of power bypass mode switches. (b) Status of receiving power mode switches. (c) Status of both charging the battery and receiving power mode switches. (d) Status of battery charging mode switches.

4.4   Mode 4: Both Charging the Battery and Receiving Power

If the insufficient battery module wants to charge its own battery and also get power simultaneously from the other module(s), for example, when the module can connect to a power supply, it can enter "both charging the battery and

Fig. 6. Communication demonstration for two SuperBot modules.

receiving power mode". When the module is in this mode, the outside power source can pass through SW1 to charge its battery and SW3 to provide its system power simultaneously as shown in Fig. 5(c).

4.5   Mode 5: Battery Charging

If the insufficient battery module only wants to charge its own battery without receiving system power, it can enter "battery charging mode". This mode is used for SuperBot in the rest state. Figure 5(d) shows the module in this mode. The outside power source can pass through SW1 to charge its battery but cannot go through SW3 to provide its system power because SW3 is open.

4.6   Dynamic Power Sharing Implementation

Before each module enters the power sharing modes, they need to communicate first to determine which power sharing mode they want to adopt. Therefore, the power sharing implementation coordinates each self-reconfigurable modular robot's power needs. Figure 6 shows two SuperBot modules exercising such communication. Figure 7 shows the power sharing implementation diagram which handles the distributed structure, as is the case for self-reconfigurable modular robots. In Fig. 7, two modules performing dynamic power sharing are demonstrated. When module 1's voltage < Vth1, it sends "Help" signal through infrared to the surrounding modules. In Fig. 7, for example, module 2 is near module 1 and receives the "Help" signal from module 1. Module 2 checks its battery voltage first and if the voltage < Vth2, it helps module 1 broadcast the "Help" signal. However if the voltage > Vth2, module 2 can decide whether it wants to help or not. If module 2 decides not to help, it checks the docking status, to see if they are already docked. Module 2 enters mode 2, acting as a bypass mode module and broadcasts the "Help" signal, however, if they are not docked yet, module 2 just helps module 1 broadcast the "Help" signal. From module 1's perspective, if it doesn't receive the confirmed signal from other modules, module 1 keeps

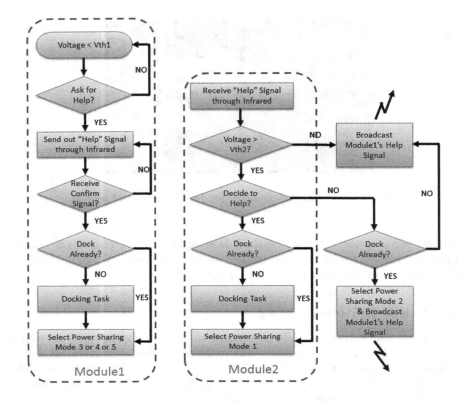

Fig. 7. Dynamic power sharing implementation

sending out the "Help" signal. Once module 1 gets the confirmed signal, it checks its docking status. If module 1 isn't docked with the module which is willing to offer power to module 1, module 1 executes the docking task and then chooses one mode from mode 3, 4 or 5 after docking. If module 1 is already docked with the module, it directly chooses one mode from mode 3, 4 or 5 to share power. If module 2 decides to help, it sends out the confirmed signal to module 1 and also checks the docking status. If they are already docked, module 2 enters mode 1 to offer power to module 1. If they are not docked yet, module 2 performs the docking task and then enters mode 2 to share power after docking.

5   Experiments and Results

Experiments have been conducted on SuperBot hardware to validate and demonstrate the functionalities of dynamic power sharing. The main idea for this experiment, as shown in Fig. 8, is to combine two SuperBot modules together to form a prototype of a robot joint under which various stress tests will be applied to show the improved energy performance of power sharing.

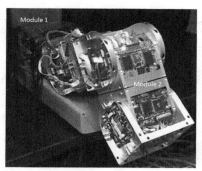

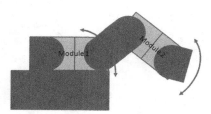

(a) The dynamic power sharing experiment setup.

(b) The side-view diagram of the dynamic power sharing experiment setup.

Fig. 8. Dynamic power sharing experiment.

5.1   Module 1: Mode 3 & Module 2: Mode 1

In this experiment, module 1 carries a heavier load than module 2. Under these conditions, module 1's power consumption is greater than module 2. Two cases were taken into consideration:

Without dynamic power sharing: the two modules' batteries were independent and the result is shown in Fig. 9(a).

With dynamic power sharing: module 1 was in mode 3 (i.e., only receiving power) and module 2 was in mode 1 (i.e., offering power) and the external power paths were connected. The result is shown in Fig. 9(b).

5.2   Module 1: Mode 4 & Module 2: Mode 1

This experiment's configuration is the same as that described in Sect. 5.1. Two cases were taken into consideration:

Without dynamic power sharing: the two modules' batteries were independent and the result is shown in Fig. 9(c).

With dynamic power sharing: module 1 was in mode 4 (i.e., both charging and powering simultaneously) and module 2 was in mode 1 (i.e., offering power) and the external power paths were connected. The result is shown in Fig. 9(d).

5.3   Module 1: Mode 3 & Module 2: Mode 1 (Same Task)

In this experiment, module 1 and module 2 performed the same task. Therefore, module 1 and module 2 had similar power consumption. Two cases were taken into consideration:

Without dynamic power sharing: the two modules' batteries were independent and the result is shown in Fig. 9(e).

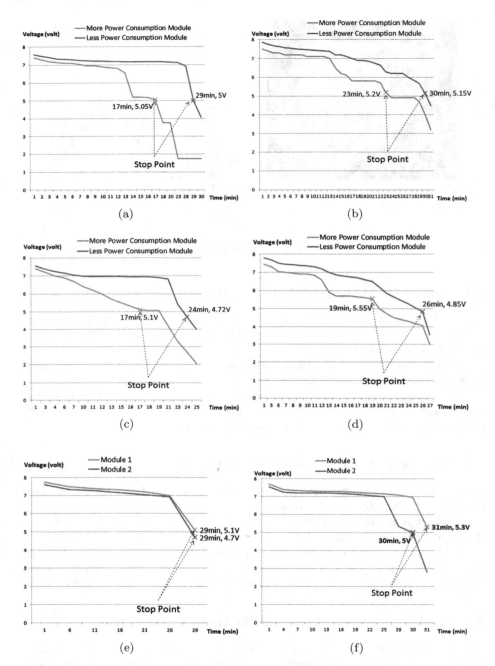

Fig. 9. Dynamic power sharing experiments results. (a) Mode 3 and mode 1 without power sharing. (b) Mode 3 and mode 1 with power sharing. (c) Mode 4 and mode 1 without power sharing. (d) Mode 4 and mode 1 with power sharing. (e) Mode 3 and mode 1 (same task) without power sharing. (f) Mode 3 and mode 1 (same task) with power sharing.

With dynamic power sharing: module 1 was in mode 3 (i.e., only receiving power) and module 2 was in mode 1 (i.e., offering power) and the external power paths were connected. The result is shown in Fig. 9(f).

6 Discussion

6.1 Summary and Contributions

According to the experimental results described in Sect. 5.1 (module 1: mode 3 & module 2: mode 1), the operation time can be increased by around 30 %. In Sect. 5.2 (module 1: mode 4 & module 2: mode 1), when module 1 is in mode 4 and module 2 is in mode 1, the operation time can be extended by almost 12 %. The difference mainly comes from charging the battery since charging the battery of the other module does not result in 100 % power transfer. However, in Sect. 5.3, if two modules do similar jobs, then the operation time with or without dynamic power sharing is almost the same. This is shown in the experimental results.

The affects of dynamic power sharing can be clearly observed when the modules have significantly different power consumption tasks because the module with higher power consumption demands can get support from other less power consumption using modules.

Through dynamic power sharing, not only do modules share power with each other, but each module can also prevent abnormal power drains from other modules, for example, in the case of a malfunctioning battery. Table 2 summarizes the results for potential battery problems. "1" means that switch is closed and "0" means the switch is open, for example, (0, 1, 0, 1) means that switch 0 is open, switch 1 is closed, switch 2 is open and switch 3 is closed.

Table 2. Solutions to potential battery problems

Solutions	M1 (SW0, SW1, SW2, SW3)	M2 (SW0, SW1, SW2, SW3)	M1 No Battery	M2 No Battery	M1 Low Battery	M2 Low Battery	M1 Battery Damaged	M2 Battery Damaged
1	(0, 1, 0, 1) Both	(1, 0, 1, 1) Offering Power	X		X			
2	(0, 1, 0, 1) Both	(1, 0, 0, 1) Bypass Power	X		X			
3	(1, 0, 1, 1) Offering Power	(0, 1, 0, 1) Both		X		X		
4	(1, 0, 0, 1) Bypass Power	(0, 1, 0, 1) Both		X		X		
5	(0, 1, 0, 0) Charging	(1, 0, 1, 1) Offering Power			X			
6	(0, 1, 0, 0) Charging	(1, 0, 0, 1) Bypass Power			X			
7	(1, 0, 1, 1) Offering Power	(0, 1, 0, 0) Charging				X		
8	(1, 0, 0, 1) Bypass Power	(0, 1, 0, 0) Charging				X		
9	(0, 0, 1, 1) Receiving Power	(1, 0, 1, 1) Offering Power			X		X	
10	(0, 0, 1, 1) Receiving Power	(1, 0, 0, 1) Bypass Power			X		X	
11	(1, 0, 1, 1) Offering Power	(0, 0, 1, 1) Receiving Power				X		X
12	(1, 0, 0, 1) Bypass Power	(0, 0, 1, 1) Receiving Power				X		X

6.2 Future Work

Power Management on SuperBot. Based on the experiments carried out on SuperBot modules, the dynamic power sharing mechanism demonstrated extended operation time. However, if the scale becomes bigger, deciding the optimized power paths becomes a complex problem.

Each module might have a different task, battery condition, level of significance, configuration, etc. Depending on the various functions, each module should coordinate to decide which mode it should be in and what its level of priority for power consumption should be. In a centralized system, it might be easier to coordinate with other modules. However, SuperBot is a decentralized system, therefore, to organize each module and determine the best configuration for optimized power consumption, further extensions will have to be made.

References

1. Fukuda, T., Nakagawa, S.: Dynamically reconfigurable robotic system. In: Proceedings of 1988 IEEE International Conference on Robotics and Automation, 1988, vol. 3, pp. 1581–1586 (1988)
2. Shen, W.M., Chiu, H., Rubenstein, M., Salemi, B.: Rolling and Climbing by the Multifunctional Superbot Reconfigurable Robotic System. Melville, New York (2008)
3. Zykov, V., Mytilinaios, E., Desnoyer, M., Lipson, H.: Evolved and designed self-reproducing modular robotics. Trans. Rob. **23**(2), 308–319 (2007)
4. Garcia, R.F.M., Lyder, A., Christensen, D.J., Stoy, K.: Reusable electronics and adaptable communication as implemented in the odin modular robot. In: Proceedings of the 2009 IEEE International Conference on Robotics and Automation, ICRA'09, Piscataway, NJ, USA, pp. 3991–3997. IEEE Press (2009)
5. Karagozler, M.E., Campbell, J.D., Fedder, G.K., Goldstein, S.C., Weller, M.P., Yoon, B.W.: Electrostatic latching for inter-module adhesion, power transfer, and communication in modular robots. In: Proceedings of the IEEE International Conference on Intelligent Robots and Systems (IROS '07), October 2007
6. Kirby, B., Aksak, B., Goldstein, S.C., Hoburg, J.F., Mowry, T.C., Pillai, P.: A modular robotic system using magnetic force effectors. In: Proceedings of the IEEE International Conference on Intelligent Robots and Systems (IROS '07), October 2007
7. Salemi, B., Moll, M., Shen, W.M.: SUPERBOT: A deployable, multi-functional, and modular self-reconfigurable robotic system. In: Proceedings of 2006 the IEEE/RSJ International Conference on Intelligent Robots and Systems, Beijing, China, October 2006
8. Chiu, H., Shen, W.M.: Concurrent and real-time task management for self-reconfigurable robots. In: Proceedings of the Third International Conference on Autonomous Robots and Agents (2006)

Heuristically-Accelerated Reinforcement Learning: A Comparative Analysis of Performance

Murilo Fernandes Martins$^{(\boxtimes)}$ and Reinaldo A.C. Bianchi

Department of Electrical Engineering – IAAA Group,
Centro Universitário da FEI, São Paulo, Brazil
murilo@ieee.org, rbianchi@fei.edu.br

Abstract. This paper presents a comparative analysis of three Reinforcement Learning algorithms (Q-learning, Q(λ)-learning and QS-learning) and their heuristically-accelerated variants (HAQL, HAQ(λ) and HAQS) where heuristics bias action selection, thus speeding up the learning. The experiments were performed in a simulated robot soccer environment which reproduces the conditions of a real competition league environment. The results clearly demonstrate that the use of heuristics substantially improves the performance of the learning algorithms.

Keywords: Reinforcement learning · Heuristics · Robot soccer

1 Introduction

In the past decades a significant amount of algorithms for Reinforcement Learning (RL) have been proposed in the literature. Amongst the proposed techniques, the class of model-free algorithms is the most widely used, since such algorithms do not require a model of the environment with which the agents interact. The Q-learning algorithm [1,2] is, perhaps, the most well-known model-free algorithm for RL. Although the implementation of the Q-learning algorithm is straightforward and its convergence to optimality has been mathematically proven [1], such convergence is rather slow, infinite time-bounded. This is due to the fact that only one state-action pair has its value updated at each iteration of the algorithm. Furthermore, the larger the state and action spaces, the more visits to the corresponding state-action pairs are necessary for the system to learn and hence the slower the learning process becomes.

Addressing the slow learning convergence of RL algorithms, many techniques have been proposed to speed up the time to converge to a (usually nearly) optimal action selection policy. This paper presents a systematic comparative analysis of performance between the Q-learning algorithm, two variants – Q(λ)-learning [1,3] (which makes use of temporal generalisations) and

---

The author acknowledges the current support of FAPESP – project n° 2012/12640-1.

A. Natraj et al. (Eds.): TAROS 2013, LNAI 8069, pp. 15–27, 2014.
DOI: 10.1007/978-3-662-43645-5_2, © Springer-Verlag Berlin Heidelberg 2014

QS-learning [4,5] (which employs spatial generalisation) – and the heuristically-accelerated variants of such algorithms, herein denoted as HAQL [6], HAQ(λ) and HAQS. The experiments were performed in a complex, dynamic, stochastic and real-time simulated robot-soccer environment called SimuroSot[1], the official FIRA Middle League Robot Soccer simulation environment.

This paper is organised as follows. Section 2 presents a formal definition of RL and the algorithms studies, including some key implementation aspects. Next, Sect. 3 introduces the mathematical formulation of the class of Heuristically-Accelerated Reinforcement Learning (HARL) algorithms, whilst details of the experimental setup are presented in Sect. 4. Then, Sect. 5 presents the results obtained, along with a comparative analysis of performance between the implemented algorithms. Lastly, Sect. 6 concludes this paper and presents ongoing and future directions.

2   Preliminaries

Usually, algorithms for RL are formally defined as Markov Decision Processes (MDP) [2]. An MDP consists of a set of states $\mathcal{S}$, a set of actions $\mathcal{A}$ (both usually finite), a reward function $\mathcal{R} : \mathcal{S} \times \mathcal{A} \rightarrow \mathbb{R}$ encoding the desired behaviour of the learning agent, and a transition function in the form $\mathcal{T} : \mathcal{S} \times \mathcal{A} \rightarrow \mathcal{S}$. The optimal solution for a deterministic MDP is defined by a policy $\pi^* : \mathcal{S} \rightarrow \mathcal{A}$ which maximises the long term delayed rewards received by the learning agent.

2.1   The Q-Learning Algorithm

Perhaps the most popular RL algorithm, the Q-learning has been extensively studied and widely used across distinct areas. This algorithm defines a function $Q : \mathcal{S} \times \mathcal{A} \rightarrow \mathbb{R}$ representing the maximum cumulative reward value that can be received by the learning agent, as shown in Algorithm 1. The current state is represented by s, while s' is the resulting state of the environment after executing the action a in state s. The value function $V : \mathcal{S} \rightarrow \mathbb{R}$, where $V(s) = \max_a Q(s, a)$, is the maximum cumulative reward the learning agent can receive from s. The discount factor $\gamma \in [0, 1)$ defines a balance between immediate and future rewards. The learning rate α determines how much the current iteration should change the values in the action-value function $Q(s, a)$. In order for the Q-learning algorithm to converge to an optimal action selection policy (in a stochastic environment), in this paper the learning rate α is updated according to Eq. 1.

$$\alpha_n = \max \left(\frac{1}{1 + visits_n(s, a)}, 0.125 \right) \tag{1}$$

where α_n is the value of α at the n-th iteration and $visits_n(s, a)$ is the number of times the learning agent has visited state s and executed action a, also at the n-th iteration. However, in this implementation, whenever the learning rate

---

[1] http://www.fira.net/?mid=simurosot

α falls below $\alpha < 0.125$, its value is kept as $\alpha = 0.125$ so that the agent never stops learning. In This can also be understood as $\lim_{n \to \infty} \alpha_n = 0.125$.

In the Q-learning algorithm, the optimal policy is defined as $\pi^*(s, a) \equiv \arg\max_{a \in \mathcal{A}} Q(s, a)$. To ensure random exploration of the environment by the learning agent, as opposed to pure exploitation, the action selection rule denoted as $\epsilon - Greedy$ is used (Eq. 2).

$$\pi(s) = \begin{cases} \arg\max_a Q(s, a) & q \leq p \\ a_{random} & \text{otherwise} \end{cases} \tag{2}$$

where $q \in [0, 1]$ is a random value sampled from a uniform distribution and $p \in [0, 1]$ is a parameter determining the exploration/exploitation rate. Also, the action a_{random} is randomly sampled from a uniform distribution of $\mathcal{A}$.

Algorithm 1. Q-learning algorithm

Input: $\mathcal{S}$: set of states, $\mathcal{A}$: set of actions, $r(s, a)$: reward function, $V(s)$: value function, $Q(s, a)$: action-value function.
1. $\forall s \in \mathcal{S}, V(s) \leftarrow 0$ and $\forall s \in \mathcal{S} \land \forall a \in \mathcal{A}, Q(s, a) \leftarrow 0$
2. Observe current state s
3. **loop**
4.   Select an action $a \in \mathcal{A}$ using the $\epsilon - Greedy$ rule
5.   Execute selected action a in current state s
6.   Receive the reward $r(s, a)$ and then observe next state s'
7.   Calculate temporal difference error TD(0): $e' = r(s, a) + \gamma * V(s') - Q(s, a)$
8.   Update $Q(s, a) \leftarrow Q(s, a) + \alpha_n * e'$ and $V(s) = \max_a Q(s, a)$
9. $s \leftarrow s'; n \leftarrow n + 1$
10. **end loop**

2.2   The Q(λ)-Learning Algorithm

The Q(λ)-learning [1, 7] is an algorithm which combines the Q-learning algorithm with temporal generalisations – the Temporal Difference method TD(λ) [2] – back-propagating the outcome of a single iteration to a history of multiple state-action pairs recently visited, known as eligibility trace [2].

In the approach presented in [1], the eligibility trace must always be reset whenever an action is randomly selected by the $\epsilon - Greedy$ rule, whereas the approach proposed in [7] does not differentiate a random action from an action that follows a greedy policy. As a consequence, for a fixed, non-greedy policy π, the Q function in the Q(λ)-learning will neither converge to Q^π nor to the optimal Q^*, but to some hybrid policy in between. However, according to [2], for a policy which becomes greedy over time the Q(λ)-learning proposed in [7] may converge to the optimal Q^* and, furthermore, result in a significantly better performance than the Q(λ)-learning algorithm proposed by [1].

In the Q(λ)-learning algorithm, the $\lambda \in [0,1]$ factor is used to discount the temporal difference error TD(λ) of the next steps when updating the action-value function $Q(s,a)$. These error values are incrementally calculated using the eligibility trace e. In this approach, a value $l(s,a)$ of eligibility trace is stored for each state-action pair. The Q(λ)-learning algorithm implemented in this paper follows the approach defined in [7], which is also detailed in [8]. The implemented algorithm is shown in Algorithm 2.

Algorithm 2. Q(λ)-learning algorithm

Input: $\mathcal{S}$: set of states, $\mathcal{A}$: set of actions, $r(s,a)$: reward function,
$\quad\quad$ $V(s)$: value function, $Q(s,a)$: action-value function,
$\quad\quad$ L: list of (s,a) pairs, that is, the eligibility trace.
1. $\forall s \in \mathcal{S}, V(s) \leftarrow 0; \forall s \in \mathcal{S} \wedge \forall a \in \mathcal{A}, Q(s,a) \leftarrow 0$ and $L \leftarrow \emptyset$
2. Observe current state s
3. **loop**
4. $\quad$ Select an action $a \in \mathcal{A}$ using the $\epsilon - Greedy$ rule
5. $\quad$ Execute selected action a in current state s
6. $\quad$ Receive the reward $r(s,a)$ and then observe next state s'
7. $\quad$ Calculate temporal difference error TD(0): $e' = r(s,a) + \gamma * V(s') - Q(s,a)$
8. $\quad$ Calculate temporal difference error TD(λ): $e = r(s,a) + \gamma * V(s') - V(s)$
9. $\quad$ **for all** $(v,u) \in L$ **do**
10. $\quad\quad$ Calculate trace decay $l(v,u) = \gamma * \lambda * l(v,u)$
11. $\quad\quad$ Update $Q(v,u) = Q(v,u) + \alpha_n * e * l(v,u)$
12. $\quad\quad$ **if** $l(v,u) < \zeta$ **then**
13. $\quad\quad\quad$ $L \leftarrow L \setminus (v,u); visited(v,u) \leftarrow 0$
14. $\quad\quad$ **end if**
15. $\quad$ **end for**
16. $\quad$ Update $Q(s,a) = Q(s,a) + \alpha_n * e$
17. $\quad$ Update eligibility trace:
18. $\quad$ $\forall u \neq a, l(s,u) \leftarrow 0; l(s,a) \leftarrow 1$
19. $\quad$ **if** $visited(s,a) = 0$ **then**
20. $\quad\quad$ $visited(s,a) \leftarrow 1; L \leftarrow L \cup (s,a)$
21. $\quad$ **end if**
22. $\quad$ $s \leftarrow s'$
23. **end loop**

In this case, α is updated using the rule defined in Eq. 1, $v \in \mathcal{S}$ is a state, $u \in \mathcal{A}$ is an action and (v,u) are the recently visited state-action pairs. Such pairs are inserted into a doubly-linked list L and, in case the eligibility trace $l(v,u) < \tau$, the corresponding (v,u) pair is removed from the list. Here, $\zeta \geq 0$ is a systematically defined threshold. Removing the (v,u) pairs from the list when their values fall below τ has been observed to considerably speed up each iteration of the algorithm. In order to ensure that the list L does not have duplicate state-action pairs, a binary function $visisted(v,u)$ is used to indicate whether a given (v,u) pair has already been visited recently and hence is in L.

By making use of replacing eligibility traces [2] (lines 24 and 25 of Algorithm 2), it is possible to define the maximum number of state-action pairs in L using the values $\gamma\lambda$ and τ. As a result, the complexity of each iteration of the algorithm does not grow linearly as the learning agent visits new state-action pairs, but it is, in the worst case, bounded to a constant length L and a manageable number of updates to the action-value function Q. Also, it is worth noticing that if $\lambda = 0$, the Q(λ)-learning algorithm becomes identical to the Q-learning.

2.3 The QS-Learning Algorithm

Whilst the Q(λ)-learning algorithm makes used of temporal generalisations, the QS-learning algorithm [4] takes advantage of *a priori* knowledge of spatial similarities employing spatial generalisation to improve the performance of the vanilla Q-learning algorithm. Depending on the similarities between state-action pairs, one single iteration of the algorithm may update more than one (s, a) pair in Q. This similarity is determined by a spreading function $\sigma(v, u, s, a) \in [0, 1]$, which may occur both in the state space $\mathcal{S}$ or the action space $\mathcal{A}$. However, in this paper, as in [4,5], only similarities in $\mathcal{S}$ are considered. The spreading function is defined in Eq. 3.

$$\sigma(v, u, s, a) = g(v, s)\delta(u, a), \quad \text{with } g(v, s) = \tau^d \tag{3}$$

where $\delta(u, a)$ is the Kronecker delta: $\delta(u, a) = 1$ if $u = a$, and $\delta(u, a) = 0$ if $u \neq a$. The function $g(v, s) = \tau^d$ defines the similarity between $v \in \mathcal{S}$ and $s \in \mathcal{S}$, where τ is a constant and d is a factor which quantifies the similarity between v and s. The proof of convergence of the QS-learning algorithm is detailed in [4], and the QS-learning algorithm implemented in this paper is described in Algorithm 3.

Algorithm 3. QS-learning algorithm

Input: $\mathcal{S}$: set of states, $\mathcal{A}$: set of actions, $r(s, a)$: reward function,
$\quad\quad V(s)$: value function, $Q(s, a)$: action-value function,
1. $\forall s \in \mathcal{S}, V(s) \leftarrow 0$ and $\forall s \in \mathcal{S} \land \forall a \in \mathcal{A}, Q(s, a) \leftarrow 0$
2. Observe current state s
3. **loop**
4. $\quad$ Select an action $a \in \mathcal{A}$ using the $\epsilon - Greedy$ rule
5. $\quad$ Execute selected action a in current state s
6. $\quad$ Receive the reward $r(s, a)$ and then observe next state s'
7. $\quad$ **for all** $v \in \mathcal{S}, u \in \mathcal{A}$ **do**
8. $\quad\quad$ **if** $\sigma(v, u, s, a) \neq 0$ **then**
9. $\quad\quad\quad$ Calculate temporal diff. error TD(0) $e' = r(s, a) + \gamma * V(s') - Q(v, u)$
10. $\quad\quad\quad$ Update $Q(v, u) = Q(v, u) + \sigma(v, u, s, a) * \alpha_n * e'$
11. $\quad\quad$ **end if**
12. $\quad$ **end for**
13. $\quad$ $s \leftarrow s'$
14. **end loop**

As with Q-learning and Q(λ)-learning, in the QS-learning algorithm the learning rate α is updated according to Eq. 1. In order to satisfy the conditions to guarantee the convergence of the algorithm, the function $\sigma(v, u, s, a)$ must decay at a faster rate than α. It is important to highlight that when $g(v, s)$ does not define any similarity between v and s, its value is thus $g(v, s) = 0$ for any state v other than s, resulting in a spreading function $\sigma(v, u, s, a) = 0$. As a consequence, the QS-learning algorithm becomes identical to the Q-learning.

3   Heuristically-Accelerated Reinforcement Learning

The use of heuristics to accelerate RL algorithms has firstly been proposed in [6], where the Q-learning algorithm was extended to take advantage of a static heuristic function defined *a priori*. This technique has also been extensively explored in goal-driven navigation tasks [9], in the multiagent robot soccer scenario [10]. The Heuristically-Accelerated Reinforcement Learning (HARL) approach has also been combined with a market-based approach applied to the Robocup 2D simulated domain [11] and with case-based reasoning [12]. In addition, in the Markov Games domain, a Heuristically-Accelerated minimax-Q algorithm (HAMMQ) was proposed in [10] and extensively analysed in domains of distinct complexity [13]. The HAMMQ algorithm is an extension of the minimax-Q algorithm proposed by [14], which is essentially the Q-learning algorithm with a *minimax* rule replacing *max* in the well-known Bellman equations [2].

In the HARL, a heuristic function is defined as $\mathcal{H} : \mathcal{S} \times \mathcal{A} \to \mathbb{R}$ and is used to bias the action selection during the learning process. This function determines how desirable would the selection of a given action $a \in \mathcal{A}$ be whilst in $s \in \mathcal{S}$. Furthermore, this function may be stationary or non-stationary, as discussed in [9]. Although the heuristic function could be extracted automatically, or from demonstrations of a teacher, it has invariably been defined *a priori* by a specialist, using the knowledge of the domain.

In order for the heuristic function to bias the action selection, the $\epsilon - Greedy$ rule must be modified, as shown in Eq. 4.

$$\pi(s) = \begin{cases} \arg\max_{a}[Q(s, a) + \xi * \mathcal{H}(s, a)] & q \leq p \\ a_{random} & \text{otherwise} \end{cases} \tag{4}$$

where $\xi \in \mathbb{R}$ is a real-valued number which weighs the influence of the heuristics in use and it is necessary to guarantee the convergence of the algorithm.

In this paper, the Q-learning variant HAQL, proposed in [6], as well as the herein proposed HAQ(λ) (Heuristically-Accelerated Q(λ)-learning) and HAQS (Heuristically-Accelerated QS-learning) – variants of the Q(λ)-learning and QS-learning, respectively – were implemented using the action selection rule defined in Eq. 4, rule which is the only modification necessary for the vanilla algorithms to take advantage of the HARL approach.

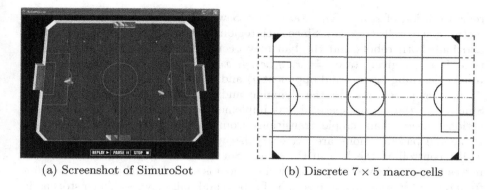

(a) Screenshot of SimuroSot   (b) Discrete 7 × 5 macro-cells

Fig. 1. The SimuroSot and diagrammatic representation of the pitch

4   Experimental Setup

A simulated robot soccer domain was proposed in [14], which consists of a two-dimensional grid world of dimensions 5 × 4, determining the state space $\mathcal{S}$, along with 5 possible actions: *Move North (N)*, *Move South (S)*, *Move East (E)*, *Move West (W)* and *Idle (I)*. In this simplistic domain there are no hidden variables, and two players (agent and opponent) compete against each other, with the ball always being in possession of one of them. This domain has been used as testbed in several studies, e.g., [6,10].

In contrast with the aforementioned domain, the FIRA SimuroSot simulator (Fig. 1a), also used as testbed in other studies (e.g., [13]), is significantly more complex. Firstly, the game is dynamic, which means the players must act upon the current state in real-time, since the ball will not stop moving whilst the players decide which action to execute next. In addition, the SimuroSot simulates cube-shaped robots with 7.5 cm sides and differential-drive kinematics, with 2 teams of 5 robots each, in a 220 × 180 cm pitch. As in the simplistic domain, only two robots are considered in the game, agent i and opponent j.

The inherent incremental error of odometry sensors is simulated, also giving the SimuroSot a stochastic characteristic. At each iteration, the simulator provides somewhat accurate information referring to the robots' position and orientation, as well as the position of the ball, playing the role of a bird's-eye view computer vision system. Also, low level position control and path planning are not readily available and had to be implemented. The whole system (RL algorithms, low level control and path planning) was implemented in C++.

4.1   Definition of Set of States $\mathcal{S}$ and Actions $\mathcal{A}$

Although the position and orientation of robots and ball are continuous variables, using such values would result in a markedly large state space, with memory requirements which are not practical when using the vanilla tabular representation of states. The state space $\mathcal{S}$ was thus discretised to 7×5 symmetric macro-cells (Fig. 1b). Since such regions are larger than the size of robots and ball, both robots and the ball may occupy the same region at the same time. Hence, a given state $s = \langle x_i, y_i, x_j, y_j, x_b, y_b \rangle$ consists of the position of the robots (learning agent i and opponent j) and the ball b.

In this paper, due to the dynamic and stochastic characteristics of the SimuroSot, the action space $\mathcal{A}$ was implemented as a set of behaviour-based actions rather than simple transitions from one discrete region to another. Behaviour-based actions are necessary because, for instance, moving regions whilst controlling the ball in the SimuroSot is far from trivial, since the ball moves freely and game only stops when a goal is scored. Furthermore, the noise from odometry sensors and position of robots and ball may result in a stochastic transition to distinct s' when repeatedly executing a given action a from state s. As a result, executing an implemented behaviour-based action $a \in \{N, S, W, E\}$ means navigating from the macro-cell at which the robot is currently located to the centre of the macro-cell corresponding to the action being executed, whilst I denotes no movement at all. Also, if the execution of an action would result in trespassing the boundaries of the pitch, the execution is considered to be a failure and the outcome is akin to executing I.

In addition, the action space $\mathcal{A}$ defined in [14] was extended with two additional actions: *Fetch Ball (F)* and *Kick to Goal (K)*. Executing F results in moving from whatever macro-cell the robot is located towards the ball (regardless of which region the ball is in). The precise desired location of the robot when executing F is defined as immediately behind the ball (considering the scoring goal side). Executing K, on the other hand, requires that the robot be in the same macro-cell as the ball, resulting in failure (i.e., no movement) otherwise. With this pre-condition fulfilled, K results in the robot hitting the ball from such an angle which allows for pushing the ball in a straight line towards the centre of the scoring goal.

Regarding the behaviour-based actions, a vanilla PID controller was implemented for the low level position control of the robots, whilst the navigation layer – which generates waypoints to move from current to desired position and orientation – was implemented using the well-known cubic Bézier curves, which presented very good results at a very low computational cost.

4.2   The Reward ($\mathcal{R}$), Spreading (σ) and Heuristic ($\mathcal{H}$) Functions

The reward function $\mathcal{R}$ was determined in such a way that whenever a goal was scored by the agent, a large positive reward value was received. Similarly, whenever a goal was suffered, the agent received a large negative reward. In addition, in order to avoid a sub-optimal, stationary behaviour by the agent, a small-valued negative reward was given for every action which did not result in a goal scored. Also, in order to discourage the agent from selecting the actions which would result in failure (representing potential collisions with the pitch boundary walls – extremely undesirable with real robots), another negative reward value was given when the outcome of an action was a failure.

The values of the reward function $\mathcal{R}$ were defined through experimentation. If $s' =$ goal scored, $r(s,a) = +1000$; when $s' =$ goal suffered, $r(s,a) = -1000$; also, $r(s,a) = 10, \forall s' \neq$ goal scored $\wedge a \neq$ failure; and if $a =$ failure, $r(s,a) = -50$.

Regarding the parameter values used in this paper, the learning rate α was initially set to 1, decaying according to Eq. 1. The exploration/exploitation ratio $p = 0.2$, as well as the dicount factor $\gamma = 0.9$ are values commonly used in the literature [10,13,14]. For the algorithms $Q(\lambda)$-learning and HAQ(λ), the factor $\lambda = 0.3$ was determined systematically, following the remarks made in [3]. Similarly, for the QS-learning and HAQS algorithms the spreading function $\sigma(v,u,s,a)$ was defined based on [5]. However, the initial value $\tau = 0.7$ decays according to the rule defined in Eq. 5a.

$$\tau_n = [0.7 - 0.1 * visits_n(v,u)]^d \tag{5a}$$

$$\lim_{n \to \infty} \tau_n = 0 \tag{5b}$$

where τ_n is the value of τ at the n-th iteration and $visits_n(s,a)$ is the number of times the learning agent has visited state v and executed action u at the n-th iteration as well. This way, the rate at which τ decays as the number of iterations n increases is faster than that of α, guaranteeing the convergence requirements, noticing Eq. 5b and recalling that α will never be smaller than 0.125 (Eq. 1).

The values of the similarity quantifier factor d are defined according to the macro-cell at which the opponent is located within v in relation to its location in s, $d = 0$ if $v = s$; $d = 1$ if v_j is adjacent in any cardinal direction of s_j; $d = 2$ if v_j is adjacent in any diagonal direction of s_j; $d = \infty$ otherwise, where $v_j \subset v$ and $s_j \subset s$ represent the position $\langle x_j, y_j \rangle$ of the robot opponent (ignoring the orientation) in states v and s, respectively.

Hence, up to 9 updates to the action-value function Q may be done per iteration. In addition, an imaginary horizontal line (axis of symmetry: $y = 2 -$ dash-dotted red line in Fig. 1b) geometrically divides the pitch in half, such that experiences of a single iteration in the upper part of the pitch can be spread out to the bottom part by mirroring states and actions, and vice-versa, thus making greater use of spatial generalisation. Mirroring states means calculating the Point Reflection, that is, the isometric involutive affine transformation in the Euclidean space $\mathbb{R}^2$ with one fixed point determined by $\rho = \langle x_k, y \rangle, \forall k \in \{i,j,b\}$ and $y = 2$, for agent i, opponent j and the ball b. Thus, for each of a maximum of 9 updates, the corresponding state v is symmetrically mirrored to $v^\rho \equiv \text{Ref}_\rho(v) = 2\rho - v$.

To illustrate this case, suppose both robots and the ball are located in the upper-leftmost macro-cell within the pitch, that is, $s = \langle x_i = 0, y_i = 4, x_j = 0, y_j = 4, x_b = 0, y_b = 4 \rangle$. By mirroring this state, both robots and the ball would be located at the bottom-leftmost macro-cell, i.e., $s^\rho = \langle x_i = 0, y_i = 0, x_j = 0, y_j = 0, x_b = 0, y_b = 0 \rangle$.

Mirroring actions, on the other hand, follows a rather simple rule: $a^\rho = S$ if $a \equiv N$; $a^\rho = N$ if $a \equiv S$; otherwise $a^\rho = a$.

In regards to the heuristic function $\mathcal{H}$, $\xi = 1$ and, as in most of the studies involving HARL algorithms, in this paper $\mathcal{H}$ was defined a priori, being as intuitive and concise as possible, as defined in Eq. 6.

$$\mathcal{H}(s, a) = \begin{cases} 100 & \text{if } s_i = s_b \land a \equiv K \\ 0 & \text{otherwise} \end{cases} \tag{6}$$

where $s_i = \langle x_i, y_i \rangle$ is the position of the learning robot in s and $s_b = \langle x_b, y_b \rangle$ is the position of the ball, also in s. Thus, the learning robot should tend to select action $a \equiv K$ whenever it is in the same macro-cell as the ball. Notice that using $\mathcal{H}$ as a controller is not a solution and would only be marginally better than random walk. This is because $\forall s_i \neq s_b$, $\mathcal{H}(s, a) = 0$ and Eq. 4 ($\arg\max_a [\mathcal{H}(s, a)]$) would often result in randomly sampling $a \in \mathcal{A}$ from a uniform distribution.

5   Results and Discussion

In order to compare the performance of the RL algorithms herein discussed, 5 trials of 500 games each were executed for each algorithm: Q-learning, Q(λ)-learning, QS-learning, HAQL, HAQ(λ) and HAQS. The virtual learning robot (the agent) always played against an opponent which chooses actions randomly.

In the SimuroSot each game consists of 5 min regardless of number of goals scored. The timer only stops when a goal is scored, then continuing once the ball and robots are automatically placed on their initial positions. As a result, each trial of each algorithm consumed 72 h ($\sim$ 42 h of actual gameplay and $\sim$ 30 h of overhead for game set up, restarting, file saving and so on). Conducting simulations in the SimuroSot requires several mouse clicks, and hence the procedure was automated using a third-party commercial software.

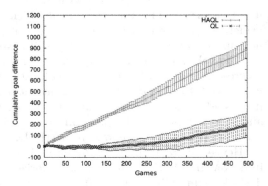

Fig. 2. Q-learning and HAQL

The results of the trials of the Q-learning and HAQL algorithms are shown in Fig. 2, whilst Fig. 3 shows the learning curves of QS-learning and HAQS and Fig. 4 presents the results of Q(λ)-learning and HAQ(λ). The graphs consist of mean and standard deviation (of the 5 trials) of the cumulative goal difference over 500 games.

By analysing Figs. 2 and 3, it is possible to notice the similarity between the Q-learning and QS-learning, as well as the HAQL and

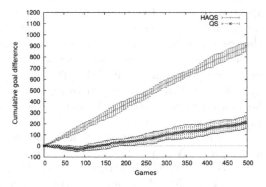

Fig. 3. QS-learning and HAQS

HAQS. This similarity, also noticed by [5], is due to the fact that spreading of experiences only occurs at the beginning of the learning, when the agent has little experience with few large rewards received (scoring or suffering a goal). Since the spreading function decays rapidly (to guarantee convergence), the QS-learning and HAQS quickly become equivalent to the Q-learning and HAQL, respectively. Despite the markedly similar performance of QS-learning and HAQS to Q-learning and HAQL, the advantage of spatial spreading of experiences can be observed by the resulting smaller standard deviation; at the early stages of learning, when visiting previously unvisited states the agent will have received rewards from the spatial spreading, thus preferring to select certain actions over others.

On the other hand, the use of temporal generalisation has a great positive impact in performance when comparing the $Q(\lambda)$-learning (Fig. 4) with Q-learning and QS-learning. This improvement is also noticeable when comparing the $HAQ(\lambda)$ (Fig. 4) with HAQL and HAQS, though not as prominent. Furthermore, two important remarks can be made from analysing the graphs in Figs. 2, 3 and 4. Firstly, the use of a good heuristic function $\mathcal{H}$ posi-

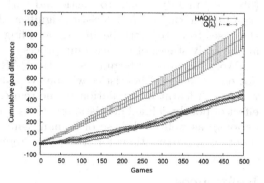

Fig. 4. $Q(\lambda)$-learning and $HAQ(\lambda)$

tively biased the action selection of the agent with little experience, thus avoiding an excessively exploratory behaviour at the beginning of the learning process. Secondly, by approximating the curves to straight lines, a considerably higher slope is noticeable with the algorithms using heuristics. The slope denotes the rate at which the agent improves its performance over time towards an optimal policy, and the difference between slopes is the result of the heuristics accelerating the learning process during the early stages. In the longer term, the influence of $\mathcal{H}$ (Eq. 4) will tend to be residual when compared to the value of Q and the slopes of the vanilla algorithms will become equivalent to their heuristically-accelerated counterparts.

In regards to the temporal generalisations in comparison with the spatial generalisations, it is worth noticing that whilst $Q(\lambda)$-learning and $HAQ(\lambda)$ spread the reward received over a trace of recently visited state-action pairs (multiple iterations) aiming at maximising the reward, the QS-learning and HAQS algorithms spread only a single position of robots and ball, and action executed (single iteration) to other states with a certain similarity (according to σ) with the original state. Therefore, propagation of high long term delayed reward values over time in the QS-learning and HAQS will be similarly slow to the propagation in the Q-learning and HAQL, respectively. On the other hand, the propagation of a trace of experiences, rather than a single experience, directs the agent towards a path of higher long term rewards, explaining why the $Q(\lambda)$-learning and $HAQ(\lambda)$ outperformed the other algorithms by a markedly long margin.

6   Conclusion and Future Work

This paper presented a comparative analysis of performance of well-known RL algorithms – Q-learning, $Q(\lambda)$-learning and QS-learning – and their heuristically-accelerated variants – the previously proposed HAQL and, to the best of the authors' knowledge, the novel algorithms $HAQ(\lambda)$ and HAQS. The results (obtained in a dynamic, stochastic and fairly realistic simulator) clearly demonstrated that the use of heuristics significantly improves the performance in all the cases from the very beginning, avoiding excessive exploration by the agent at the early stages of the learning process.

Work has already begun testing the HARL approach with real robots, where not only the use of heuristics will be evaluated, but also the transfer of policies $\pi : S \to A$ learnt in simulation to the real robots. Future work will explore the effect of poorly defined heuristic functions (potentially leading to poorly selected actions), as well as the use of human demonstrations as heuristics. The use of function approximations for state space representation will also be addressed.

References

1. Watkins, C.: Learning from delayed rewards. Ph.D. thesis, University of Cambridge, England (1989)
2. Sutton, R.S., Barto, A.G.: Reinforcement Learning: An Introduction. Adaptive Computation and Machine Learning. MIT Press, Cambridge (1998)
3. Wiering, M., Schmidhuber, J.: Fast online q(lambda). Mach. Learn. **33**(1), 105–115 (1998)
4. Ribeiro, C., Szepesvári, C.: Q-learning combined with spreading: convergence and results. In: ISRF-IEE International Conference on Intelligent and Cognitive Systems (Neural Networks Symposium), pp. 32–36 (1996)
5. Ribeiro, C., Pegoraro, R., Costa, A.: Experience generalization for concurrent reinforcement learners: the minimax-qs algorithm. In: Proceedings of the First International Joint Conference on Autonomous Agents and Multiagent Systems, pp. 1239–1245. ACM, NY (2002)
6. Bianchi, R.A.C., Ribeiro, C.H.C., Costa, A.H.R.: Heuristically accelerated Q–learning: a new approach to speed up reinforcement learning. In: Bazzan, A.L.C., Labidi, S. (eds.) SBIA 2004. LNCS (LNAI), vol. 3171, pp. 245–254. Springer, Heidelberg (2004)
7. Peng, J., Williams, R.: Incremental multi-step q-learning. Mach. Learn. **22**(1–3), 283–290 (1996)
8. Wiering, M., van Hasselt, H.: Ensemble algorithms in reinforcement learning. IEEE Trans. Syst. Man Cybern. Part B **38**(4), 930–936 (2008)
9. Bianchi, R., Ribeiro, C., Costa, A.: Accelerating autonomous learning by using heuristic selection of actions. J. Heuristics **14**(2), 135–168 (2008)

10. Bianchi, R., Ribeiro, C., Costa, A.: Heuristic selection of actions in multiagent reinforcement learning. In: Proceedings of the 20th International Joint Conference on Artifical Intelligence, pp. 690–696. Morgan Kaufmann Publishers Inc. (2007)
11. Gurzoni Jr, J.A., Tonidandel, F., Bianchi, R.A.C.: Market-based dynamic task allocation using heuristically accelerated reinforcement learning. In: Antunes, L., Pinto, H.S. (eds.) EPIA 2011. LNCS, vol. 7026, pp. 365–376. Springer, Heidelberg (2011)
12. Bianchi, R.A.C., Ros, R., Lopez de Mantaras, R.: Improving reinforcement learning by using case based heuristics. In: McGinty, L., Wilson, D.C. (eds.) ICCBR 2009. LNCS, vol. 5650, pp. 75–89. Springer, Heidelberg (2009)
13. Bianchi, R., Martins, M., Ribeiro, C., Costa, A.: Heuristically-accelerated multiagent reinforcement learning. IEEE Trans. Cybern. **44**(2), 252–265 (2013)
14. Littman, M.L.: Markov games as a framework for multi-agent reinforcement learning. In: Proceedings of the 11th International Conference on Machine Learning (ML-94), pp. 157–163. Morgan Kaufmann, New Brunswick (1994)

TREEBOT: Tree Recovering Renewable Energy Robot

Giovanni Gerardo Muscolo[1,2(✉)] and Rezia Molfino[2]

[1] Creative and Visionary Design lab, Humanot s.r.l., Prato, Italy
muscolo@dimec.unige.it
[2] PMAR lab, Department of Applied Mechanics and Machine Design,
Scuola Politecnica, University of Genova, Genova, Italy

Abstract. The authors aim at designing, prototyping and validating a new generation of ICT hardware and software technologies bio-inspired from trees, called TREEBOT, endowed with distributed sensing, alternators and intelligence for tasks of recovering renewable energy. TREEBOT takes inspiration from, and aim at reproducing, the flexibility, resistance and adaptation capabilities of trees. The tree robot will be composed of a network of sensorized and actuated roots, branches and leaves, displaying rich sensing and coordination capabilities as well as energy-efficient actuation and high sustainability, typical of the trees. Each tree branch and leave will consist of a robotic artefact that comprises sensors, actuators, alternators, control units, and by an elongation zone that mechanically connects the apex and the trunk of the robot.

Keywords: Robot autonomy including energy self-sufficiently · Robot-environment interaction · Bio-inspired robot · Advanced sensors and actuators

1 Concept and Research Objectives: Beyond the State of the Art

Trees have evolved very robust growth behaviours to respond to changes in their environment and a network of highly sensorized tree roots, branches and leaves to efficiently explore the air and soil volume in order to up-take water and to search solar light. The TREEBOT proposal has three major goals: (1) to abstract and synthesize with robotic artefacts the principles that enable trees to effectively and efficiently explore and adapt to ground and underground environments; (2) to analyze the interaction between the environment and trees using new methodologies to recovery renewable energy by means of the oscillation of the branches and of the leaves during the interaction between the wind and the tree, by means of the accumulation of the solar light in the leaves in a process similar to the photosynthesis and by means of the interaction between tree roots and the soil. (3) to formulate scientifically testable hypotheses and models of some unknown aspects of trees, like the role of local communication among branches during adaptive growth and the combination of rich sensory information to produce collective decisions. The new technologies expected to result from TREEBOT concern new typologies of alternators, energy-efficient actuator systems, chemical and physical micro-sensors, sensor fusion techniques, and

A. Natraj et al. (Eds.): TAROS 2013, LNAI 8069, pp. 28–29, 2014.
DOI: 10.1007/978-3-662-43645-5_3, © Springer-Verlag Berlin Heidelberg 2014

Fig. 1. TREEBOT: Tree Recovering Renewable Energy roBOT

distributed, adaptive control in networked structures with local information and communication capabilities. The TREEBOT robot is conceived as an autonomous system to be deployed in the environment by a network of sensorized and coordinated roots, branches and leaves, steered by osmotic-based actuators [1] that can collectively and adaptively explore the environment, using decentralized and local communication and control. From a long-term perspective, the key characteristics of tree roots, branches and leaves are: (1) Exploration capabilities: advancing in the air in a coordinated way based on the concept of a network of many growing branches. (2) Actuation: exploiting a principle inspired by trees (osmosis) to develop a new generation of actuators, characterized by low power consumption and high force which can bring new insights to longer-term research in robotic actuation solutions. (3) Sensing: Trees embed tens of sensor typologies (e.g. touch, humidity, gravity, ions, etc.) thanks to which the various parts of the tree can collectively elaborate information for implementing complex, adaptive behaviours. They fuse their multiple sensory information at a very low level and with very limited, if any, computing resources. These features represent a source of inspiration for developing innovative sensing systems and new sensor fusion techniques [2]. (4) Collective adaptive behaviour: capabilities inspired from the experimental evidence of coordinated behaviour of the branch apices to maximize air exploration will drive the development of novel methods of collective sensor fusion and decision making in decentralized structures with local computation and simple communication. (5) Recovering renewable energy: the capabilities to recover energy using, inside the branches and leaves, new alternators activated by the interaction with the wind, and using new methodologies to accumulate the solar energy and to up-take water from the soil, will be investigated. These features represent a source of inspiration to research novel renewable energy robotic systems (Fig. 1).

References

1. Burgert, I., Fratzl, P.: Actuation systems in plants as prototypes for bioinspired devices. Phil. Trans. R. Soc. A **367**, 1541–1557 (2009). doi:10.1098/rsta.2009.0003
2. Fratzl, P., Barth, F.G.: Biomaterial systems for mechanosensing and actuation. Nature **462** (2009). doi: 10.1038/nature08603

Coy-B, an Art Robot for Exploring the Ontology of Artificial Creatures

Paul Granjon[✉]

Cardiff School of Art and Design, Cardiff Metropolitan University,
Cardiff, UK
pgranjon@cardiffmet.ac.uk

Abstract. The author is a performance and visual artist whose interest lies in the co-evolution of humans and machines, a subject he explores with self-made machines. The paper describes the aims, method, and context of *Coy-B*, a robot designed for a performance art experiment in human-robot interaction loosely based on Joseph Beuys' *I Like America and America Likes Me* (1974) where the German artist shared a gallery space in New York for several days with a wild coyote. *Coy-B* will feature in a series of durational performances for an autonomous mobile robot and a human, where the robot will take the role occupied by the coyote in Beuys' piece. Diametrically opposed to the coyote who symbolised a natural instinctual dimension, the *Coy-B* robot is a representative of contemporary techno-scientific achievements, a fully artificial creature.

Keywords: Autonomous · Self-motivated · Human-robot interaction · Art · Performance · Beuys · Johnston · Bio-inspired · Machinic life

In *The Allure of Machinic Life* John Johnston originates in the work of W. Grey Walter and other early British cyberneticians the emergence of a 'machinic life' where the development of A-life, AI, robotics and digitisation reaches a critical complexity that strongly undermines the differentiation between living and non-living things, enabling the emergence of complex and adaptive 'liminal machines' [1]. In Johnston view the new 'sciences of the artificial' (artificial intelligence, artificial life, robotics) 'have been able to produce [...] a completely new kind of entity'. The new machines require a new ontology which the *Coy-B* performance proposes to empirically explore. At the core of the project is the construction of an intrinsically motivated learning robot capable of interacting with its environment and with a human in a life-like manner.

Robotic creatures have roamed in art galleries since the 1960s, generating a great variety of responses from audiences and art critics. In the article *Robot and Cyborg Art*, art critic Jack Burnham envisions that the 'cultural tradition with the art object is slowly disappearing and being replaced by what might be called "systems consciousness"'. Based on scientific-technological evolution, 'these new systems prompt us not to look at the skins of objects but at those meaningful relationships within and beyond their visible boundaries' [2]. In many ways, the *Coy-B* performance is an experiment in relationships and exploration of boundaries, with two main aims:

A. Natraj et al. (Eds.): TAROS 2013, LNAI 8069, pp. 30–33, 2014.
DOI: 10.1007/978-3-662-43645-5_4, © Springer-Verlag Berlin Heidelberg 2014

– To generate an experiment in human-robot interaction with a metaphorical dimension that will provide material for reflection, dialogue and analysis on the ontology of artificial creatures.
– To truly experience the unfolding of a relation between an intelligent mobile machine and a human sharing a common territory over a set duration.

An abundant documentation was recorded during Beuys' performance, including black and white cine footage which I used to identify aspects of the interaction between the artist and the animal. The overall volume and weight of the final robot will be similar to those of a coyote, but the body, motor system and appearance will be structurally different. The design will favour functionality and avoid gratuitous zoomorphic aspects such as fur, tail, eyes or ears. The body will likely be based on a wheeled platform equipped with an extending, rotating neck/arm (Fig. 1). A set of jaws mounted on the neck will be fitted with pointed teeth strong enough to pull at things and to provide a bite of adjustable power.

One of the most prominent aspects of the coyote's behaviour is his determined avoidance of physical touch with the human. The machine will implement the basic avoidance drive at the hardware level, through implementation of a hard-wired behavioral layer. The other prominent, instinctual or physiological aspects of the coyote's behaviour are constant awareness and monitoring of the animal's environment, resting and feeding. These will not be hard-wired, but operate at a very high priority level within the software of the machine. The machine will extract information from its environment with a comprehensive array of sensors. A vision and depth sensor will be used to navigate the space, differentiate the human from other features, and locate objects. An array of microphones will allow acoustic source

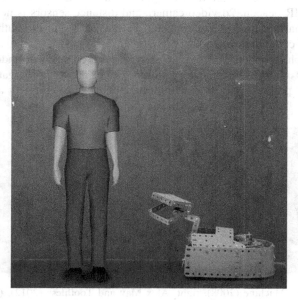

Fig. 1. Preparatory model simulation of Coy-B robot (image by Paul Granjon)

localisation, sound processing and recognition. The robot will detect touch on key parts of its body with contact sensors. It will also feature an olfactory organ and will be able to mark the territory in a similar way to a mammal, Finally the machine might be given the ability to detect some of the human's cerebral activity with a brainwave sensor system, enabling it to react to variations in mood or peaks in cerebral activity triggered by fear or surprise.

Coy-B will be an intrinsically motivated learning robot, able to develop and adjust idiosyncratic behaviours according to its interactions with the environment, including the human. Its behavioral design draws from several examples in cybernetics and robotics sciences. Bottom-up approaches such as the non-representational, hard-wired navigation of W. G. Walter's cybernetic turtles, the adaptive capabilities of R. Brook's behavior-based robots and the emerging fitness aspect of evolutionary robotics are complemented with more computing-heavy functions inspired by MIT's Kismet's synthetic nervous system [3], the curiosity function of Frédéric Kaplan's Aibos [4] and the use of adaptive resonance theory (ART) neural networks for implementing associative memory predictors as seen in the motivated reflex agents developed by Rob Saunders [5]. The starting point of the project was the *Biting Machine*, a simple automaton built by the author in 2008 [6] (Fig. 2). A more complex prototype platform was built in August 2012 in collaboration with artist programmer Alex May in order to test the suitability of a Microsoft Kinect three-dimensional vision sensor for differentiating a human figure from other objects. An on-board Linux machine and an Arduino board were used for processing the data and interfacing with the hardware.

The prototype robot called *Toothless* (Fig. 2), is able to locate and approach a (slow-moving) human in its environment. The experiment demonstrated that the Kinect was not sufficiently effective for human detection when mounted on a mobile platform and too power-hungry for the application. A simpler solution involving a combination of IR beacon, 2d video camera and distance sensors is now considered. A fully functional machine is not expected before 2015 and the author is open to suggestions and collaborations.

As well as philosophers and sociologists who are engaged in understanding the changes brought upon by technical evolution, visual artists investigate the field of techno-scientific progress and its dynamic interaction with humanity and the world,

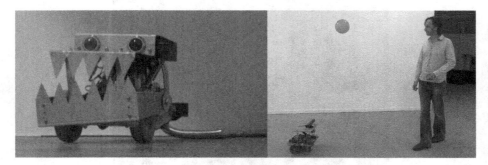

Fig. 2. left: Biting Machine (2008), right: Alex May and Toothless (2012), (photographs by Paul Granjon)

using tools and methods derived from science. Aspects of the *Coy-B* performance can be compared to a scientific experiment in human-robot interaction. A key difference is that the experiment's main aim is not to test a novel techno-scientific development but to produce a empirically based metaphor for the relationship between humans and artificial creatures in the 21st century. Electronic arts specialist Stephen Wilson states that 'In a techno-scientific culture, artistic probing of the world of research is a critical, desperate need. We need people looking at these fields of inquiry from many frames of reference, not just those sanctioned by academia or commerce' [7]. At a time when machinic life becomes a tangible possibility it is my ambition that probing with the *Coy-B* project will uncover ground for reflection and insight relevant for both artistic and scientific contexts.

References

1. Johnston, J.: The Allure of Machinic Life. MIT Press, Boston (2008)
2. Burnham, J.: Robot and Cyborg Art in Beyond Modern Sculpture, pp. 68–77. George Brazillier Inc, New York (1968)
3. Breazeal, C.L.: Designing Social Robots. MIT Press, Boston (2002)
4. Kaplan, F.: Les Machines Apprivoisées Vuibert, Paris (2005)
5. Saunders, R., et al.: Curious places, proactive adaptive built environments. In: Proceedings of AISB'07 Symposium on Agent Societes for Ambient Intelligence, Newcastle, UK (2007)
6. Granjon, P.: http://www.zprod.org/PG/machines/bitingMachine.htm (2013). Accessed 4 September 2013
7. Wilson, S.: Art and Science. Thames & Hudson, London (2010)

Time Preference for Information in Multi-agent Exploration with Limited Communication

Victor Spirin$^{(\boxtimes)}$, Stephen Cameron, and Julian de Hoog

University of Oxford, Oxford, UK
victor.spirin@cs.ox.ac.uk

Abstract. Multi-agent exploration of unknown environments with limited communication is a rapidly emerging area of research with applications including surveying and robotic rescue. Quantifying different approaches is tricky, with different schemes favouring one parameter of the exploration, such as the total time of exploring 90 % of the environment, at the expense of another parameter, like the rate of information update at a base station. In this paper we present a novel approach to this problem, in which agents choose their actions based on the time preference of the base station for information, which it encodes as the desired minimum ratio of base station utility to total agent utility. We then show that our approach performs competitively with existing exploration algorithms while offering additional flexibility, and holds the promise for much improvement regarding incorporation of various information preferences for the base station.

1 Introduction

Multi-agent exploration of environments where communication between agents is limited has been a rapidly emerging area of research in recent years. Several approaches have been suggested, ranging from those that aim to always maintain a communication link between the agents and the base station, ensuring that any new information gets to the base station as soon as it is discovered [1,8], to approaches where agents are allowed to explore the environment without putting any effort to bring the information back to the base station until the exploration effort is over [13], to strategies that lie in between [3,6,9,10]. For simplicity of modelling most of the work has assumed a two-dimensional environment where the aim is to provide information at a single base station, but the approaches have natural extensions to more complicated domains.

Each of those approaches has their strengths and weaknesses; they all work best in different specific scenarios. From looking at their performance, we can see that there is one major factor that allows us to decide which approach to choose in favour of another in any given situation - and that factor is the *time preference* of the base station for information about the environment. The base station may value information obtained sooner higher than information obtained later - for example, when rescuing people from a building that is on fire,

A. Natraj et al. (Eds.): TAROS 2013, LNAI 8069, pp. 34–45, 2014.
DOI: 10.1007/978-3-662-43645-5_5, © Springer-Verlag Berlin Heidelberg 2014

the human rescue team is likely to prefer to have information about *some* of the environment sooner, than have a complete map of the environment later, when it may already be too late. In other scenarios, however, we may prefer to have information about the whole of the environment sooner, and not be too interested in getting a more constant flow of information. We may also have preferences that lie somewhere in between.

In this paper we present a novel approach that attempts to solve the problem by providing the human operator controlling a team of robots with an intuitive way of specifying what their time preference for new information is, and have the agents automatically adapt their cooperative behaviour accordingly. We describe our approach in Sect. 3, and in Sect. 4 we present results obtained in simulation that show how with our approach, agents can adapt their behaviour to changes in the time preference. We also compare our results with other approaches, and analyse how the performance of the system changes as the time preference is changed. In Sect. 5 we discuss ways in which this approach can be further developed and extended.

2   Related Work

Since our approach was developed to primarily deal with scenarios where a team of agents has a goal of exploring an unknown environment, while having limited communication between each other and the base station, in this section we will give an overview of some of the existing approaches to the same problem. Of course, when exploring unknown environments with limited communication, there is always going to be a tradeoff between the speed and frequency of getting new information to the base station, and the time it takes to explore the whole environment. To the best of our knowledge, all existing approaches to the problem optimise for a particular tradeoff between the two. Here we will look at an approach that aims to explore all of the environment as soon as possible, and an approach that aims to minimise the latency in getting new information to the base station while still being able to explore all of the environment in reasonable time. All of the approaches we describe here are built upon the frontier-based exploration framework as described in [11].

2.1   Frontier Exploration

Using frontiers to distribute the exploration task among multiple agents is a common approach to multi-agent exploration. A frontier is a boundary between the explored and the unexplored parts of the map [13]. Agents can then allocate frontiers among themselves by estimating the path costs of themselves and other agents in their vicinity to the frontiers, so as to maximise overall exploration utility. However, a frontier on its own gives no information on the potential information gain of exploring the area that lies behind the frontier, which can lead to inefficient allocations. The concept of *frontier polygons* was introduced in [12] to deal with that problem; a frontier polygon is the polygon that is formed

Fig. 1. An example of a map in simulation with visible frontier polygons. A robot is shown in blue, green areas are 'safe space', while white areas need to be further explored and are bounded by 'frontier polygons', shown in red (Color figure online).

between a frontier and the boundaries of *safe space*. Safe space is comprised of the areas of the map that the agent has been closer to than the full range of his sensor (normally, it is set to be around half the sensor's full range) (Fig. 1).

We can then estimate the potential information gain from a frontier by using the area of the frontier polygon as the estimate. This can be especially important for our proposed approach, as having an estimate of potential information gain from a frontier is crucial when deciding whether to continue to explore the frontier, or to return to the base station; but it can enhance the frontier exploration approach in general by allowing agents to better allocate frontiers among themselves. In particular, we can control the exploration behaviour by setting the constant n in the equation which estimates the potential information gain when calculating the utility of a frontier

$$U(p_i) = A(p_i)/C^n(p_i) \qquad (1)$$

where $U(p_i)$ is the utility of the frontier polygon p_i, $A(p_i)$ is the area of the frontier polygon, and $C(p_i)$ is the cost of the path to the frontier polygon's centre. Low values of n mean that agents will favour the exploration of larger frontier polygons, such as corridors or halls, and higher values of n mean that the agents will more often tend to examine nearby smaller areas, such as rooms [11]. In most experiments in this paper we used the value $n = 2$, which tends to provide a good balance between the two in practice [5].

2.2 Return When Done

This strategy is the most straightforward, and is the direct application of frontier exploration. All agents continue to explore the environment without having to return to the base station at regular intervals, and only once all the frontiers have been explored do they return to base. In practice, this often ensures that the whole of the environment is explored in shortest time, however that may not always be the case due to the fact that only returning to the base station after the exploration is finished means that not only is the base station not getting

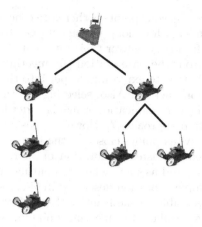

Fig. 2. An example of the agent hierarchy tree. The base station is at the root of the tree; relays are shown in red, and explorers are shown in blue (Color figure online).

frequent updates on the exploration effort, but information sharing between agents may also be reduced, which may lead to the same areas getting explored more than once.

2.3 Role-Based Exploration

With this approach, agents are divided into two groups: explorers and relays. The task of the explorers is to explore as much of the environment as possible and return it back to pre-agreed rendezvous points at pre-arranged times. The task of relays is to communicate information between rendezvous points (where they communicate with explorers or other relays), or between a rendezvous point and the base station.

Here, teams of agents have a rigid hierarchy tree which is manually selected before the agents enter the environment; however, agents may switch positions in the tree throughout their mission. However, the shape of the tree itself does not change (Fig. 2).

When an explorer meets its parent relay, they exchange information about the environment, ensuring that they both have the same knowledge about it. Then, the explorer suggests a rendezvous point, normally near a frontier that it plans to explore next, and a fallback rendezvous point in case the primary one cannot be reached, which is especially useful in dynamic environments (if the explorer and the relay happen to meet before they both reach their meeting point, they act as if they have reached it and proceed to exchange information and to replan their next meeting). Because the explorer and the relay share the same information, the explorer can predict how long it will take for the relay to get back to its own parent, and then travel to the new meeting point, and can therefore decide when it should stop exploring the new frontiers and get back to the agreed rendezvous point [6].

The selection of the rendezvous points is therefore crucial for the performance of the algorithm, as it affects how long the agents get to spend exploring the environment, and how often they deliver their information back to the base station. The further away from the base station the meeting points are, the more biased the exploration effort is to exploring deeper into the environment instead of relaying the information back [4]. Also, selecting meeting points at junctions and in corridors where the communication range is wider leads to increased performance of the exploration approach [7]. However, this approach does not allow to easily move the rendezvous points closer to the base station/deeper into the environment to favour either faster exploration or more frequent communication with the base station, and as such, when comparing our proposed approach against it, we used the implementation described in [4], where rendezvous points are selected as close as possible to the frontier that the explorer plans to investigate next, while trying to place the meeting point in a junction or an open space, which maximises communication range.

3   Proposed Method

3.1   Overview

When a team of autonomous agents is sent to explore an unknown environment where communication is limited, resources can generally be allocated in two different ways: collecting new information about the environment; and keeping the base station (and other agents) updated of the current progress made by the agents. Usually, an agent would have to allocate its resources to only one of the above tasks at any given time. As a result, the human operator commanding a team of robots would need to somehow specify how the team resources should be allocated between the two tasks.

There are several straightforward ways of doing that. The operator may want to specify maximum latency of information propagation in the team - that is, the maximum time between information exchanges of an agent with the base station. While at first it seems like a useful way of specifying the desired team behaviour and resource allocation, there are several problems with it. How can the operator know apriori what maximum latency is appropriate for a given scenario? It is likely to be desirable for the latency to be very low at the start of the exploration, while the agents are exploring parts of the environment close to the base station, and to increase as the agents progress deeper into the environment. But how should the latency increase, to ensure that agents do not waste their resources communicating with the base station when no new information has been discovered to communicate, and that all of the environment gets explored? The operator may not have the answers to these questions without some information about the environment, and by the time that information is obtained it could be too late to set the behaviour of the team.

Another way of specifying the desired team behaviour could be by setting the rate of information update at the base station to be a function of utility gathered by the agents that has not yet been delivered to the base station. For example,

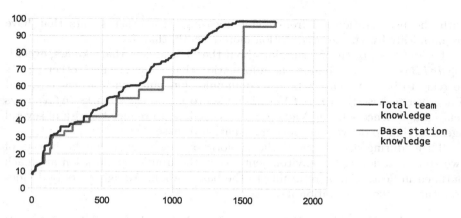

Fig. 3. A graph showing combined team knowledge and base station knowledge changes with time over the course of a simulated exploration mission using role-based exploration strategy.

if we are only interested in building a map of the environment, the operator might specify the desired team behaviour by setting the target minimum ratio of information about the environment known by the base station, to the amount of information known by all the agents combined (Fig. 3).

That single parameter, the target ratio, would then be a real number ranging from 0 to 1. Setting it to 0 would result in greedy exploration behaviour, where all of the team resources are used to gather new information; while setting it to 1 would ensure maximum connectivity to the base station. By setting the ratio to a value between 0 and 1, it is possible for the operator to specify how they want the team resources to be allocated between discovering new information and maintaining communication with the base station in a meaningful way. Of course, as the agents in the team are operating with limited information about the state of other agents in the team and about the environment, their behaviour will not match the target ratio precisely, but as we show in the section describing our simulation results, even with the team using a simple heuristic that crudely approximates the resulting ratio, it provides a promising way of specifying the desired team behaviour, with the team behaviour changing accordingly with the target ratio changes.

3.2   Implementation

For this paper, we used a simple implementation of the approach described above. Before the start of the exploration, the user sets a target information ratio $targetInfoRatio \in [0; 1)$, which gets propagated to all the agents in the team before the exploration begins.

For each agent i, let $infBase_i$ be the information i believes the base station to have at the current time. $infBase_i$ is obtained directly from communicating with the base station, or from communicating with other agents that have communicated with the base station more recently than i.

Let $infNew_i$ be the information about the environment that i knows, excluding $infBase_i$, and excluding the information that i has given to other agents to relay to base. When i gets into communication range with an agent j, and j is closer to base than i, $infNew_i$ is added to $infNew_j$, i marks $infNew_i$ as relayed and hence sets $infNew_i$ to $\varnothing$. This is done to reduce the risk of several agents trying to deliver the same information to base.

Then, during each cycle of the exploration, each agent can be in one of two states: exploring the environment (using the frontier exploration approach outlined in Sect. 2.1), or returning to the base station. An agent i only decides to return to the base station if

$$|infBase_i|/(|infBase_i| + |infNew_i|) < targetInfoRatio, \qquad (2)$$

where $|infBase_i|$ and $|infNew_i|$ are the utilities of $infBase_i$ and $infNew_i$ accordingly. Otherwise agent i continues to explore the environment.

4 Simulation Results

4.1 Simulator

We used the MRESim simulator [6] to evaluate our approach. MRESim simulates sensor data, communication, movements and collisions of multiple agents in a 2D environment consisting of free space and obstacles. The actions performed by the simulator at each time step are shown in Algorithm 1.

Actions taken at each time step by MRESim

```
foreach agent do
   nextLoc = requestDesiredLocation(agent);
   if isValid(nextLoc) then
      move(agent, nextLoc);
      sensorData = simulateSensorData(agent, nextLoc);
      sendData(agent,sensorData);
   end
end
foreach agent do
   foreach agent2, agent2 != agent do
      if isInRange(agent, agent2) then
         communicateData(agent, agent2);
      end
   end
end
updateGUI();
```

The simulator assumes perfect localisation and sensor data. While this assumption is unrealistic, it still allows us to get a good idea about how different agent cooperation strategies perform against each other and see what their strengths and limitations are.

For all of the experiments, we used a standard path loss communication model with a wall attenuation factor [2].

4.2 Set Up

We used 4 different maps for our experiments, as shown in Fig. 4: a small room-based map that consists of corridors and a number of rooms to be explored; a cluttered environment; a large "library" map, consisting of many rooms and corridors to be explored; and a large outdoor environment. We initially did 4 runs on each of the maps using 6 different exploration strategies, a total of 96 runs: using our approach with target ratios of 0.95, 0.90, 0.75, 0.50 and 0.30 and using role-based exploration. For each of the runs on the first and second maps, we had a total of 4 agents navigating the environment; for the library map and for the outdoors map we used a total of 8 agents. In the runs where we used role-based exploration, half of those agents were assigned the roles of "relays", each of them relaying information for one other agent.

The results obtained from doing the runs on the 4 maps appeared to be similar to each other, so we decided to focus on running a larger number of

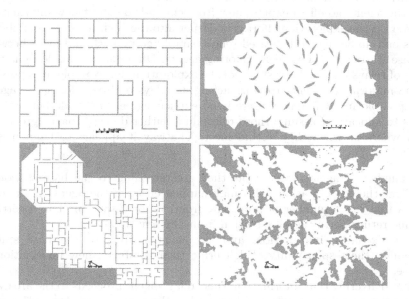

Fig. 4. 4 maps used in the simulations: rooms, cluttered, library and outdoors (starting from top left, clockwise)

simulations on the "rooms" map. We ended up doing 48 runs of each type on the "rooms" map, for a total of 288 runs. We present the results of those simulations below.

4.3   Results

The results of the simulation runs on the "rooms" map are shown in Fig. 5.

As we can see, our approach with a target ratio of 0.95 manages to explore 98 % of the environment faster than role-based exploration, while the average ratio of total agent knowledge to base station knowledge is very similar to that of role-based exploration. This was an expected result, as our approach does not designate a number of agents to be used only as relays throughout the simulation runs, which should result in a more efficient use of resources to reach a particular target ratio.

Another interesting observation is that as we decrease the target ratio from 0.95 to 0.3, reducing the average and minimum actual observed ratios between total agent and base station utilities accordingly, it has less and less of an effect on increasing the overall speed of exploration.

4.4   Emergent Behaviour

During our simulations, we found that with higher values for the target ratio and with higher number of agents (4 or more), at the start of the exploration, all of the agents go off to explore new frontiers and to collect new information. However, as the exploration effort gets deeper into the environment, a number of agents end up acting as dedicated relays, simply going back and forth between the base station and the other exploring agents. Often, agents would behave as chains of relays - at the later stages of the exploration, for example, it is possible for the majority of agents to start acting as relays and only for a few agents to keep exploring. Of course, there is no explicit agreement made between the agents to allocate or assume those roles, and neither do they make agreements about where or when they should meet. The way it appears to happen is as follows:

1. A number of agents meet while they are returning to the base station to deliver their information. This may happen either if the agents flock to an area that has a number of promising frontiers, or if they meet in a corridor while returning to the base station from different areas.
2. They exchange information, and the agent nearest to the base station assumes the responsibility of delivering their combined new information to base.
3. After delivering that information, this "relay" agent proceeds to the area with the most promising frontiers. Since he has the same knowledge as the other agent had at the time of meeting, he will likely go to the same "promising" frontier as the other agent did, meeting him - or another agent relaying for him and returning to base - on the way.

Mean and std. deviation of the number of cycles to explore
98% of the environment and deliver the information
to the base station

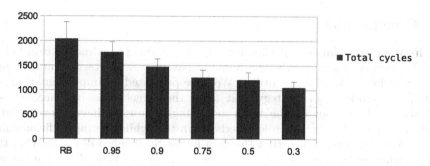

Mean and std. deviation of the average ratio between
total agent and base station utility
during a simulation run

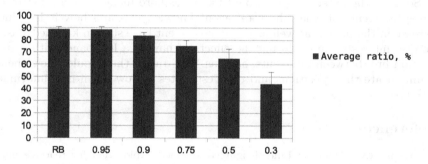

Mean and std. deviation of the minimum ratio between
total agent and base station utility
during a simulation run

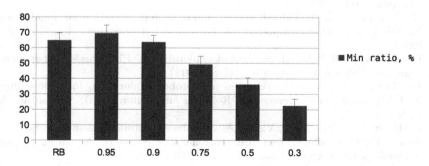

Fig. 5. Graphs showing mean and standard deviation values obtained from the simulation runs using role-based exploration, and our approach using target ratios of 0.95, 0.9, 0.75, 0.5 and 0.3.

The cycle above results in the emergent behaviour of chains of relays being formed as they are needed to keep delivering information to the base station at a frequency that is appropriate for the target ratio set up by the operator.

5   Conclusions

We have shown a simple, but effective way of specifying the desired team behaviour by means of setting a single numeric parameter, the target ratio of base station utility to total agent utility. We have presented an implementation of a distributed exploration strategy that takes the target ratio into account and adjusts the behaviour of the team accordingly, and we have shown that it performs competitively with role-based exploration while offering additional flexibility. We have also shown that the gain in the total speed of exploration that we get when we reduce the target ratio seems to get a lot smaller than the corresponding increase in the cost (reduced rate of base station updates) as the target ratio gets closer to 0 - which may be useful when deciding which target ratio should be used for any particular situation.

Some of the extensions we would like to explore include having better estimates by agents of what the total agent knowledge is and how it is going to increase in the future, as well as what the current base station knowledge is. We are also interested in exploring the effects of having a low-bandwidth communication link between all agents, such as VHF radio, that would allow them to communicate their positions and their estimates of how much new information they possess.

References

1. Arkin, R.C., Diaz, J.: Line-of-sight constrained exploration for reactive multiagent robotic teams. In: 7th International Workshop on Advanced Motion Control. Proceedings (Cat. No.02TH8623), pp. 455–461. IEEE (2002)
2. Bahl, P., Padmanabhan, V.N.: RADAR: an in-building RF-based user location and tracking system. In: Proceedings of the IEEE Infocom 2000, vol. 2, pp. 775–784. Tel-Aviv, Israel (2000)
3. Balch, T., Powers, M.: Value-based communication preservation for mobile robots. In: 7th International Symposium on Distributed Autonomous Robotic Systems (2004)
4. De Hoogm, J.: Role-based multi-robot exploration. D.Phil Thesis, Department of Computer Science, University of Oxford (2011)
5. De Hoog, J.: Using mobile relays in multi-robot exploration. In: Proceedings of ACRA (Australian Conference on Robotics and Automation). Melbourne, Australia (2011)
6. De Hoog, J., Cameron, S., Visser, A.: Role-based autonomous multi-robot exploration. In: Proceedings of the International Conference on Advanced Cognitive Technologies and Applications (COGNITIVE) (2009)
7. De Hoog, J., Cameron, S., Visser, A.:. Selection of rendezvous points for multirobot exploration in dynamic environments. In: Proceedings of AAMAS (Workshop on Agents in Realtime and Dynamic Environments, International Conference on Autonomous Agents and Multi-Agent Systems). Toronto, Canada (2010)

8. Howard, A., Mataric, M.J., Sukhatme, G.S.: An incremental deployment algorithm for mobile robot teams. In: IEEE/RSJ International Conference on Intelligent Robots and System, vol. 3, pp. 2849–2854. IEEE (2002)
9. Mosteo, A.R., Montano, L., Lagoudakis, M.G.: Multi-robot routing under limited communication range. In: 2008 IEEE International Conference on Robotics and Automation, pp. 1531–1536. IEEE (2008)
10. Vazquez, J., Malcolm, C.: Distributed multirobot exploration maintaining a mobile network. In: 2nd International IEEE Conference on 'Intelligent Systems'. Proceedings (IEEE Cat. No.04EX791), pp. 113–118. IEEE (2004)
11. Visser, A., Slamet, B.: Balancing the information gain against the movement cost for multi-robot frontier exploration. In: European Robotics Symposium (2008)
12. Visser, A., Van Ittersum, M., Jaime, L.A.G., Stancu, L.A.: Beyond frontier exploration. In: Proceedings of the 11th Robocup International Symposium (2007)
13. Yamauchi, B.: Frontier-based exploration using multiple robots. In: AGENTS '98: Proceedings of the Second International Conference on Autonomous Agents, pp. 47–53. ACM, New York (1998)

Discrimination of Social Tactile Gestures Using Biomimetic Skin

Hector Barron-Gonzalez(✉) and Tony Prescott

Sheffield Centre of Robotics, University of Sheffield,
Western Bank, Sheffield S10 2TN, UK
{hector.barron,t.j.prescott}@sheffield.ac.uk

Keywords: Social robotics · HRI · Tactile gesture recognition

The implementation of novel tactile sensors has yielded original mechanisms for human-robot interaction that support the interpretation of complex social scenarios. For instance, the recognition of social tactile gestures is an important requirement in the design of robot companions because it enables the android to engage with human drives. We are interested on implementing such a functionality upon the biomimetic skin of the iCub android [1].

The iCub robot has been provided with a capacitive-based artificial skin in order to augment its sensorial capacities. The robot has 384 tactile sensors (*taxels*) that acquire force data during contact over arms and torso, with a sample rate of $50\,Hz$. Additional sensors are located in fingertips, whose hyperacuity allows shape recognition.

Other recent studies have also been focused on tactile gesture recognition using different types of artificial skin. The work in [2] utilizes photoreceptors to extract information about force, position and frequency. Models are generated using *Support Vector Machines* (SVM) and *k-Nearest Neighbour* to classify different affective gestures. The classification rate obtained with only tactile information was 67 %, requiring of visual information to increase reliability up to 90.5 %. On the other hand, Silvera et.al. [3] employed an impedance-based skin to recognize gestures using spatio-temporal features such as pressure, contact area and duration. A boosting-based classifier produced a classification rate up to 90 %, using tactile data generated from only one participant. The classification accuracy decreased up to 70 % when tactile data was generated by several participants.

In order to evaluate the feasibility of using the artificial skin for gesture discrimination, eight tactile gestures with high social content (*C1 = poke, C2 = caress, C3 = grab, C4 = stroke, C5 = tickle, C6 = pat, C7 = slap* and *C8 = pinch*) were selected from the state of art [3], considering that these gestures are also

H. Barron-Gonzalez—The research leading to these results has received funding from the European Union Seventh Framework Programme FP7/2007-2013, under grant agreement No 270490-EFAA.

A. Natraj et al. (Eds.): TAROS 2013, LNAI 8069, pp. 46–48, 2014.
DOI: 10.1007/978-3-662-43645-5_6, © Springer-Verlag Berlin Heidelberg 2014

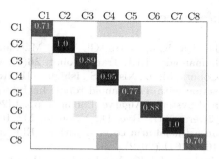

a. Tactile gesture over the forearm. b. Confusion matrix.

Fig. 1. Tactile gesture discrimination

well discriminated by the human being. Several participants were instructed to perform similarly the tactile gestures over the skin (see Fig. 1a).

The extraction of tactile features involved three steps. First, contact areas for each frame were computed using hierarchical clustering with cutoff at $2.5\,cm$. Second, static tactile features such as contact force, area and position were extracted from each contact region. And third, a set of temporal features were computed for each gestures, such as magnitude of displacement and duration. Finally, we gathered a set of 301 instances.

The classification was achieved using an ensemble of classifiers based on Least Squares Support Vector Machines [4]. Each classifier was trained to recognize one type of gesture. The type of each instance was determined by the classifier with the highest ranking. The classification rate is illustrated in Fig. 1b by the confusion matrix with a k-fold cross validation ($k = 4$). The total classification rate was of 86.2 %, which is comparable with respect to the state of art. Other experiments involving less classes (e.g. caress, poke, grab and stroke) generated a classification rate up to 97 %.

On the contrary to other works that require either additional information about the contact body part or visual support, this is the first work that demonstrates successfully that the iCub skin provides sufficient information to discriminate tactile gestures. The extracted tactile features produced a consistent classification, although it is not possible to define simple rules directly from them. These features are easily identified by the human being because they are programmer-based, but it might be difficult to provide an intuitive description of each tactile gesture. Our future work involves exploring other more bioinspired strategies to generate more grounded tactile features.

References

1. Dahiya, R.S., Metta, G., Valle, M., Sandini, G.: Tactile sensing from humans to humanoids. IEEE Trans. Robot. **26**(1), 1–20 (2010)
2. Cooney, M.D. , Nishio, S., Ishiguro, H.: Recognizing affection for a touch-based inter-action with a humanoid robot. In: International Conference on Intelligent Robots and Systems, Algarve, Portugal, pp. 1420–1427 (2012)
3. Silveram, D., Rye, D., Velonaki, M.: Interpretation of the modality of touch on an artificial arm covered with an EIT-based sensitive skin. Int. J. Robot. Res **31**, 1627–1641 (2012)
4. Suykens, J.A.K., Vandewalle, J.: Least squares support vector machine classifiers. Neural Process. Lett. **9**(3), 293–300 (1999)

Bio-inspired and Aerial Robotics

A Personal Robotic Flying Machine with Vertical Takeoff Controlled by the Human Body Movements

Vittorio Cipolla[1,2], Aldo Frediani[1,2], Rezia Molfino[3],
Giovanni Gerardo Muscolo[3,4(✉)], Fabrizio Oliviero[1],
Domenec Puig[5], Carmine Tommaso Recchiuto[6], Emanuele Rizzo[2],
Agusti Solanas[5], and Paul Stewart[7]

[1] Aerospace Division, Department of Civil and Industrial Engineering,
University of Pisa, Pisa, Italy
[2] SkyBox Engineering s.r.l, Pisa, Italy
[3] PMAR lab, Department of Applied Mechanics and Machine Design,
Scuola Politecnica, University of Genova, Genoa, Italy
info@humanot.it
[4] Creative and Visionary Design lab, Humanot s.r.l, Prato, Italy
[5] University Rovira i Virgili, Tarragona, Spain
[6] Electro-Informatic lab, Humanot s.r.l, Prato, Italy
[7] The University of Lincoln, Lincoln, UK

Abstract. We propose a cooperative research project aimed at designing and prototyping a new generation of personal flying robotic platform controlled by movements of the human body using a symbiotic human-robot-flight machine interaction. Motors with ducted fun propulsion and power supply, and a VSLAM system will be integrated in the final flight machine with short and vertical takeoff and landing capability and composite (or light alloy) airframe structure for low speed and low altitude flight. In the project, we will also develop a flight simulator to test the interaction between the flying machine and the human body movements. In this first step, for human safety, the flying machine will be controlled by an autopilot colligated in a closed-loop control with the simulator.

Keywords: Autonomous vehicles · Personal robotics · Human-robot interaction and interfaces · Aerial robots · VSLAM · Navigation

1 Concept and Research Objectives: Beyond the State of the Art

Many personal flying machines have been conceived (e.g. Paragliding, Deltaplane, Jet pack, Rocket belt, Backpack helicopter, Wing suit, Flying Wingpack, etc.). All these solutions have been only developed for experts in aircraft piloting or for athletes able to face great challenges. As a matter of fact, the safety problems of these machines are very serious. The basic idea of the present project is to advance the research in the opposite direction: to allow everyone to fly quietly, slowly, safely and simply

A. Natraj et al. (Eds.): TAROS 2013, LNAI 8069, pp. 51–52, 2014.
DOI: 10.1007/978-3-662-43645-5_7, © Springer-Verlag Berlin Heidelberg 2014

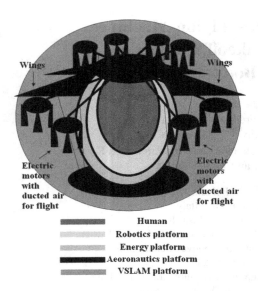

Human
Robotics platform
Energy platform
Aeoronautics platform
VSLAM platform

on personal machines designed to become a common way of transport. The perspectives for future applications include also the possibility of achieving a fully automatic vehicle to be used to rescue people during catastrophes or, in practice, a general purpose flying robot. The main novelty of this research is that the personal robotic flying machine combines the properties of a helicopter and a fixed wing aircraft with high efficiency. The properties of vertical take-off and landing as a helicopter and horizontal cruise as a fixed wing aircraft are made possible by a proper aerodynamic and propulsion system allowing us to modify the configuration of the vehicle. The concept of the proposed flying machine is totally innovative in comparison to the previously enumerated examples. In particular, the flying machine will consist of the following main elements: an upper lifting system with tilting wings, a propulsion system and a pilot housing requiring an unprecedented 120 kW power output peak to the fan(s) with an estimated mass of 15 kg. The ducted fan is the preferred technology for this application due to its ability to have more, and shorter blades and reduced velocities that result in lowered aerodynamic losses at the blade tips, allowing quieter and safer operation with thrust vector capability. The whole system will be powered with Li-Ion polymer and Li-Iron phosphate chemistries. These battery types require tight control of charging voltages and monitoring of the cells that make up a vehicle battery pack. To improve the safety during landing, a VSLAM [2] system, endowed with advanced cameras, will map the environment so as to detect obstacles and will analyze the characteristics of the surface in which the robot is willing to land. Finally a screw theory approach [1] will be used in order to design the tilting system on the flying machine and to control the machine oscillations generated by aerodynamic forces and human movements. However in this first implementation, an autopilot will provide the input to control the tilting wings of the flying machine in relation to the human movements captured on the simulator. Anyway, safety issues will be already taken into account: multiple aerodynamics controls and fractioned propulsion will minimize the risks.

References

1. Ball, R.S.: The Theory of Screws: A study in the dynamics of a rigid body. Hodges, Foster & Co., Dublin (1876)
2. Eade, E., Drummond, T.: Scalable monocular SLAM. In: CVPR, vol. 1, pp. 469–476 (2006)

Developing the Cerebellar Chip as a General Control Module for Autonomous Systems

Emma D. Wilson[1][(✉)], Sean R. Anderson[1], Tareq Assaf[2],
Jonathan M. Rossiter[2], Martin J. Pearson[2], and John Porrill[1]

[1] Sheffield Centre for Robotics (SCentRo), University of Sheffield, Sheffield, UK
e.wilson@sheffield.ac.uk
[2] Bristol Robotic Laboratory (BRL), University of the West of England
and University of Bristol, Bristol, UK

Abstract. Biological systems have evolved robust, adaptive control strategies to deal with a wide range of control tasks in time varying systems and environments. The cerebellum is the brain structure particularly associated with the control of skilled movements, the advantageous properties of the cerebellum can be exploited for robotic control applications. In this contribution we present a bioinspired cerebellar control algorithm. We extend the existing cerebellar inspired adaptive filter control algorithm, previously applied to plants of specific order, to the control of general n^{th} order plants. This is done by augmenting the existing cerebellar algorithm with a reference model, a technique used in model reference adaptive control. This augmented cerebellar controller is applied successfully to the simulated control of a general plant, and to the real time control of a dielectric electroactive polymer actuator. This augmented biomimetic control strategy has promise for the control of human-centred robots operating in unstructured environments.

1 Introduction

A central goal of bioinspired design is to translate and exploit the advantageous features of biological systems into their engineered counterparts [1]. Biological systems have evolved highly successful solutions that are robust and can handle a hugely diverse range of control tasks [2,3]. Hence bioinspired adaptive control strategies may provide solutions to the new control challenges posed by using robots in unstructured, human environments [4].

In humans, the cerebellum plays an important role in the fine-tuning of motor control tasks [5,6]. An intriguing feature of the cerebellum is the repeating architecture [7–9], implying that the same architecture and algorithm can be used in the control of a myriad of tasks [10,11]. In the cerebellum [11], individual cerebellar microcircuits have similar internal structure, but unique external connections. Hence an exciting possibility for the future control of autonomous systems is the development of a 'cerebellar chip' that can be plugged into existing control systems, augmenting performance to fine-tune the control of any task.

A. Natraj et al. (Eds.): TAROS 2013, LNAI 8069, pp. 53–63, 2014.
DOI: 10.1007/978-3-662-43645-5_8, © Springer-Verlag Berlin Heidelberg 2014

The cerebellar microcircuit has similar structure to an adaptive filter [6,12]. The adaptive filter based cerebellar algorithm has been successfully applied to other robotic control applications: the control of a simulated robot arm [13], and of a robot eye actuated by pneumatic artificial muscles [3]. These applications have been to systems in which the choice of position or velocity control (inspired by biology) lead to a plant with equal numbers of poles and zeros. However, to provide a modular, generally applicable solution that could be used in human centred robotics in a wider range of tasks, the algorithm must be extended to be applicable to a range of plants of more general orders.

The aim of this paper is to build on biological models of cerebellar function in specific control tasks, such as the vestibulo-ocular reflex, and extend them to develop an adaptive control module that can be used in a wider rage of contexts. The approach we take here is to describe the performance required from the cerebellar controller by using ideas from model reference adaptive control (MRAC) [14].

In this contribution a 'cerebellar chip' controller that augments the cerebellar adaptive filter algorithm with MRAC is developed. We demonstrate the potential of the cerebellar chip algorithm by applying it to the real time displacement control of a soft actuator. We show that the cerebellar chip is able to track the actuators displacement response accurately, despite changes to the dynamics over time.

The paper is organised as follows. The original cerebellar algorithm is described in Sect. 2. Section 3 develops the novel cerebellar and MRAC hybrid controller. Details of the experiment for the real time control of a DEA are given in Sect. 4. Section 5 presents the results, these are discussed in Sect. 6.

2   Biological Background

The structure of the cerebellar microcircuit is uniform, suggesting that there is a single 'cerebellar algorithm' with a general signal processing capacity. This gives rise to the chip metaphor of cerebellar organisation (see Fig. 1A) [11]. The Vestibular Ocular Reflex (VOR), the reflex that stabilises visual images on the retina during head movements [5], is the experimental model that has been used previously to investigate cerebellar function [3,18]. In this section the cerebellar algorithm, previously developed in the context of the VOR, is described.

It has been shown that the cerebellar microcircuit can be mapped onto an adaptive filter structure [6], this structure is shown in Fig. 1B. In the adaptive filter model initial processing by the granule cells is modelled as bank of filters,

$$p_j = G_j(\mathbf{u}) \qquad (1)$$

where G_j is the transfer function of the jth filter, $\mathbf{u}$ the motor command signal, and p_j the filter output. For commonly used filters and input statistics these filter outputs are highly correlated, therefore a matrix Q is used to decorrelate these signals and speed up learning, giving the signals q_i. This decorrelation

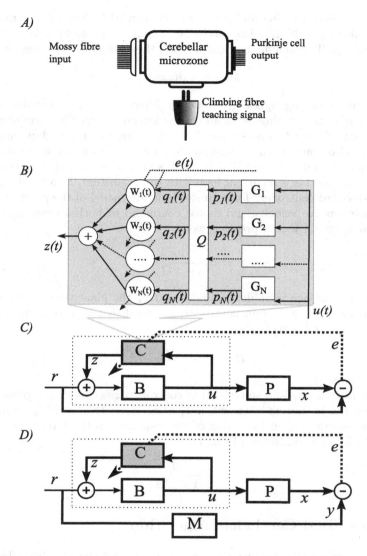

Fig. 1. Cerebellar based control schemes. (A) The 'cerebellar chip'. Each cerebellar microzone has a similar internal organisation, but its own set of connections. (B) Interpretation of cerebellar microcircuit as adaptive filter. Processing by the granular layer is modelled as a filter bank $p_j = G_j(\mathbf{u})$, followed by signal decorrelation using a matrix Q. The Purkinje cell output z is a weighted sum of the inputs, q_j, where the filter weights are adjusted according to the covariance learning rule (C) Control system based on linearised VOR (ignoring the vestibular system). The motor command u is generated by a fixed brainstem element B, and adaptive cerebellar filter C connected in a recurrent loop. (D) Adapted control for generalised plant model. The control diagram has been extended to include a reference model M specifying how the controlled system should behave.

stage is key to fast learning and may be implemented biologically by plasticity in the granular layer [15,16]. The cerebellar output z is given by a weighted sum of the signals q_i. The weights are learnt using the covariance learning rule

$$\delta w_{ij} = -\beta \langle e_i q_j \rangle \tag{2}$$

where e_i is the teaching signal, given as the difference between the desired and actual output at the ith time step, and β the learning rate. The form of Eq. (2) is equivalent to the least mean square (LMS) learning rule from adaptive control theory [17]. This learning rule approximately minimises the sensory error.

A linearised model of cerebellar control is shown in Fig. 1B. The cerebellar filter C is embedded in a recurrent loop with the brainstem, B, a fixed, approximate feed-forward controller. The transfer function model of the plant P for the relation between eye velocity and motor command in both an eye model [18], and a robotic eye [3] was in previous work,

$$P = \frac{ks}{s + 1/T_p} \tag{3}$$

where s is the Laplace variable, k a gain, and T_p the plant time constant. The corresponding fixed brainstem controller was modelled as a combination of a leaky integrator and scalar gain

$$B = g_s + \frac{g_i}{s + 1/T_i} \tag{4}$$

The brainstem provides some level of control (perfect plant compensation is achieved by the brainstem alone when $T_i = \infty$, $g_s = 1$, $g_i = 1/T_p$), which the cerebellum improves upon by tuning of the response via the filter weights. The overall feed-forward controller of Fig. 1B is

$$K = \frac{B}{1 - CB} \tag{5}$$

3   Generalised Cerebellar Algorithm

This section presents the modifications made to the cerebellar algorithm described above to make it generally applicable to control plants of any order. Firstly, the control system modifications are described, and then the corresponding modifications to the learning rule.

When testing the described cerebellar control algorithm, previously the controlled plant has always had equal poles and zeros (velocity control is used in the VOR which adds a zero to the plant transfer function, and ensures the transfer function of the controlled plant has equal poles and zeros). This simplifies the inverse feed-forward control of the plant as the inverted plant model (controller) also has equal poles and zeros so is proper. In this section we describe how the cerebellar algorithm can be extended to control general plants of the form

$$P = \frac{\prod_{i=1}^{m} s + b_i}{\prod_{j=1}^{n} s + a_j} \tag{6}$$

where $n \geq m$. If $n > m$ then the inverse compensator being learnt will have an improper transfer function $K = P^{-1}$. Calculating the output of a improper transfer function requires differentiators, is difficult to realise, and leads to noisy high frequency performance; it can also lead instabilities in the learning rule [19]. To ensure that the desired controller is well-behaved a reference model is used which effectively specifies a realistic response for the controlled plant [14,20]. We will use a reference model of the form

$$M = \frac{1}{(\tau s + 1)^{m-n}} \tag{7}$$

where M specifies the desired response of the controlled system, τ is a time constant, and $n - m$ the reference model order. The time constant τ can be chosen to ensure response roll-off at frequencies above the operating range of the plant, hence the overall response is not effected much [20]. With this reference model the required controller is $K = P^{-1}M$, giving a proper controller of the form

$$K = \frac{\prod_{j=1}^{n} s + a_j}{\prod_{i=1}^{m} (s + b_i)(\tau s + 1)^{n-m}} \tag{8}$$

When the controlled system has no reference model and the form of Fig. 1C the recurrent loop means that the error signal itself is a suitable training signal [13]. With the inclusion of a reference model the error signal used in the LMS rule must first be filtered through a copy of M (Fig. 1D) to ensure stable learning. In biological systems such pre-filtering is sometimes called an eligibility trace.

4   Experimental Setup

The cerebellar algorithm with the MRAC extension was applied to the experimental control of a one degree-of-freedom dielectric electroactive polymer actuator (DEA). DEAs respond with large displacements to suitable electrical stimuli [21], they are compliant actuators with high energy density, large strain capability, low mass, a relatively fast response and the capability for self-sensing [22,23].

This section details the experimental set up used to test one degree-of-freedom control of a DEA. Our DEA is comprised of a thin passive elastomer film, sandwiched between two compliant electrodes. In response to an applied voltage the electrodes squeeze the film in the thickness direction, resulting in biaxial expansion. In order to constrain the control signal to be 1-D, a spherical load was placed at the centre of a circular DEA and the motion of this in the vertical plane (i.e. vertical displacement) measured.

The DEA was comprised of acrylic elastomer 3 m VHB 4905, with an initial thickness of 0.5 mm. This elastomer was pre-stretched biaxially by 350 % (where 100 % is the unstretched length) prior to being fixed on a rigid Perspex frame, with inner and outer diameters of 80 mm and 120 mm respectively. A circular conductive layer of carbon grease (MG Chemicals) was brushed on both sides of the VHB membrane and a 3 g sphere was placed on the actuator.

For control hardware, a CompactRio (CRIO-9014, National Instruments) platform, with input module NI-9144 (National Instruments) and output module NI-9264 (National Instruments) was used in combination with a host laptop computer. Input and output signals were sampled simultaneously at 50 Hz. A laser displacement sensor (Keyence LK-G152) was used to measure the vertical movement of the mass sitting on a circular DEA. This signal was supplied to the input module of the CRio. From the output module of the CRio, voltages in the range 1.1 V–3.75 V were passed through a potentiostat (HA-151A HD Hokuto Denko) and amplified (EMCO F-121 high voltage module) with ratio of 15 V:12 kV and applied to the DEA. During control experiments adaptive feed-forward control was used to adjust the voltage input to the DEA to track a desired reference signal.

5   Results

This section presents results from combining MRAC with the cerebellar algorithm to control plants with more poles than zeros. Initially the effect of the reference model parameters on the response, and the control of a simulated plant is considered. Next the algorithm is applied to the real time control of a DEA.

5.1   Behaviour of Reference Model

The reference model M determines the behaviour of the controlled system [20]. The effect of changing the model order, and the time constant τ on the response of the reference model is shown in Fig. 2. The smaller τ, the higher frequency at which the response starts to roll off. The greater the model order the faster the roll off, and larger the phase shift. If the value of $1/\tau$ is large in comparison to the maximum frequency in the signal the original signal is not altered much by the inclusion of a reference model.

5.2   Adaptive Cerebellar Control of Simulated Plant Using MRAC

The performance of the original cerebellar algorithm, and that augmented with a reference model, were compared when applied to the control of a simulated plant.

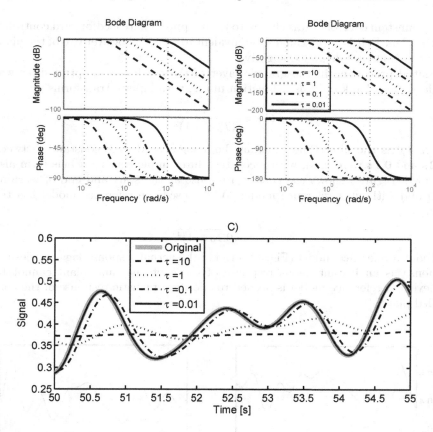

Fig. 2. Effect of reference model parameters on the response, where the reference model is described as in Eq. (7). (A) Bode plot for a first order reference model for a range of time constants. (B) Bode plot for a second order reference model for a range of time constants. (C) Time domain plot for a first order reference model of how the time constant effects the desired response. The original signal is 0–1 Hz band limited white noise, constrained to range from 0.2 to 0.6.

The simulated plant used has four poles and no zeros, this demonstrates the applicability of the control method to general, high order plants. The simulated plant is described as

$$P = \frac{1}{(s-8)(s-2)(s^2+20s+60)} \tag{9}$$

In each case the corresponding fixed brainstem controller was described as

$$B = \frac{(s-8.4)(s-2.1)(s^2+20s+60)}{(0.1s+1)^4} \tag{10}$$

The brainstem controller was chosen to be an approximate feedforward controller for the plant P. It is approximately equivalent to MP^{-1}, however two of the plant time constants are degraded slightly.

Initial processing in the granule layer of the cerebellar adaptive filter was modelled as a bank of basis α-function filters, with laplace transforms

$$G_N = \frac{1}{(T_N s + 1)^2} \tag{11}$$

where T_N is the filter time constant. Four log-spaced time constants between $0.02\,\mathrm{s}$ and $0.1\,\mathrm{s}$ were used, with a constant filter implementing a bias term also included. Figure 3 shows the tracked response at the start and end of learning with and without a reference model. Where used the reference model has the form

$$M = \frac{1}{(0.1s + 1)^4} \tag{12}$$

Without a reference model (Fig. 3A - right) the actual response lags the desired response as an instantaneous response of the controller and plant cannot be achieved. A reference model is needed to specify a realistic response of the controlled plant.

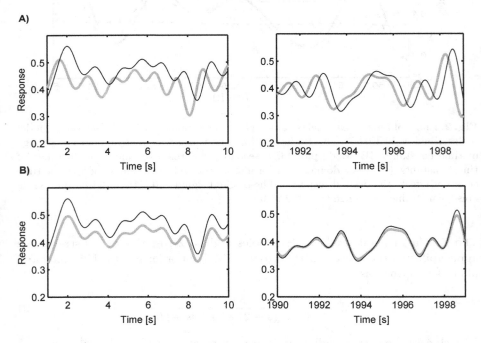

Fig. 3. Cerebellar inspired control of a fourth order simulated plant described by Eq. (10). (A) Desired (—) and actual (—) response at the start and end of learning when no reference model is used. (B) Desired (—) and actual (—) displacement response at the start and end of learning when a second order reference model with $\tau = 0.1$ is used.

5.3   Adaptive Control of DEA Using MRAC

In this experiment the cerebellar algorithm, augmented with a reference model, was applied to the real time control of the DEA actuator. If the input range is constrained, the displacement response of the DEA can be described as a first order linear system with one pole and no zeros. A first order reference model with $\tau = 0.1$ was used to define the desired system response. The brainstem was a sloppy compensator with input (v) to output (u) model equations,

$$\tau \dot{u}_t = a_0 \dot{v} + v \tag{13}$$

$$u = (u_t - c_0)/b_0 \tag{14}$$

where $a_0 = 0.087$, $b_0 = 0.331$, $c_0 = -0.317$, and as in the reference model $\tau = 0.1$. The cerebellar filters consisted of four α-function filters, with log spaced time constants between $0.02\,$s and $0.5\,$s, with an additional constant filter implementing a bias term.

The algorithm was tested for its ability to learn the dynamics in order to track the vertical displacement of the DEA. Figure 4 shows the results from a displacement tracking run. The learning rate, β was chosen to give robust stable learning on a time-scale that allowed tracking of variations in model parameters. The algorithm produced accurate displacement tracking, which remained stable over the length of the experiment.

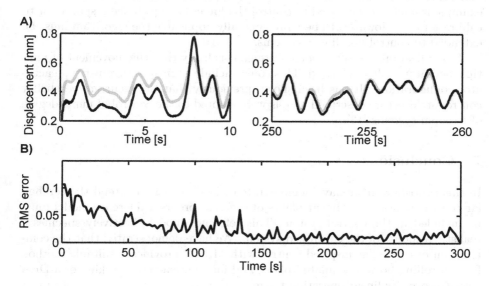

Fig. 4. Experimental control of DEA using cerebellar inspired adaptive controller synthesised with reference model. (A) Desired (—) and actual (—) response at the start and end of learning. (B) Windowed RMS errors during learning

6   Discussion

As the domain of operation of autonomous robots increases, new control challenges arise [4]. Algorithms capable of controlling systems in unstructured, changing environments are required. The 'cerebellar chip' provides a potential solution. This study has demonstrated that augmenting a bio-inspired control strategy (cerebellar control) with MRAC provides a potential general solution for robotic control.

The reference model determines how the controlled plant should behave. A controlled plant with more poles than zeros cannot respond instantaneously, if a reference model is not used it is constrained to try to do so. If the time constant of the reference model is small, the roll off in response occurs at high frequency and low frequency signals are not affected. Augmenting the cerebellar algorithm with MRAC means that the controlled plant is driven to match a realistic reference signal.

In experiments the DEA dynamics changed over time. These changes are likely to be due to stress and strain relaxation (creep), fatigue, aging and trauma. These highlight the need for an adaptive control algorithm. The cerebellar algorithm with reference model was able to track the displacement response of the DEA effectively despite changes to the dynamics.

The cerebellar algorithm described here has the potential to provide a modular controller for soft robots, and can be applied to generic control tasks. For example it could also be used to control the force, or impedance response, or to calibrate tasks. However, to be more generally applicable the algorithm must be extended to control non-linear systems.

In applications to soft robotic systems it is likely that the movement of multiple actuators will be coupled. This could include agonist-antagonist type actuator configurations. It has been shown previously that the recurrent cerebellar control architecture described here is well adapted to the control of multi-degree of freedom systems [13].

7   Conclusions

In this contribution we have demonstrated that in order to extend the cerebellar control algorithm to linear plants of different orders a reference model must be included in the control system. This reference model effectively specifies a realistic response of the controlled system. We have demonstrated that the combination of cerebellar inspired control with MRAC provides a suitable method for controlling both a simulated plant, and for the real-time tracking of a DEA actuator over its linear operating range.

Acknowledgements. This was supported by an EPSRC grant no. EP/IO32533/1, *Bioinspired Control of Electro-Active Polymers for Next Generation Soft Robots.*

References

1. Pfeifer, R., Lungarella, M., Iida, F.: Self organization, embodiment, and biologically inspired robotics. Science **318**, 1088–1093 (2007)
2. Javaherian, J., Huang, T., Liu, D.: A biologically inspired adaptive nonlinear control strategy for applications to powertrain control. In: 2009 IEEE International Conference on Systems, Man and Cybernetics (2009)
3. Lenz, A., Anderson, S.R., Pipe, A.G., Melhuish, C., Dean, P., Porrill, J.: Cerebellar-inspired adaptive control of a robot eye actuated by pneumatic artificial muscles. IEEE Trans. Syst. Man. Cybern. B **39**(6), 1420–1422 (2009)
4. De Santis, A., Siciliano, B., De Luca, A., Bicchi, A.: An atlas of physical human-robot interaction. Mech. Mach. Theory **43**(3), 253–270 (2008)
5. Ito, M.: The Cerebellum and Neural Control. Raven, New York (1984)
6. Dean, P., Porrill, J., Ekerot, C.F., Jörntell, H.: The cerebellar microcircuit as an adaptive filter: experimental and computational evidence. Nat. Rev. Neurosci. **11**(1), 30–43 (2010)
7. Eccles, J.C., Ito, M., Szentgothai, J.: The Cerebellum as a Neuronal Machine. Springer, Berlin (1967)
8. Marr, D.: A theory of cerebellar cortex. J. Physiol. **202**, 437–470 (1969)
9. Albus, J.S.: A theory of cerebellar function. Math. Biosci. **10**, 25–61 (1971)
10. Ito, M.: Control of mental activities by internal models in the cerebellum. Nat. Rev. Neurosci. **9**(4), 304–313 (2008)
11. Porrill, J., Dean, P., Anderson, S.R.: Adaptive filters and internal models: Multilevel description of cerebellar function. Neural networks. http://dx.doi.org/10.1016/j.neunet.2012.12.005. 28 Dec 2012
12. Fujita, M.: Adaptive filter model of the cerebellum. Biol. Cybern. **206**, 195–206 (1982)
13. Porrill, J., Dean, P.: Recurrent cerebellar loops simplify adaptive control of redundant and nonlinear motor systems. Neural Comput. **19**(1), 170–193 (2007)
14. Landau, Y.D.: Adaptive Control: The Model Reference Approach (Control and System Theory). Marcel Dekker, New York (1979)
15. Schweighofer, N., Doya, K., Lay, F.: Unsupervised learning of granule cell sparse codes enhances cerebellar adaptive control. Neuroscience **103**(1), 35–50 (2001)
16. Coenen, O.J.D., Arnold, M.P., Sejnowski, T.J.: Parallel fiber coding in the cerebellum for life-long learning. Auton. Robot. **11**, 291–297 (2001)
17. Widrow, B., Stearns, S.D.: Adaptive Signal Processing. Prentice Hall, Upper Saddle River (1985)
18. Dean, P., Porrill, J., Stone, J.V.: Decorrelation control by the cerebellum achieves oculomotor plant compensation in simulated vestibulo-ocular reflex. Proc. R. Soc. B **269**(1503), 1895–1904 (2002)
19. Morari, M., Zafiriou, E.: Robust Process Control. Prentice-Hall, Englewood Cliffs (1989)
20. Kaufman, H., Itzhak, B., Sobel, K.: Direct Adaptive Control Algorithms: Theory and Applications, 2nd edn. Springer, New York (1998)
21. Bar-Cohen, Y.: Electroactive polymer (EAP) actuators as artificial muscles: reality, potential, and challenges. SPIE Press, Bellingham (2001)
22. Pelrine, R., Kornbluh, R.D., Pei, Q., Stanford, S., Oh, S., Eckerle, J., Full, R.J., Rosenthal, M.A., Meijer, K.: Dielectric elastomer artificial muscle actuators: toward biomimetic motion. Proc. SPIE **4695**, 126–137 (2002)
23. OHalloran, A., OMalley, F., McHugh, P.: A review on dielectric elastomer actuators, technology, applications, and challenges. J. Appl. Phys. **104**(7) 071101 (2008)

UAV Horizon Tracking Using Memristors and Cellular Automata Visual Processing

Ioannis Georgilas[1,2]($\boxtimes$), Ella Gale[1,2], Andrew Adamatzky[1,2], and Chris Melhuish[2]

[1] International Centre for Unconventional Computing,
University of the West of England, Frenchay Campus, Bristol BS16 1QY, UK
giannis.georgilas@brl.ac.uk
http://uncomp.uwe.ac.uk
[2] Bristol Robotics Laboratory, University of Bristol and
University of the West of England, T Block, Frenchay Campus,
Bristol BS16 1QY, UK
http://www.brl.ac.uk

Abstract. Unmanned Aerial Vehicles (UAV)s can control their altitude and orientation using the horizon as a reference. Typically this task is performed via edge-detection vision processing techniques implemented in a computer or digital electronics. We demonstrate a proof-of-principle for a memristive cellular automata (CA) system which can simply interface with an analog electronic control system. Our aim is a cheaper, lighter and more robust low-level system. Low-quality, noisy and wide-angle images consistent with cheap cameras have been tested and, even with these issues, the system can recognise the tilt angle and express it as relative activation of cells at the edge of a CA which could be used to drive motors to right the aircraft.

Keywords: Cellular automata · Memristors · Image processing · UAV

1 Introduction

Unmanned Aerial Vehicles (UAVs) use a variety of sensor systems in order to establish their location and control their attitude [1,2]. One of the main methods is the use of horizon detection in order to establish the altitude and attitude of the vehicle and produce the necessary corrections [3,4]. Most of those methods are based on edge-detection of the horizon and are performed by computation with a digital computer [5]. In this paper we propose using a controller based on analog electronics, namely a combination of memristors and Cellular Non-linear Networks (CNN). Such an unconventional computing based electronic system obviates the need for a digital computer, decreasing the cost of the system and providing a continuous and seamless operation of the horizon tracking system which allows for more reactionary and fast control.

A. Natraj et al. (Eds.): TAROS 2013, LNAI 8069, pp. 64–75, 2014.
DOI: 10.1007/978-3-662-43645-5_9, © Springer-Verlag Berlin Heidelberg 2014

Memristors are the recently discovered [6] 4th fundamental circuit element [7] which are essentially resistors with memory. They are of interest as they are non-linear (in fact the simplest non-linear circuit element [7]) and they have been suggested as possible components for neuromorphic (brain-like) computers [6] – an exciting development given Memristor theory has recently been suggested to explain neuronal action [8,9]. Memristors are also low power consumption and can be made on the nanoscale, making them attractive for green computing.

Memristors are usually considered as a.c. components and are identified by a distinctive pinched Lissajous I-V curve and its response to changing voltage frequency response [10]. However memristor operation under d.c. power supply can be useful for some applications and has been quantified [12] and investigated for application to logic [13], chaotic circuits [14] and time-perception in robotic control structures [15].

The intriguing d.c. characteristics of the memristor can be used to pre-process the output of a photosensor and before the signal is input to a CNN lattice. CNN is a parallel processing paradigm [16] with the advantages that its continuous nature offers extremely fast computation and scalable implementation. CNN have been proposed as image processing methods [17] and their cellular structure works well with the rectangular format of image pixels.

CNNs can be implemented using any non-liner function. For the specific application and proof-of-concept implementation investigated here, we are going to use Cellular Automata (CA) [18] a discrete finite-state approach that functionally resembles some aspects of CNNs. A light-sensitive CA resembles the human eye in that it is distributed and relies on spatial organisation. CA have been extensively investigated for their computation abilities [19], as controllers for automation applications [20] and in image processing [21].

Memristive cellular automata where the cells are joined by memristive links have been used for image processing [22]. Memristive grids have been shown to be capable of identifying moving aspects of a scene [23]. In this paper we look at a standard 2D CA with a memristor input layer for edge detection and horizon location and deviation measurement.

Our proposed integration of the CA and memristors technologies is tested on an image processing task crucial to autonomous systems, specifically location of the horizon. We shall describe the steps of our proposed analog model that combines memristors and CA and demonstrate how the continuous nature of the former can be coupled with the finite state nature of the latter: this combination gives interesting spatial-temporal behaviour which is well-suited for image-processing.

We will use the simplest possible implementation with the aim of interfacing it with a control system that could be implemented using only electronic circuits instead of computers. This could lead to cheaper, lighter and more robust UAV control systems. We will also use test images of a quality consistent with cheap, low quality optics to allow the resulting UAVs to be built cheaply.

The paper is organised as follows. First, the model of the sensing system will be describe in Sect. 2. The proof-of-concept application of edge-detection

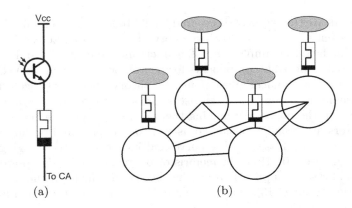

Fig. 1. Proposed System. (a) Detail of the sensor-memristor part of the circuit. (b) Grey circles represent the photosensors with gray which are connected to cellular automata cells (represented as spheres) via memristor driven circuits.

will be analysed in Sect. 3 and the sample images will be presented. In Sect. 4 the application of the proposed method for horizon sensing control will be demonstrated. The experiments contacted involve both artificial horizons and real-world examples.

2   The Model

This section describes the elements of the analog processing methodology and their interconnection. We are modelling a memristor connected to a light-sensor as shown in Fig. 1(a), so that the memristor's phase change can excite the connected 2D CA below it. The hardware is not a true 3D CA and is better conceived as an extended 2D (or (2+1)D) CA as shown in Fig. 1(b). This architecture allows the best spatial distribution of the system and is also scalable. Potentially, an analog vision system can be created based on this design using Integrated Circuits (IC) fabrication technologies.

2.1   Memristor d.c. Response

The memristor responds to the application of a constant d.c. voltage (caused by light falling on the light sensor) with the graph shown in Fig. 2 as discussed in—[12]. The system is excited into a high current ON phase and then decays into a low current OFF phase. We have decided to discretise this process as the decay from phase A (ON) to phase B (OFF). The device is taken to be fully discharged (i.e. have lost the short-term memory of the spike) at the time taken to lose 99 % of the difference between the peak height (i_{max}) and the equilibrated value (i_∞): this time is τ_{99}. The switch from A $\rightarrow$ B is taken to be half the difference between τ_{99} and $t(i_{max})$, i.e. the half way point. Note that the precise times

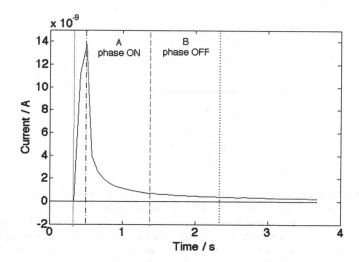

Fig. 2. An example memristor spike response to constant voltage. The rising edge between the orange line and the black dot-dashed line is the excitation. Phase A is the memristor in the ON phase, phase B is the memristor in the OFF phase. The black dotted line is the point at which the memristor has decayed to 99 % of its peak value.

for this process can be tuned by using different materials and different types of Memristors. The rising edge of the spike is often very fast, and this is the gap between the orange line and the peak as shown in Fig. 2.

We are using a highly simplified, 2-state course-grained model of the memristor where the decay from i_{max} to i_{99} is modelled as the phase transition from $A \rightarrow B$. The memristor has some internal phase (specifically the resistance of the material) which decays; τ_{99} is a measure of the time taken for this decay to happen to the point where, for practical purposes, the B phase is indistinguishable from the fully equilibrated state. This is a measure of the short-term memory of the memristor [12], after τ_{99} you can not tell if the Memristor had been charged, before that you can.

2.2 Cellular Automata Excitable Lattice

As the CA implemented in this system we are using an extension of the 2^+-medium. Specifically we are including four more states corresponding to the different areas of the memristor curve. These extra states are provide a form of short-term memory to the system.

The simple form of the rules for the 2^+-medium is given in Eq. 1. In words, for a two-dimensional lattice A, each cell x takes three states, resting(0), excited (2) and refractory(1). A resting cell is excited if the number of excited neighbours is exactly two. At the next time step, an excited cell will then takes a refractory state and a refractory cell will then takes a resting state unconditionally.

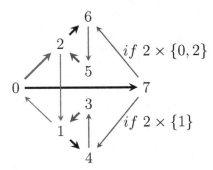

Fig. 3. States transition diagram. Red arrows indicate CA rules transitions, **bold red** is the usual excitation in 2^+-medium. **Black** arrows indicate memristor excitations. Blue arrows indicate state transition based on the memristor model. The transition from *state-7* to *states-6* or *state-4* happens based on the neighbourhood of the cell. Specifically 7 to 6 if there are two *state-2* cells or all cells are *state-0*, and 7 to 4 if there are two *state-1* cells. **Bold blue** arrows indicate the transition from memristor space to CA space which is controlled by $\chi(y, \{3,5\}) = 3$ (Color figure online).

$$
x^{t+1} = \begin{cases} 2, & x^t = 0 \text{ and } \sum_{y \in u(x)} \chi(y, 2) = 2 \\ 1, & x^t = 2 \\ 0, & \text{otherwise} \end{cases} \tag{1}
$$

where $\chi(y, 2) = 1$ if $y = 2$, and 0 otherwise.

The original excitation of the lattice is done by the transition from the resting state, *state-0*, to *state-7*. Intermediate excitation can be performed from *state-1* and *state-2*. The full state transition diagram is shown in Fig. 3. *States-0,1,2* represent the standard 2^+-medium CA model. *State-0* is the resting state. *State-7* is the rising edge of the d.c. memristor response if the CA is resting and only excited by the light input. The memristor then moves to the ON phase A, causing the CA below it to move to either *state-6* or *state-4*, depending on the CA neighbourhood . The memristor then decays from the ON to OFF (phase B) and this causes the CA below it to move to either *state-5* or *state-3*. Finally, the CA reverts back to CA rules *state-1* or *state-2*. This last transition can also happen to adjacent cells if $x^t = 0$ *and* $\chi(y, \{3,5\}) = 3$. From *state-2* and *state-1* the lattice can be re-excited by Memristors to *state-6* and *state-4* respectively.

3   Edge-Detection Examples

To demonstrate the usefulness of this system as a image processing method, we shall give two examples of edge-detection. We use a CA set up in *state-0* which is excited to *state-7* by application of light, we then investigate the dynamics of the CA. Note that we are only flashing light on the first step, not after, and

Table 1. Population Progression & Compression Rate in generation 3.

	G0	G3	G4	Compression (%)
Flower	58,240	3,276	4,572	94.6
Waterfall	30,651	11,455	18,413	62.6

thus the transitions from *state-2*→*state-6* and *state-1*→*state-4* will not happen in this experiment.

Pictures were selected to be differing in subject, 500 × 500 pixels in size, and in RGB colour-map. The blue channel was selected to be used as the greyscale representation of the image (because the blue channel of RGB photos shot under normal light usually contains the highest contrast and darkest shadows). The greyscale images then had their histograms equalised such that the histogram covered the entire range from 0:255 using MATLAB's *histeq* function. The equalised version was converted to black and white (monochrome) with MATLAB's *im2bw* function using the default value of 128 as threshold (so exactly half the tones are converted to black, and half to white). The binary image was then converted into a pixel-map of values 0 (black) and 7 (white) to match the states of the CA.

The CA rules were implemented using the software package Golly [24] with one cell representing one pixel, the initial state of all cells was *state − 0*. The pixel-map of the black and white image was then 'flashed' on the lattice to model the CA retina being flashed with light. After the original excitation the CA automata rules were allowed to normally evolve as per the transition diagram of Fig. 3.

As Fig. 4 shows after only 3 generations, the memristor-excited CA lattice has found the edges of the objects. This suggests a novel algorithm for edge-detection. This worked best in pictures with high contrast and a contiguous, simple shape, such as the flower, and less well in pictures with low contrast. The latter will be of minor importance in the case of horizon detection since sky and ground can have a significant contrast. Also, in the case of sky and sea, coloured filters (such as a red filter) or infrared images could be used and give a substantial contrast.

3.1   Compression

As shown in Table 1 the population of non resting cells in the CA lattice for three generations, zero (initial excitation), three (edge detection) and four (pattern dissolving). Since generation three has a direct connection to initial excitation generation, the decrease in excited-cell population can be conceived as compression and the respective compression rates can be also seen in the table. Also, in generation 4 the population goes up hence the optimum compression rate is achieved in generation 3. This suggests that the first 3 steps of a CA algorithm like this could be used as a novel compression or encryption routine.

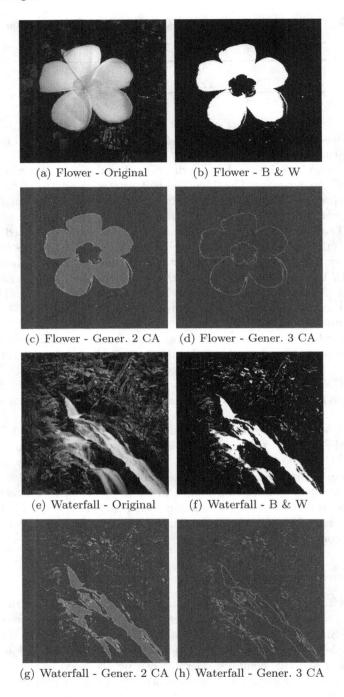

(a) Flower - Original   (b) Flower - B & W

(c) Flower - Gener. 2 CA   (d) Flower - Gener. 3 CA

(e) Waterfall - Original   (f) Waterfall - B & W

(g) Waterfall - Gener. 2 CA (h) Waterfall - Gener. 3 CA

Fig. 4. Images of a flower and a waterfall. The different phases of the processing method. Image copyright Ella Gale.

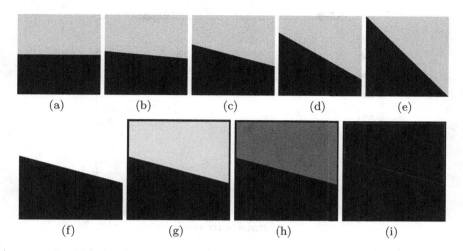

Fig. 5. (a)–(j) Artificial horizons in different angles (0, 5, 15, 30, 45 degrees). (k) Black and white version of 15 degrees horizon, (l) exposure to CA via Memristors (*state-7*), (m) Generation 2 of CA and (n) Generation 3 of CA.

4   Horizon Detection

Using the technique described in the previous section, here we shall demonstrate how horizon detection could be performed for the stabilisation of UAVs. We will investigate the most simple case where the horizon is visible and vertical. Our intention is to demonstrate the feasibility of the system as a potential controller. The control signal will be generated in a similar manner as for phototaxis of a light-excited robot in [25]. The control signal will be proportional to the number of non-resting cells on the border of the lattice. For our example 500×500 matrices this will be columns and rows 0 and 501.

A set of experiments using an ideal horizon, Fig. 5, was performed to evaluate the number of none resting cells in the aforementioned borders. The black and white version of the artificial horizon is 'exposed' as the light-map applied to the memristor-CA lattice, as in Fig. 4b and f. The CA evolve normally for 3 generations, Fig. 5(h), and then the border cells are calculated in generation 3, Fig. 5(i). Specifically, because artificial horizon angles are in the range $[0, 45°]$, only the left (column 0) and the right (column 501) were evaluated. Figure 6 depicts the number of non-resting cells in left and right borders for horizons exposed to the CA lattice. The numbers diverge the greater the horizon angle of the horizon, making it useful for control feedback.

4.1   Real Image Example

To investigate the generalisation of the proposed method, a test example of an image with a real horizon is given, Fig. 7. This image is taken from a wide-angle camera (so the horizon may be distorted), the contrast had been pushed

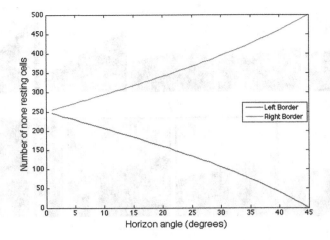

Fig. 6. The number of none resting cells in rows 0 and 501 of the CA lattice in relation to horizon angle.

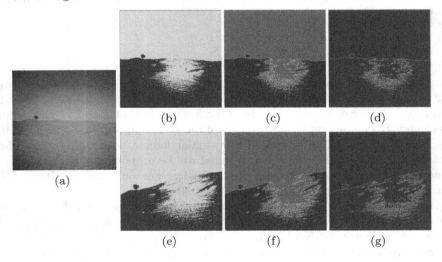

Fig. 7. Image of real horizon. (b)–(d) Processing performed in zero rotation, and (e)–(g) processing performed in 15° rotation. Image copyright Ella Gale.

leading to an increased about of colour noise, has strong vignetting (which makes the edges darker and the centre brighter) common to cheaper cameras (such as those found in web, phone and surveillance cameras) and an exceptionally bright foreground (over-exposed desert) making it a difficult test case. The experiment was evaluated at both 0° and 15° rotation (the latter performed in software). The results are shown in Table 2 and indicate that there are more non-resting cells on the left border than the right (44.8 %) and thus a motor driven by the proportion of activated cells would correct the roll of the UAV. The differences in the zero angle case (−5 %) are significantly smaller and are attributed to

Table 2. Number of none resting cells in real horizon images.

	Left border L	Right border R	Difference L-R	Mean (L+R)/2	% Difference ((L-R)/Mean)×100
Zero rotation	247	260	−13	253.5	−5.13
15° rotation	314	199	115	256.5	+44.8

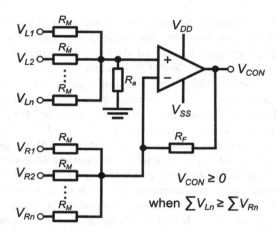

Fig. 8. Analog comparator of the voltage outputs of the left and right sides of the lattice. V_{CON} range and relation to $\sum V_{xn}$ is affected by the values of resistors R_M, R_a and R_F.

the non-perfect shape of the horizon, as expected in real-world situations. By using a photograph with high vignetting, the middle of the picture is lighter and picked up as an edge, even with this erroneous edge detection, this has not prevented the system from identifying the tilt away from a level horizon. This result demonstrates that robustness of the system and this indicates that low-quality cameras (such as a phone or web camera) would be sufficient for this task.

4.2 Analog Control Signal Generation

In order to maintain the analog nature of the proposed approach the control signal is generated using analog addition and comparison of the states (voltages) at the borders of the lattice. The circuit we propose to use can be seen in Fig. 8. It consists of an operational amplifier that acts as a summation for the voltages and comparator of the sums. The output of the amplifier will take values between V_{DD} (positive supply voltage) and V_{SS} (negative supply voltage) as a function of the values of $\sum V_{xn}$ and resistors R_M, R_a and R_F. V_{L1} to V_{L500} is the voltage from the left border of the lattice and V_{R1} to V_{R500} from the right side of the lattice. The output voltage V_{CON} can be fed as signal to other control circuits depending on the hardware selected.

Since the findings described here are mainly simulated, the inherent problems of analog electronics (i.e. noise, ubiquity, size) can be ignored. Nonetheless, in the case where actual prototypes are going to be constructed, special care will have to be taken to counter the before mentioned issues.

5   Conclusions and Further Work

The system implementation proposed here combines the benefits of both memristors and CA technologies to achieve an analog horizon detection method. The selection of analog processing instead of digital processing was based on the continuous and nature-like operation of the former.

In order to prove the feasibility of the method a simple edge-detection sample has been initially performed and the findings where implemented for the detection of horizon. The method worked well for both artificial and a simple real horizon test. In order to extract information to be used for the control of the vehicle an analog method has been used. The states of the image, CA lattice borders has been evaluated using an analog summation and comparison circuit and the output voltage could be used to control the inclination of the vehicle.

Memristors are low power consumption devices and, as we've shown both that digital computer control hardware is unnecessary and that the horizon detection works with the type of optics found in cheap lenses, we expect that such a control system would be economically attractive. We anticipate that interaction of memristor spikes (caused by flashing the network at some point during the 1st and 3rd generations) could allow a way of picking up on moving edges. As we have prototype memristors, future work will involve the building of a prototype control for UAV based on continuous image analysis. Also, more complicated Memristor-CA interactions will be investigated, based on a gradient approach than a border case. This way any edge effects and noise present in real images will be compensated for.

References

1. Ettinger, S., Nechyba, M., Ifju, P., Waszak, M.: Vision-guided flight stability and control for micro air vehicles. Adv. Robot. **17**(7), 617–640 (2003)
2. Shabayek, A., Demonceaux, C., Morel, O., Fofi, D.: Vision based uav attitude estimation: progress and insights. J. Intell. Rob. Syst. Theory Appl. **65**(1–4), 295–308 (2012)
3. Todorovic, S., Nechyba, M., Ifju, P.: Sky/ground modeling for autonomous mav flight. In: Proceedings of the IEEE International Conference on Robotics and Automation, ICRA'03, vol. 1, pp. 1422–1427 (2003)
4. Dusha, D., Boles, W., Walker, R.: Attitude estimation for a fixed-wing aircraft using horizon detection and optical flow. In: IEEE 9th Biennial Conference of the Australian Pattern Recognition Society on Digital Image Computing Techniques and Applications, pp. 485–492 (2007)
5. Chen, Y., Abushakra, A., Lee, J.: Vision-based horizon detection and target tracking for UAVs. In: Bebis, G., et al. (eds.) ISVC 2011, Part II. LNCS, vol. 6939, pp. 310–319. Springer, Heidelberg (2011)

6. Strukov, D.B., Snider, G.S., Stewart, D.R., Williams, R.S.: The missing memristor found. Nature **453**, 80–83 (2008)
7. Chua, L.O.: Memristor - the missing circuit element. IEEE Trans. Circuit Theory **18**, 507–519 (1971)
8. Chua, L., Sbitnev, V., Kim, H.: Neurons are poised near the edge of chaos. Int. J. Bifurcat. Chaos **11**, 1250098 (2012). (49pp)
9. Chua, L., Sbitnev, V., Kim, H.: Hodgkin-huxley axon is made of memristors. Int. J. Bifurcat. Chaos **22**, 1230011 (2012). (48pp)
10. Chua, L.: Resistance switching memories are memristors. Appl. Phys. A Mater. Sci. Process. **102**, 765–782 (2011)
11. Georgiou, P., Yaliraki, S., Drakakis, E., Barahona, M.: Quantitative measure of hysteresis for memristors through explicit dynamics. Proc. R. Soc. A **468**, 2210–2229 (2012)
12. Gale, E., de Lacy Costello, B., Adamatzky, A.: Observation and characterization of memristor current spikes and their application to neuromorphic computation. AIP Conf. Proc. **1479**, 1898 (2012)
13. Gale, E., de Lacy Costello, B., Adamatzky, A.: Boolean logic gates from a single memristor via low-level sequential logic. In: Mauri, G., Dennunzio, A., Manzoni, L., Porreca, A.E. (eds.) UCNC 2013. LNCS, vol. 7956, pp. 79–89. Springer, Heidelberg (2013)
14. Gale, E., de Lacy Costello, B., Adamatzky, A.: Observations of bursting spike patterns in simple three memristor circuits. arXiv (2012) (preprint). http://arxiv.org/pdf/1210.8024v1.pdf
15. Gale, E., de Lacy Costello, B., Adamatzky, A.: Does the d.c. response of memristors allow robotic short-term memory and a possible route to artificial time perception? In: ICRA 2013 Workshop - Unconventional Approaches to Robotics, Automation and Control, Inspired by Nature (UARACIN) (2013)
16. Chua, L.O., Yang, L.: Cellular neural networks: theory. IEEE Trans. Circuits Syst. **35**(10), 1257–1272 (1988)
17. Crounse, K.R., Chua, L.O.: Methods for image processing and pattern formation in cellular neural networks: a tutorial. IEEE Trans. Circuits Syst. I Fundam. Theory Appl. **42**(10), 583–601 (1995)
18. Wolfram, S.: Theory and Applications of Cellular Automata. World Scientific, Singapore (1986)
19. Adamatzky, A.: Computing in Non-linear Media and Automata Collectives. Institute of Physics Publishing, Bristol (2001)
20. Georgilas, I., Adamatzky, A., Melhuish, C.: Towards an intelligent distributed conveyor. In: Herrmann, G., Studley, M., Pearson, M., Conn, A., Melhuish, C., Witkowski, M., Kim, J.-H., Vadakkepat, P. (eds.) TAROS-FIRA 2012. LNCS, vol. 7429, pp. 457–458. Springer, Heidelberg (2012)
21. Rosin, P.: Training cellular automata for image processing. IEEE Trans. Image Process. **15**(7), 2076–2087 (2006)
22. Adamatzky, A., Chua, L.: Memristive excitable cellular automata. Int. J. Bifurcat. Chaos **21**(11), 3083–3102 (2012)
23. Lim, C., Prodromakis, T.: Computing motion with 3d memristive grids. arXiv (2013) (preprint). http://arxiv.org/pdf/1303.3067v1.pdf
24. Golly http://golly.sourceforge.net/. Accessed 29 April 2013
25. Adamatzky, A., Melhuish, C.: Phototaxis of mobile excitable lattices. Chaos. Soliton. Fract. **13**(1), 171–184 (2002)

Sensor Integrated Navigation for a Target Finding UAV

Michael J. Park$^{(\boxtimes)}$ and Charles Coldwell

M.I.T. Lincoln Laboratory, 244 Wood Street, Lexington, MA 02420, USA
{michael.park,coldwell}@ll.mit.edu

Abstract. Unmanned Aerial Vehicle (UAV) flight control and sensor data acquisition are typically decoupled in intelligence-related missions. In the archetypal case, a UAV flies pre-programmed GPS waypoints while an on-board sensor streams data to a ground control station observed by a human or post-processed for analysis. In this paper, we present work on a simple UAV system that incorporates sensor data to statistically minimize the time to autonomously locate an RF emitter on the ground. The demonstrated system uses a two-stage approach to finding the RF target: (1) randomized Lévy flight to search the ground space and, (2) simplex minimization to home in on the target. Flight tests showed that the UAV consistently located the target within 10 m accuracy in a large outdoor field. Algorithm pseudocode, heat map plots, and supplementary videos of the UAV successfully executing its search task are presented.

Keywords: Autonomous systems · UAV autopilot · Lévy flight · Simplex minimization

1 Introduction

Sensor integrated path planning has been a central area of research for autonomous robotic systems for many decades. In recent years, goal-oriented path planning with obstacle avoidance has gained widespread exposure with the 2004 and 2007 DARPA Unmanned Ground Vehicle (UGV) challenges. The motivation for sponsoring autonomous UGV missions is understandable, given that such competitions push the limits of what is capable today in numerous fields in robotics, from advanced sensor capabilities to artificial intelligence.

In more recent years, due in part to improved lithium polymer batteries, cheaper brushless motors, and rapid dissemination of technological developments through online videos, task planning for small quadrotor platforms has become an active area of autonomous systems research [1,2]. While many of these efforts have paved the way for 3D motion planning and deterministic control in a mapped indoor positioning environment, work remains ahead for increasing the capability of these novel systems to exploit their own sensors to make navigation decisions autonomously. In particular, an Unmanned Aerial Vehicle (UAV) that

A. Natraj et al. (Eds.): TAROS 2013, LNAI 8069, pp. 76–89, 2014.
DOI: 10.1007/978-3-662-43645-5_10, © Springer-Verlag Berlin Heidelberg 2014

incorporates its own sensors to complete some pre-determined mission more efficiently would improve upon the pre-programmed flight paths that are decoupled from potentially useful sensor data.

In this paper, we present work on a self-navigating UAV that uses GPS and an onboard RF receiver to locate an RF transmitter target within a designated area of interest. The UAV locates its target by flying a randomized path over its search region and homing in over the target using sensor inputs to its real-time updated flight plan. The search strategy used is a two-step approach: (1) Lévy flight over the area until the required RF sensor detections are acquired; (2) Simplex Minimization to home in over the RF transmitter where the received signal strength is near its maximum. In this work, we describe the search strategy methodology, UAV system hardware, software algorithm implementation, and flight demonstration results.

2   Search Strategy for Finding the Target in the Minimum Time

The predominant method for aerial Intelligence, Search, and Reconnaissance (ISR) missions over a designated area of interest is to use pre-programmed flight paths that cover the specified area. "Lawnmower" path or polygonal flight paths are common examples. In UAV testing situations, the acquired on board data is typically post-processed and analyzed; in real-time operations the data is streamed back to a human operator that can update the flight plan or manually fly to home in on an area of interest. We demonstrate an alternate approach of using useful onboard sensor data to guide the aerial flight path that statistically minimizes the overall flight time to discover the target.

In essence, the problem we are trying address is to *find a known target located in an unknown place within a region of known bounds in the minimum time possible*. This problem assumes that the sensors range to the target is smaller than the overall search space, hence the need to move the UAV throughout the search space to find it. A paper by Viswananthan et al. [3] shows that statistically, the optimal flight path strategy is a generalized random walk called Lévy flight. In Lévy flight, the directions of flight are random and isotropic (as in a usual random walk) but the step sizes are chosen from a probability distribution that is typically $\frac{1}{r}$ (Cauchian) or $\frac{1}{r^2}$ (Gaussian) where r is the radius step size (as opposed to a usual random walk where all of the step sizes are the same). Figure 1 shows an example of a Lévy flight with 100 steps drawn from a Cauchian step size distribution. Note that using an inverse power distribution yields step sizes that more often small and clustered and only occasionally far and distant. It has been shown that predators such as sharks and bird of prey naturally adopt this search strategy when hunting [4].

Since Lévy flights are theoretically unbounded, we impose a search limit of 90 m around a designated ground origin for implementation and testing purposes. The RF transmitter of interest is assumed to be within this search radius. During the course of its search flight, the UAV ignores and does not fly to points outside

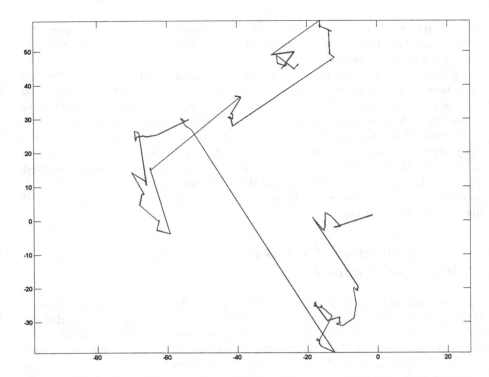

Fig. 1. An example 100-step 2-dimensional Lévy flight path with step sizes chosen from a Cauchian $\frac{1}{r}$ probability distribution.

of this bounded radius. Additionally, the UAV is constrained to fly at a fixed altitude of 20 m over the search area.

During the course of its flight, the UAV collects Received Signal Strength Indication (RSSI) data emitted from the RF transmitter on the ground. As the UAV flies closer to the transmitter, the detected RSSI signal increases. If the RSSI values are above a pre-defined threshold amplitude, the UAV transitions from the random Lévy flight mode, to a flight optimization mode that directs the UAV over the region where the RSSI strength is at a maximum. The method we implemented for this optimization is simplex minimization and was introduced by Nelder and Mead in 1965 [5]. Our applied method essentially takes 3 RSSI points in the flight plane and uses the values to select and fly to a 4th point which is expected to have a stronger signal. When the 4th point is reached, the UAV uses this as a vertex for the next iteration of the optimization scheme. Figure 2 shows an example of this algorithms progression toward a point source through iterative measurements of a scalar value (analogous to our physical implementation where the scalar value is the RSSI of the RF receiver). The algorithm implementation will be discussed further in the pseudocode section of this paper.

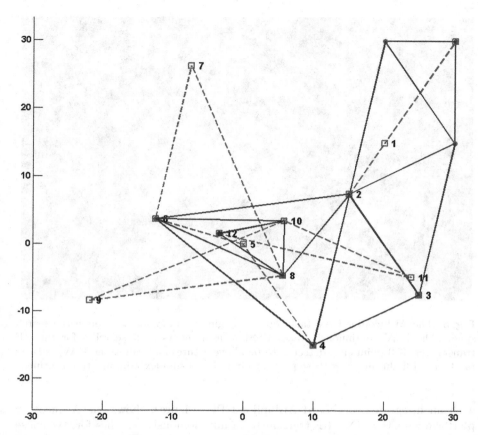

Fig. 2. Example of simplex minimization algorithm. Algorithms starts with initial "seed" triangle at the top right corner and iteratively progresses toward the point source at $(0,0)$. The blue solid lines indicate the simplex triangles and the red dotted lines indicate the flight paths that the algorithm requires for measurement to create the next simplex triangle. Details on these paths described later in this paper and in Nelder and Meads 1965 paper [5].

Figures 3, 4, and 5 illustrate our hybrid approach to finding the RF transmitter target within a bounded region in the minimum time possible. In the first step, the UAV flies randomly in the Lévy fashion until at least 3 RSSI points were above a threshold value. Next, the UAV initiates a simplex minimization routine that iteratively flies toward the target. Lastly, the UAV autonomously lands and reports its location back to an operator.

3   UAV System Architecture

For this work, we used and built upon various commercially available equipment and software. For the UAV platform, we used the 3D Robotics Arducopter

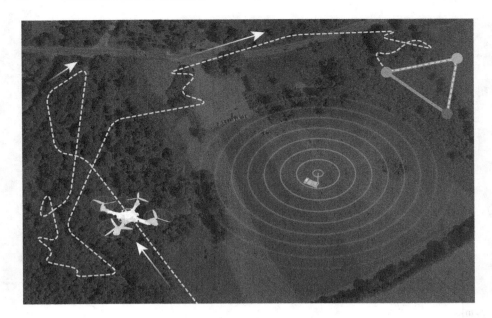

Fig. 3. The Arducopter UAV takes off and begins its Lévy flight through the search space. The UAV continuously takes RSSI measurements as it searches for the RF transmitter. If 3 points are detected to be above a threshold value, the UAV ends its randomized flight and uses these points to initialize a simplex minimization routine.

with an integrated, on-board Ardupilot CPU. Figure 6 shows the Arducopter platform with the LINX RF transmitter it autonomously searches for. We chose this platform primarily for its ease of development with its open source software and configurable hardware system. We leveraged the Arducopters existing flight dynamics control, GPS-guided flight capability, altimeter sensor, and basic communication structure. In addition, we added various specialized communication and monitoring messages for flight tests. We will discuss the software changes in the next section of this paper.

The primary hardware modification made to the Arducopter was the integration of a LINX RSSI receiver with analog input to the central Ardupilot controller. Figure 7 shows this hardware along with other peripheral system devices. The LINX receiver measures physical proximity to the transmitter that is placed in a field as the searched object. The RSSI output voltage amplitude decreases as $\frac{1}{r^2}$, where r is the distance between the transmitter and receiver. The receiver was also calibrated to detect the transmitter only within a limited radius (roughly 20 m), such that the UAV would generally need to search throughout the area of interest to detect and locate the transmitter target. Figure 8 shows an example RSSI amplitude versus distance measurement.

Other hardware peripheral devices were integrated into our system for various communication and control capabilities. A Spektrum DX8 900 MHz manual

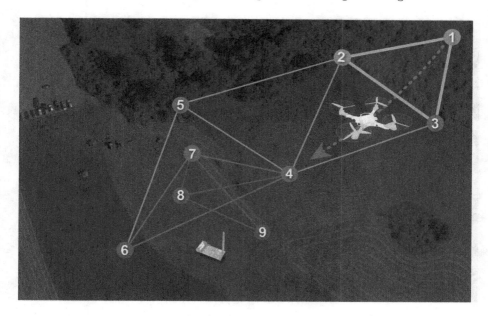

Fig. 4. The UAV uses the initially defined simplex triangle to iteratively move itself toward the RF transmitter. Each subsequently defined simplex triangle brings the UAV closer to the RF transmitter source.

radio controller was used for flight take-offs and manual override to the Ardupilot controller for testing purposes. In cases when the UAV would complete a successful geo-location run, the Spektrum controller was used only for take-off purposes. The UAV is capable of automatic take off (it is a built in feature in the Ardupilot CPU), but for our safety purposes, our take offs were manually operated. Control was switched to fully autonomous when the UAV was a safe distance away from our operator. The Spektrum manual override functionality also was useful in the algorithm development stages when we would be testing a single feature of the autonomy algorithm and then switch to manual operation to bring the UAV back to its launch point. A Dell notebook computer with MATLAB 2012a operated as our ground station controller to the UAV. A script in MATLAB would communicate to the Arducopter via a 900 MHz DIGI XBee module and receive telemetry data such as GPS, altitude and RSSI strength. In return, the MATLAB script would process this data and send back Lévy flight and simplex minimization generated GPS points to the Ardupilot CPU. In this manner, the RSSI sensor data would guide the Arducopter toward autonomous geo-location in real-time. In addition, the Dell PC was connected to the internet via a Samsung Android 4G connection. With this data link, MATLAB would query a United States Geological Survey (USGS) server and build maps for an overview of the region of interest. In this way, our UAV system is applicable to fly anywhere there is GPS coverage and an internet connection.

Fig. 5. After a series of simplex minimization sequences has completed, the UAV autonomously lands near the RF transmitter and broadcasts its geographic location to the ground station.

4   Software Algorithm Implementation

The autonomous control scheme for the target finding UAV is a finite state machine that takes as input the aircrafts current GPS location, RSSI sensor values, and target area domain boundaries to continuously search for the RF transmitter. The pseudocode below outlines the important elements of the algorithm we implemented.

Note that the UAV reports GPS and RSSI data to the ground station PC which in return sends algorithm-generated GPS waypoints and flight instructions to the UAV. The algorithm script runs on MATLAB which interfaces with both the UAV and a USGS map server to construct a visual overview of the flight path and RSSI detections.

MATLAB pseudocode used to direct an Arducopter UAV to autonomously find and report the location an RF transmitter of interest using RSSI amplitude data. Flight data is monitored in real-time and in successful runs, the UAV automatically lands near the transmitter at the end of its search.

```
RETRIEVE map from USGS web server using latitude, longitude &
radius inputs SET exitLevy = 0
WHILE exitLevy = 0
  // Generate Levy flight:
  CHOOSE random direction
```

Fig. 6. 3D Robotics Arducopter with LINX RF receiver input (attached to bottom right arm) and associated RF transmitter search target.

```
  CHOOSE step size from a 1/x (Cauchian) distribution
  UPLOAD GPS coordinate waypoint to Arducopter
  PLOT received location & RSSI telemetry data on map
  IF at least 3 cumulative RSSI values are above a given threshold
    SET exitLevy = 1
  END IF
END WHILE
SORT telemetry data in order of decreasing RSSI
ENUMERATE all possible triangles with vertices at points
where RSSI was measured
CHOOSE the first triangle with good geometric spread to seed the
simplex algorithm
// Begin simplex minimization
LOOP simplex
  REFLECT the vertex with the lowest RSSI through the line defined
  by the other two vertices
    DEFINE this point to be the Candidate Vertex
    FLY to Candidate Vertex & RECORD RSSI
    IF (Candidate Vertex RSSI > previous highest RSSI)
      DEFINE Candidate Vertex to be another step of the same size
      in the same direction
      FLY to Candidate Vertex & RECORD RSSI
```

Fig. 7. Electronic/computational hardware for the UAV system. The Ardupilot CPU (*a*) receives signal strength data from the LINX receiver (*b*). The CPU also has a GPS antenna (*c*), 900 MHz XBee serial data link to a PC (*d*), and 2.4 GHz manual override signal input (*e*). The LINX RF transmitter (*f*) and USB XBee PC serial data link (*g*) are also shown.

```
  ELSE IF (Candidate Vertex RSSI < than the previous second
  lowest RSSI)
     DEFINE Candidate Vertex to be half the distance toward the
     line defined by the other two points
     FLY to Candidate Vertex & RECORD RSSI
  END IF
 DEFINE new simplex triangle consisting of the previous two
 highest RSSI vertices & the Candidate Vertex
ITERATE simplex UNTIL step < epsilon
LAND Arducopter & report final GPS location
```

The main structural feature of the algorithm finite state machine is that it is divided into Lévy flight and the simplex minimization sections. The UAV will only enter the homing simplex stage after at least 3 RSSI telemetry points are above a defined threshold. The other requirements of these 3 initializing points above the RSSI threshold are that they must define a triangle with points that are not close to collinear and the must not be too close together. This ensures that the "seed" triangle for the subsequent simplex optimization is adequately chosen for recursively homing in on the target.

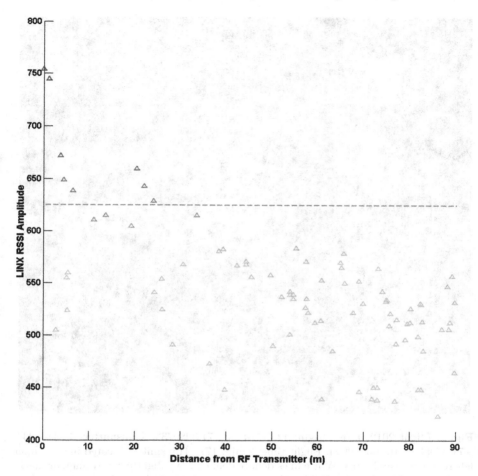

Fig. 8. An example scatter plot of Receive Signal Strength Indicator (RSSI) RF receiver amplitude versus radial distance from corresponding transmitter. The green dashed line indicates the defined threshold amplitude needed to transition from a random Lévy search mode to a simplex minimization search mode. The data corresponds to the flight data from Fig. 9.

5   Flight Test Results

A series of flight tests were conducted at the Davis RC airfield in Sudbury, MA in the early fall of 2012. Figures 9, 10, and 11 show sample flight test plots that display the flight telemetry of the autonomous UAV as it searched for and homed in on its RF transmitter target. Starting from the origin marked at the center of each plot, the UAV autonomously flew and reported its geographic and RSSI measurements as it flew within the designated 90 m radius. For these examples, the RF transmitter is located in the upper left region of the circular area, as

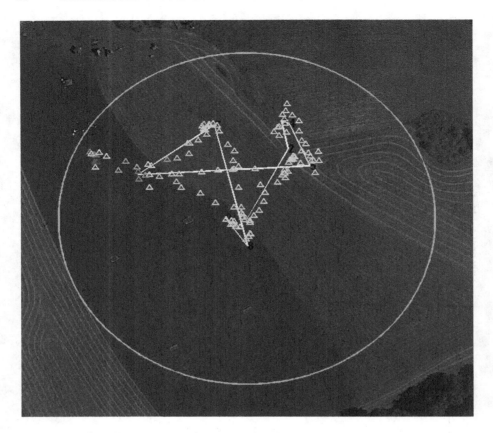

Fig. 9. Flight 20120927e. Origin: 42.415745°, −71.398073°. Transmitter: 42.416023°, −71.398960°. Radius: 90 m. Flight time: 2 min, 17 s. Transmitter located in the upper left region of search area. White lines denote defined Lévy flight path, triangular markers denote UAV telemetry RSSI and coordinate readings. Color scale goes from yellow (low RSSI) to red (high RSSI). Flight data collection initialized at origin (center of the circle) (Color figure online).

marked by a red 'X' in the plots. These plots were produced in real-time as the UAV flew over the region and reported its location to a MATLAB PC console. The plots were produced using MATLAB Mapping Toolbox 2012a.

We conducted a total of 20 fully autonomous search flights at the Davis RC airfield, 17 of which successfully located and landed within 10 m of the target RF transmitter. The 3 unsuccessful flights failed due to inconsistent RSSI measurements that resulted in a simplex convergence time that was greater than the flight time allowed by the Arducopters battery power source. In these cases, our test pilot noted the low power indicator and implemented the manual override control to safely land the Arducopter before the battery power completely expired.

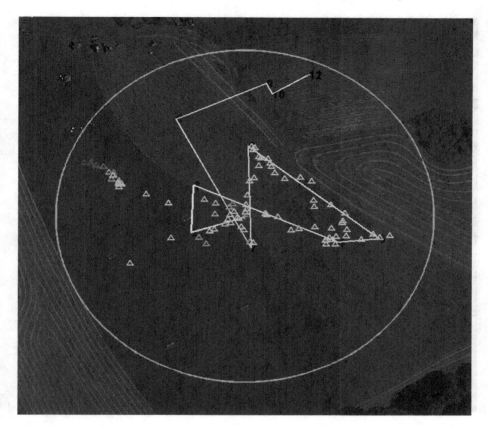

Fig. 10. Flight 20120927b. Same origin, transmitter location, and search radius as for Fig. 9. Flight time: 1 min, 32 s. For this flight, note that the UAV departs from its Lévy flight path toward the end to home in on the target with the simplex optimization routine.

Additional tethered flight tests were conducted at the MIT Briggs Field in Cambridge, MA and at the Air Force Research Laboratory field on Hanscom Air Force Base. These flights were for development purposes and only components of the algorithm were tested at these locations.

6   Supplemental Video Material

Annotated flight test videos and a hardware description video are available as supplementary material for this paper. Also available is a MATLAB simulation of a simplex minimization routine.

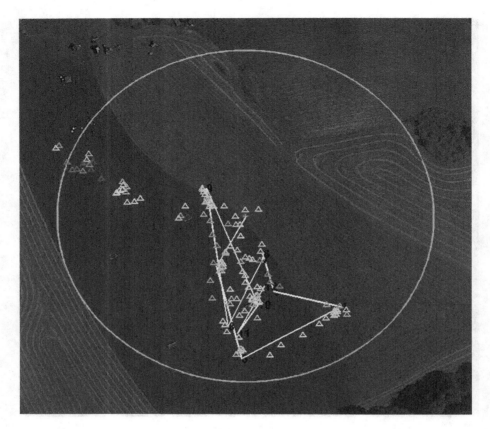

Fig. 11. Flight 20120927a. Same origin, transmitter location, and search radius as for Fig. 9. Flight time: 2 min, 8 s.

Acknowledgments. The authors thank Michael Klinker and Will Bartlett for field testing support; Daniel Jih for early simplex development; Paul Huckfeldt for graphics; Jon Barron for photography; Michael Boulet, Andy Vidan and Bernadette Johnson for program support and funding; and Spring Berman for an insightful discussion. This work is sponsored by the Department of the Air Force under Air Force Contract #FA8721-05-C-0002. Opinions, interpretations, conclusions and recommendations are those of the authors and are not necessarily endorsed by the United States Government.

References

1. Mellinger, D., Michael, N., Kumar, V.: Trajectory generation and control for precise aggressive maneuvers with quadrotors. Int. J. Robot. Res. **31**(5), 664–674 (2012)
2. Schwager, M., Julian, B.J., Rus, D.: Optimal coverage for multiple hovering robots with downward facing cameras. In: 2009 IEEE International Conference on Robotics and Automation, Kobe, Japan, pp. 3515–3522 (2009)

3. Viswanathan, G.M., Buldyrev, S.V., Havlin, S., da Luz, M.G.E., Raposo, E.P., Stanley, H.E.: Optimizing the success of random searches. Nature **401**, 911–914 (1999)
4. Sharks Have Math Skills. http://news.discovery.com/animals/sharks/sharks-math-hunt.htm
5. Nelder, J.A., Mead, R.: A simplex method for function minimization. Comput. J. **7**(4), 308–313 (1965)

Inventing a Micro Aerial Vehicle Inspired
by the Mechanics of Dragonfly Flight

Nina Gaissert[(⊠)], Rainer Mugrauer, Günter Mugrauer,
Agalya Jebens, Kristof Jebens, and Elias Maria Knubben

Bionic Learning Network, Festo AG & Co. KG,
Ruiter Strasse 82, 73734 Esslingen, Germany
{niga,kknb}@de.festo.com,
{r.mugrauer,g.mugrauer}@airstage.de,
{agalya.jebens,kristof.jebens}@jntec.de

Abstract. Dragonfly flight is unique: Dragonflies can manoeuvre in all directions, glide without having to beat their wings and hover in the air. Their ability to move each of their four wings independently enables them to slow down and turn abruptly, to accelerate swiftly and even to fly backwards. We looked into the mechanics of the dragonfly flight and managed to transfer its flight dynamics into an ultralight flying object: the BionicOpter. With a wingspan of 63 cm and a body length of 44 cm, the model dragonfly weighs just 175 g. A brushless motor actuates the four wings and is used to alter the flapping frequency. Eight servo motors allow the amplitude and the twisting angle of each wing to be changed independently making the BionicOpter almost as agile and fast as its natural role model. Here we present how dragonfly flight dynamics can inspire future design of MAVs.

Keywords: Bio-inspired engineering · Biologically-inspired robots · Aerial robotics · Unmanned aerial vehicles · Dragonfly flight

1 Introduction

Almost 300 million years ago dragonflies shared the land with early amphibians and the first reptiles. Some of the ancient dragonflies were enormously large. *Meganeuropsis permiana* reached a wingspan of up to 75 cm [2]. However, due to the decline in atmospheric oxygen levels during the Upper Permian Period and due to the evolution of birds, these giant dragonflies disappeared [2] and today's smaller but fast and agile dragonflies evolved. For the last 150 million years the body of the dragonfly has not changed much as shown in Fig. 1. This consistency is only possible because dragonflies are optimally adapted to their environment and life conditions.

Dragonflies are predators that catch their prey during flight and some even mate during flight. They live close to water which they need for egg deposition. There are about 6000 species of dragonflies with wingspans ranging from 18 mm to 190 mm. Their segmented body weighs less than 0.2 g, consisting of the head, the thorax with its six legs and a long abdomen. They have complex eyes that are perfectly adapted to fast visual perception during flight. The ability to move each of its four wings

A. Natraj et al. (Eds.): TAROS 2013, LNAI 8069, pp. 90–100, 2014.
DOI: 10.1007/978-3-662-43645-5_11, © Springer-Verlag Berlin Heidelberg 2014

Fig. 1. Fossil of *Aeschnogomphus intermedius*. This dragonfly had a wingspan of 19 cm and was conserved in Solnhofen limestone for the last 150 millions of years. It was found in Blumenberg, Eichstaett, Germany.

independently enables the dragonfly to fly in all directions and to execute the most complicated flight manoeuvers. Astonishingly, a wing weighs only about 0.002 g and has a thickness of 3 μm at its thinnest point [3]. Yet dragonflies can reach a maximum speed of 54 km/h and with some tailwind they can fly up to 1000 km [4]. Some dragonflies even have been observed to fly with half a wing missing.

Dragonflies can slow down and turn abruptly during flight. They can accelerate swiftly and even fly backwards. Some dragonflies (Anisoptera) have broader hind-wings that allow them to glide through the air. Finally and most importantly, drag-onflies can hover in the air. Thus dragonflies can master all flight conditions of a helicopter, a plane and even a glider. This makes them a very interesting and unique biological role model for bio-inspired engineering. Hovering in particular is important for many real-world applications of micro aerial vehicles (MAVs), such as data collection in research, exploration of wind turbines or rescue missions.

Most interestingly, the dragonfly cannot only hover in the horizontal plane but is also able to hover in the vertical plane. With this function the dragonfly exceeds all common aerial vehicles including customary quadrocopters. Hovering in several positions allows for more degrees of freedom for the flying object when interacting with a second object in its surroundings and thus will lead to new applications of MAVs in the future.

The uniqueness of the dragonfly flight inspired us to look into the flight mechanics of real dragonflies and to turn their complexity into a technical reality. With Bioni-cOpter we addressed this challenge. In this paper we describe the technical setup and give an outlook on how the dragonfly flight can inspire future MAVs (Fig. 2).

2   Lightweight Design with Intelligent Kinematics

2.1   The Natural Role Model and Bio-inspired Engineering

The mechanics of dragonfly flight are unique: Dragonflies can manoeuvre in all directions, glide without having to flap their wings and hover in mid-air. Most

Fig. 2. BionicOpter and its natural role model. On the left a natural dragonfly of the infraorder Anisoptera is displayed. These insects are characterized by the large multifaceted eyes that dominate the head, the four strong transparent wings and the elongated body. On the right the BionicOpter is displayed. BionicOpter was mostly inspired by the physiology and flight dynamics of Anisoptera, since these special dragonflies are not only able to fly extremely well, they can also hover in mid-air and glide.

interestingly, they can hover in the horizontal plane and in the vertical plane. This becomes possible since they can control each of the four wings independently in amplitude and frequency. Further, they can twist their wings. To do this they use direct flight muscles, i.e. muscles that are directly attached to the joint of the wings. Each wing is moved by at least five distinct muscles. To control these muscles actively and precisely, and thus to define the position of the wing exactly, three to fifteen neurons innervate each muscle [5]. Finally, numerous sensors distributed across the whole body of the dragonfly deliver input about the position of each wing and the corresponding state of the environment.

We took inspiration from this highly complex system. However, we did not copy it one-to-one. Instead we combined the knowledge obtained from the real dragonfly, with newest materials, the latest technical inventions and with knowledge gained from other MAVs such as quadrocopters, and thus transferred the dragonfly dynamics to the technical world. This was only possible with a consistently lightweight design and by integrating the functions in the smallest spaces: sensors, actuators and mechanical components as well as communication and open- and closed-loop control systems are installed in a very small space and connected to one another.

The result is an MAV with flapping wing drive, encompassing thirteen degrees of freedom. A brushless motor in the bottom part of the housing provides the drive for the common beat frequency of the four wings. Each wing is individually twisted by one servo motor. One additional servo motor at the joint of each wing controls the amplitude. A linear movement in the wing root continuously adjusts the integrated crank mechanism to vary the deflection of the wing. The swiveling of the wings determines the thrust direction. The thrust intensity can be regulated using the amplitude controller. The combination of both enables the dragonfly to hover on the spot, manoeuvre backwards and transition smoothly from hovering to forward flight. The last four degrees of freedom are in the head and tail. The body of the dragonfly is fitted with four flexible muscles made of nitinol. These shape memory alloys (SMAs) contract when exposed to heat and expand when they cool down. Passing an electric current through the SMAs produces ultralight actuators that move the head horizontally and the tail vertically (See Fig. 3).

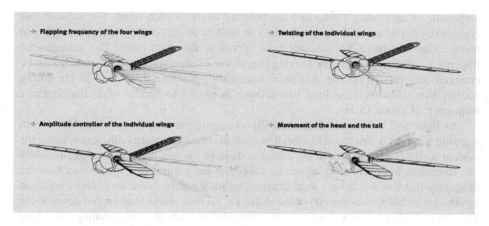

Fig. 3. Individually controlled wings. Just like the real dragonfly BionicOpter can control each of the four wings independently. One servo motor per wing controls the amplitude, another controls the twisting of the wing. Furthermore, the flapping frequency of all four wings can be controlled by the central motor and can be varied between 15 and 20 Hz.

All of the motors, as well as the SMAs, the on-board electronics and the communication systems are actuated by two LiPo cells with 7.4 V. While the motors and mechanic components are positioned within the housing the batteries are positioned in front of the housing to allow an easy and quick exchange. The batteries are covered by the head that nicely resembles the design of a dragonfly head. Furthermore, the whole design of the BionicOpter, especially the long and fragile tail closely follows the shape and design of the natural dragonfly.

In spite of its technical complexity the BionicOpter weighs just 175 g because of a consistently lightweight design and because of the fact that the housing, the tail, as well as the mechanical system were generated using laser-sintered polyamide and aluminium. Besides the low weight of the laser-sintering material, this technology allowed us to quickly generate new prototypes and to vary the wing position, the wing design etc. in just a few days. Besides polyamide and aluminium, other lightweight materials were used. Deep-drawn ABS terpolymer was used for the head and to cover the housing, while carbon-fibres and polyester membrane were used for the wings.

The final BionicOpter with a wingspan of 63 cm and a body length of 44 cm is almost as agile and fast as the natural dragonfly, not only because of its smart mechanics but also because of the on-board electronics and the accompanying software, which control the flapping frequency, amplitude and twisting angle. This way it becomes possible to control the BionicOpter with a digital spectrum transmitter via wireless remote control or even via a smartphone.

2.2   Actuation of Wings

As described above, dragonflies are such agile fliers because they can move their wings independently from each other. They can alter the frequency of the flapping, the amplitude of each wing stroke and the twisting angle of each wing. We reduced this complexity for space and weight reasons as well as to create a system that is easy to handle. One brushless VS external rotor serves as the main motor that actuates the flapping of all four wings. By altering the power of the motor, the frequency can be changed between 10 Hz the minimum necessary for stable flight and 20 Hz during acceleration. However, the final BionicOpter is easiest to handle when flapping at a frequency of about 15 Hz.

As Ratti and colleagues described [6], changing the frequency of an MAV with flapping wing drive can result in vibrations that decrease the energy efficiency, make the system hard to control and can even lead to damage the mechanical components when the vibrations build up. Here again we looked at the natural role model and how the dragonfly handles this issue. Real dragonflies only rarely move their forewings and hindwings in full synchrony (0° phase shift), i.e. all four wings beat up and down at the same time. Full synchrony actually only occurs during short phases of mating or hunting and is highly energy consuming [4, 7]. Moreover, dragonflies only rarely move their forewings and hindwings in an opposing fashion (180° phase shift), i.e. the forewings strike upwards while the hindwings strike downwards. Mostly, the wing pairs move in a phase shift of 50° to 100°, which decreases energy consumption especially during forward flight. Most interestingly for natural dragonflies the hindwing leads this movement and is thus moved upwards slightly before the forewing follows.

Although the phase shift was an initial source of inspiration for the design of the BionicOpter, for our system we found that vibrations were reduced best when the forewing leads the movement. A phase shift of ~40° allows a stable flight both indoors and outdoors.

The main motor actuates the four wings using a crankshaft for each wing. This crankshaft is used to convert the rotating movement actuated by the motor into an oscillating movement of the wing. The interesting aspect of the crankshaft is that the proximal part of the crankshaft leaves the housing containing the motor in a straight line. However, the distal part is tilted. Since the position of the mechanical crosspoint can be shifted along the crankshaft from a more distal to a more proximal position the offset of the crankshaft changes and by this the amplitude of the wing movement changes (See Fig. 4). The movement of the crosspoint is actuated by a servo motor. The resulting deflection varies between 80° and 130°.

A second servo motor actively changes the twisting angle of the wing by up to 90°.

Taken together, we can control the frequency of all four wings together, and the amplitude and twisting angle of each wing independently, resulting in 9 degrees of freedom. This way the direction of thrust and the intensity of thrust for all four wings can be adjusted individually, thereby enabling the remote-controlled dragonfly to move in almost any orientation in space.

2.3 Wing Design

The movement of the wings of the BionicOpter resembles the wing movement of the natural dragonfly very nicely. Furthermore the design of the wings was inspired by the shape and structure of the wings of the real dragonfly.

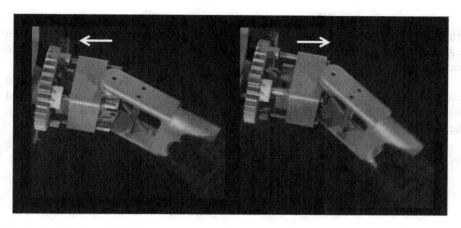

Fig. 4. Amplitude modulation. A crankshaft is used to convert the rotating movement of the motor (positioned outside of these image sections on the left) into an oscillating movement of the wing (lower right-hand corner of these image sections). The crankshaft leaves the housing containing the motor in a straight line. However, the distal part is tilted. The mechanical crosspoint can be shifted along the crankshaft as marked by the arrows using a servo motor. This enables the amplitude of the wing stroke to be modulated continuously (see animation for further details).

Wings of dragonflies are highly fascinating. They only weigh about 0.002 g although they span a distance of up to 10 cm. The wings consist of a thin transparent membrane strengthened by a number of longitudinal veins. Together with the cross-connections the veins form a beautiful pattern of cells. The front edge of the wing consists of much thicker veins with a thickness of up to 220 μm. The rear edge measures only 3 μm in thickness [3]. The front edge is thus much more rigid and stable than the flexible rear edge.

The same is true for the wings of the BionicOpter. They consist of a carbon-fibre frame enclosing a thin polyester membrane. Although the thickness of the carbon-fibre is the same across the wing, the width, and thus the stiffness, varies. The front edge has an increased width that decreases towards the tip of the wings and further decreases towards the rear edge leading to a stable and rigid front edge and a flexible rear edge. Based on this structure the wings show a passive-torsion effect during flapping. While flapping up and down the base of the front edge leads the movement but the tip of the wing and the rear edge can move like the sail of a sailing boat and thus form a profile as known for rotor blades. This increases the thrust during flying. However, the point of the passive-torsion has to be located along the outer part of the wing. When the length of the wing is increased or when the frequency of flapping is increased this point of passive-torsion moves along the front edge of the wing towards the centre of the MAV. As soon as the point of passive-torsion passes the middle of the wing, the wing stops flapping up and down. Instead, it shows an effect comparable to a double pendulum and all of the thrust suddenly ceases. The same occurs when the stiffness of the wing is decreased. Thus the wings were optimized for stiffness to allow passive torsion but to prevent a double pendulum effect.

In addition we found that shorter wings made the BionicOpter easier to handle during flight manoeuvres since shorter wings can flap at a higher frequency. With a higher frequency tilting of the wings and amplitude changes of each wing stroke become more immediate.

We tested several wing lengths at different flapping frequencies. Furthermore, we stabilized the wing by inserting cross-sections as apparent in the wings of natural dragonflies. Different patterns of cross-stabilizations can be seen in Fig. 5 as well as the final wing layout.

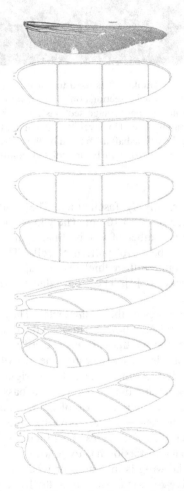

Fig. 5. Wing design. The drawing at the top is of a forewing of *Megatypus schucherti* [1] by Günter Bechly showing the structure and the vein patterning of a dragonfly wing. We tested several different designs and lengths of wings, four of these designs are displayed here. The drawing at the bottom represents the final wing design resulting in a wingspan of 63 cm.

2.4 Wing Positioning

When the first demonstrators of the dragonfly inspired MAV were designed and tested, we positioned the wings in correspondence to the natural role model, i.e. parallel to each other, at a 90° angle relative to the sagittal plane of the body. However, this design of a MAV was hard to control during flight. Here we decided to take inspiration from the technical world and looked at customary quadrocopters. Quadrocopters have four rotors. Each rotor generates a distinct momentum, which then sum up to yield the final force that makes the MAV fly. These four rotors are positioned at the edges of a square, since this positioning facilitates the corresponding calculations as well as the final handling of the MAV.

According to the design of quadrocopters we changed the positioning of the four wings of the BionicOpter away from a parallel position, by shifting the wings by 30°. Main forces generated by the four wings are thus better distributed and almost span across a square plane as valid for quadrocopters. When hovering with this wing positioning all four wings move forward and backwards in a horizontal plane instead of flapping up and down. The wings almost touch each other in front and behind the body, closing a 360° circle. This circle shows similarities in thrust generation with the circle spanned by the rotors of a helicopter. This clever design change made the MAV much easier to handle during flight (Fig. 6).

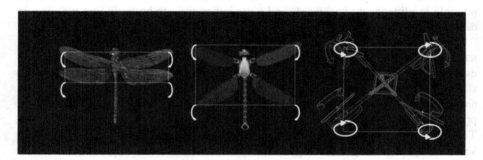

Fig. 6. Force distribution. On the left the natural dragonfly is displayed. The wings are almost parallel and the points of the main forces are positioned on the edges of an elongated rectangle. On the right a quadrocopter is displayed (picture from [8] adapted). The main forces are positioned at the edges of a square. We took inspiration from both and arranged the wings in an X-shape forming an angle of 30°. This made the final MAV easier to handle during flight.

3 Open- and Closed-Loop Control on Board

BionicOpter is a highly complex system. Yet, it is easy to operate using a digital spectrum transmitter or a smartphone. The flapping frequency, amplitude and installation angle are controlled by software and electronics; the pilot just has to steer the dragonfly. There is no need to coordinate the complex motion sequences. The signals from the remote control are transmitted via wireless modules.

During operation, the remote-control system transfers the signals that tell the object which direction to fly in and at what speed. A high-performance ARM microcontroller calculates all the parameters that can be adjusted mechanically based on the recorded flight data and the pilot's input. The processor actuates the nine servo motors to translate these parameters into movement using beat frequency, a swivel device and the amplitude controller.

In order to stabilize the flying object, data on the position and the twisting of the wings is continuously recorded and evaluated in real time during the dragonfly's flight. The acceleration and tilting angle of the BionicOpter in space can be measured using the inertia sensors. Furthermore, integrated position and acceleration sensors detect the speed and spatial direction of the dragonfly's flight.

4   Findings for Future MAV Design

After bird flight had been deciphered with the SmartBird in 2011, dragonflies were the next-bigger challenge for us. Dragonflies exceed birds especially in terms of agility. Fascinated by this agility we studied dragonfly flight and managed to transfer the dynamics into an ultralight flying object: the BionicOpter. Like its natural role model BionicOpter can manoeuvre in all directions, accelerate swiftly and turn abruptly. Furthermore we were able to demonstrate that with such an MAV design it becomes possible to glide, to fly backwards and most importantly to hover in mid-air, even in different orientations.

Only with a consistently lightweight design and by integrating all functions in the smallest space, it was possible to make BionicOpter fly. Continuous recording and evaluation of the position and twisting of the wings in real time was necessary to stabilize the flying object. The findings about lightweight design, function integration and especially about continuous diagnostics to guarantee operational reliability and process stability will be transferred to industrial use in factory automation by Festo. The BionicOpter itself, however, will not be turned into a product but will be presented as a fascinating technology carrier. In this sense it can inspire researchers in two ways: in a technical sense BionicOpter can inspire future design of MAVs that are able to hover in different positions as its natural role model. In a biological sense, it can be used to better understand dragonfly flight.

Like a helicopter dragonflies can hover on the spot, they can fly up and down, forward and even backwards in this flight conditions. Unlike a helicopter, however, dragonflies do not need to tilt forwards to generate forward thrust. This means that they can accelerate while keeping a horizontal position. Further, they can hover in the horizontal plane but also in the vertical plane. With these functions dragonflies exceed every customary quadrocopter and can only be compared to the newest quadrocopters with slewing rotators (see [8] for an interesting technical setup).

Hovering in several positions allows for more degrees of freedom for the flying object when interacting with a second object in its surroundings and thus will lead to new applications of MAVs in the future. In research it could be used for data and sample collection. Because of the extreme agility such an MAV could easily pass between the narrow branches of a tree and reach e.g. a bird's nest. The MAV could quickly fly towards the desired position in the horizontal, plane-like mode then move into the vertical and hover in this position in front of the interesting spot, circle around it and take pictures from different angles and possibly even take samples. In line with this example, a possible industrial application is the exploration of wind turbines. The blades and the rest of the turbines have to be checked for damage on a regular basis. MAVs make it easy to approach the blades at high altitudes. A dragonfly-inspired MAV however, could examine the damage from different positions and different angles. Furthermore, with an attached tool, it could even repair or exchange small parts.

These are just some possible applications which can be solved with an MAV able to hover in different orientations. Dragonflies can do so very easily. Furthermore they can even fly in different orientations. This will open up further ideas for applications. With BionicOpter we showed that it is possible to transfer the flight dynamics of a dragonfly to the technical world. In this sense we hope that dragonfly flight will inspire the research community to think about the design of future MAVs in a new way.

Finally, and most interestingly, by studying BionicOpter further and by comparing it to real dragonflies, we can learn a lot about the incredible dynamics, the mechanics and especially the behaviour of the fascinating natural role model: the dragonfly (Fig. 7).

Fig. 7. Different positions in space while hovering. With dragonfly-inspired MAVs it becomes possible to take different orientations in space while flying or hovering by twisting the four wings in regards to the body plane. Here the different wing positions during hovering are shown, when BionicOpter hovers in the vertical (upper picture) or horizontal (lower picture) plane. An intermediate state is shown as well.

Acknowledgment. Numerous people contributed by adding important knowledge regarding flight dynamics, electronics, 3D printing, etc. and helped with generating pictures, animations and videos. The authors want to thank all of these helpful colleagues.

Appendix

Videos showing BionicOpter's flight behaviour and an animation describing the actuation of the BionicOpter can be found here: www.youtube.com/user/FestoHQ.
Real-world recordings: http://youtu.be/nj1yhz5io20.
Animation: http://youtu.be/JUAD7nhyzhU.

References

1. Bechly, G., et al.: New results concerning the morphology of the most ancient dragonflies (Insecta: Odonatoptera) from the Namurian of Hagen-Vorhalle (Germany). J. Zool. Syst. Evol. Res. **39**, 209–226 (2001)
2. Probst, E.: Wer war der Stammvater der Insekten?: Interview mit dem Stuttgarter Biologen und Paläontologen Dr. Günter Bechly. Grin Verlag, Santa Cruz (2012)
3. Zhao, H.X., Yin, Y.J., Zhong, Z.: Assembly modes of dragonfly wings. Microsc. Res. Tech. **74**(12), 1134–1138 (2011)
4. Silsby, J., Trueman, J.: Dragonflies of the World. CSIRO Publishing, Collingwood (2001)
5. Simmons, P.: The neuronal control of dragonfly flight. I. Anatomy. J. Exp. Biol. **71**, 123–140 (1977)
6. Ratti, J., Jones, E., Vachtsevanos, G.: Fixed frequency, variable amplitude (FiFVA) actuation systems for micro aerial vehicles. In: IEEE International Conference on Robotics and Automation, Shanghai, China (2011)
7. Wakeling, J., Ellington, C.: Dragonfly flight. II. Velocities, accelerations and kinematics of flapping flight. J. Exp. Biol. **200**(Pt 3), 557–582 (1997)
8. Ryll, M., Bülthoff, H.H., Giordano, P.R.: First flight tests for a quadrotor UAV with tilting propellers. In: IEEE International Conference on Robotics and Automation, Karlsruhe, Germany (2013)

Computer Vision

Sensor Data Fusion Using Unscented Kalman Filter for VOR-Based Vision Tracking System for Mobile Robots

Muhammad Latif Anjum[1](✉), Omar Ahmad[1], Basilio Bona[2],
and Dong-il "Dan" Cho[3]

[1] Department of Mechanical and Aerospace Engineering (DIMEAS),
Politecnico di Torino, Corso Duca degli Abruzzi 24 10129 Turin, Italy
{muhammad.anjum, omar.ahmad}@polito.it
[2] Department of Control and Computer Engineering, (DAUIN),
Politecnico di Torino, Corso Duca degli Abruzzi 24 10129 Turin, Italy
basilio.bona@polito.it
[3] Department of Electrical Engineering and Computer Science/ASRI/ISRC,
Seoul National University, 1 Gwanak-ro, Gwanak-gu, Seoul 151-742, Korea
dicho@snu.ac.kr

Abstract. This paper presents sensor data fusion using Unscented Kalman Filter (UKF) to implement high performance vestibulo-ocular reflex (VOR) based vision tracking system for mobile robots. Information from various sensors is required to be integrated using an efficient sensor fusion algorithm to achieve a continuous and robust vision tracking system. We use data from low cost accelerometer, gyroscope, and encoders to calculate robot motion information. The Unscented Kalman Filter is used as an efficient sensor fusion algorithm. The UKF is an advanced filtering technique which outperforms widely used Extended Kalman Filter (EKF) in many applications. The system is able to compensate for the slip errors by switching between two different UKF models built for slip and no-slip cases. Since the accelerometer error accumulates with time because of the double integration, the system uses accelerometer data only for the slip case UKF model. Using sensor fusion by UKF, the position and orientation of the robot is estimated and is used to rotate the camera mounted on top of the robot towards a fixed target. This concept is derived from the vestibulo-ocular reflex (VOR) of the human eye. The experimental results show that the system is able to track the fixed target in various robot motion scenarios including the scenario when an intentional slip is generated during robot navigation.

Keywords: Sensor fusion · Unscented Kalman filter · VOR · Vision tracking systems

1 Introduction

Applications of robotics in various fields of science and engineering are increasing. Robots need a set of accurate and reliable sensors and control techniques to perform various tasks assigned to them. Accurate vision systems are especially crucial for

A. Natraj et al. (Eds.): TAROS 2013, LNAI 8069, pp. 103–113, 2014.
DOI: 10.1007/978-3-662-43645-5_12, © Springer-Verlag Berlin Heidelberg 2014

robots involved in target tracking, visualization and vision based decision making. Accurate localization of mobile robots is an essential requirement for mobile robots navigation and target tracking. Various sensors have been utilized to achieve an accurate positioning system for mobile robots. The system proposed in this paper estimates the robot motion information and further uses this information to implement vision tracking system by rotating the camera mounted on the mobile robot towards the target. This concept is based on the Vestibulo-Ocular Reflex (VOR) of the human eye. The system can be divided into two blocks. Block 1 estimates robot motion information using sensor data fusion and block 2 uses this information to rotate the camera mounted on the robot to track the fixed target.

Shim, E.S. *et al.* presented a stable vision system for mobile robots using encoder data [2]. It detects a target within a few meters and maintains it fixed in the center of the image frame during locomotion. The proposed system uses encoder data to calculate robot motion and periodically uses vision sensor signals to compensate for the errors. Their system is highly dependent on vision sensor information in the slip case and therefore performance deteriorates.

Jaehong Park *et al.* developed a high performance vision tracking system for mobile robots using sensor data fusion via Kalman filter [1]. The robot motion information is computed by low cost accelerometer data, gyroscope data, and encoder data. The vision information is obtained by camera images of the object on locomotion during vision tracking. Researchers have also used fuzzy controller [10] and double Kalman filters [9, 11] to obtain vision tracking systems for mobile robots.

We are using UKF which is an advanced filtering technique as compared with EKF. Furthermore, we are no longer using vision sensor (camera) image information which makes the system computationally heavy causing problems for tracking the target in continuous real time environment. UKF is much more efficient to deal with systems with severe nonlinearities because it is not dependent on first order linearization. The concept of unscented Kalman filter, also known as Sigma Point Kalman Filter (SPKF), has been studied by many researchers. Julier and Uhlmann compared the performance of this new filter (UKF) with the previously used EKF [6]. They argued that because EKF is based on the first order linearization of the nonlinear systems, its performance deteriorates especially when the systems are highly nonlinear. They introduced this new filter which is based on selecting a deterministic set of points, called the sigma points. These sigma points capture the true mean and covariance of the system when propagated through the nonlinear systems.

2   The Vision Tracking System

The vision tracking system implemented in this paper is inspired by Vestibulo-Ocular Reflex (VOR) of the human eye. The vestibulo-ocular reflex is the eye reflex movement that stabilizes images on the retina during head movement, by producing an eye movement in the direction opposite to head movement, thus preserving the image on the center of the visual field. For example, when the head moves to the right, the eyes move to the left, and vice versa. Since slight head movement is present all the time, the VOR is very important for stabilizing vision. Patients whose VOR is impaired find it difficult to read, because they cannot stabilize the eyes during small head tremors. The VOR does not depend on visual input and works even in total darkness or when the eyes are completely closed [4].

The VOR has both rotational and translational aspects. When the head rotates about any axis (horizontal, vertical, or torsional), distant visual images are stabilized by rotating the eyes about the same axis, but in the opposite direction. When the head translates, for example during walking, the visual fixation point is maintained by rotating the gaze in the opposite direction, based on the translational distance covered by the head.

The vestibulo-ocular reflex needs to be fast. For clear vision, head movement must be compensated almost immediately; otherwise, vision corresponds to a photograph taken with a shaky hand. To achieve clear vision, signals from the semicircular canals are sent as directly as possible to the eye muscles. The connection involves only three neurons, and is correspondingly called the three neuron arc [4]. Using these direct connections, eye movement lags the head movement by less than 10 ms, and thus the vestibulo-ocular reflex is one of the fastest reflexes in the human body.

The system proposed in this paper uses data from various sensors onboard the robot to detect the robot motion, and then rotates the camera based on the robot motion information. The whole concept can be partitioned into two blocks:

- Robot motion information using sensor fusion by UKF
- Camera rotation in opposite direction of the robot motion (VOR based concept)

Low cost inertial sensors (MEMS based gyroscope and accelerometer) are used along with robot wheel encoders to implement the robot motion information algorithm. The motion information obtained from the motion information block is used to rotate the camera towards the target based on the VOR concept of the human eye.

3   The Mathematical Modeling

The robot position and orientation can be estimated based on the encoder measurements using the following state model. The linear and angular velocities v, w can be obtained from differential drive kinematics.

$$p_k = \begin{bmatrix} x_k \\ y_k \\ \theta_k \end{bmatrix} = \begin{bmatrix} x_{k-1} + v_{k-1} \cos(\theta_{k-1})\Delta t \\ y_{k-1} + v_{k-1} \sin(\theta_{k-1})\Delta t \\ \theta_{k-1} + w_{k-1}\Delta t \end{bmatrix} \tag{1}$$

Odometry is based on the assumption that wheel revolutions can be translated into linear displacement relative to the floor. This assumption is only of limited validity. One extreme example is wheel slippage: if one wheel was to slip on, say, an oil spill, then the associated encoder would register wheel revolutions even though these revolutions would not correspond to a linear displacement of the wheel. To avoid this error, our proposed system does not use encoder data in the case of slippage.

Gyroscopes measure the angular velocities which can be integrated to give the orientation. Gyroscopes have bias and drift errors which should be properly tackled to avoid the unbounded accumulation of errors in position and orientation [3]. The data from gyroscope can be used to calculate robot orientation as follows.

$$\theta_k = \theta_{k-1} + w_k \Delta t \tag{2}$$

The accelerometer measures the linear acceleration of the robot which can be integrated to give velocity and position for mobile robots. Accelerometer data is very noisy because it naturally incorporates the gravity vector. Linear position estimation with information from accelerometers is more susceptible to errors due to the double integration process. The following equations are used to obtain linear position from accelerometer data.

$$v_{k+1} = v_k + a_k \Delta t$$
$$x_{k+1} = x_k + v_k \Delta t \tag{3}$$

We have used a two-wheeled robot with 3 degrees of freedom containing motion in x/y axis and rotation in the z-axis. Two coordinate frames are used to model the robot navigation; the earth coordinate frame (X, Y, Z), and the robot coordinate frame (x, y, z) as shown in Fig. 1.

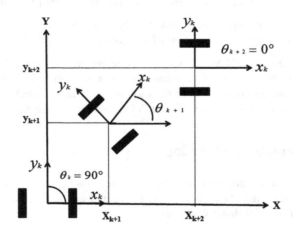

Fig. 1. The coordinate system used for robot navigation

4 The Unscented Kalman Filter

The unscented Kalman filter (UKF) is a recursive minimum mean square estimator (MMSE) based on the optimal Gaussian approximate Kalman filter framework that addresses some of the approximation issues of the EKF. Unlike the EKF, the UKF does not explicitly approximate the nonlinear process and observation models; the state distribution is still represented by a Gaussian random variable (GRV), but it is specified using a minimal set of deterministically chosen sample points called the sigma points. These sample points completely capture the true mean and covariance of the GRV, and when propagated through the true nonlinear system, captures the posterior mean and covariance accurately to the 2nd order for any nonlinearity, with errors only introduced in the 3rd and higher orders [6].

An unscented transformation is based on two fundamental principles. First, it is easy to perform a nonlinear transformation on a single point (rather than an entire *pdf*). Second, it is not too hard to find a set of individual points (sigma points) in state space whose sample *pdf* approximates the true *pdf* of a state vector. These sigma points $x^{(i)}$ can be calculated from the mean $\bar{x}$ and a deviation from the mean $\chi^{(i)}$ obtained from the square root decomposition of covariance matrix P. For an n-element state vector x, the sigma points can be calculated as follows:

$$x^{(i)} = \bar{x} + \chi^{(i)} \qquad i = 1, \ldots, 2n$$

$$\chi^{(i)} = \left(\sqrt{nP}\right)^T_i \qquad i = 1, \ldots, n \tag{4}$$

$$\chi^{(n+i)} = -\left(\sqrt{nP}\right)^T_i \qquad i = 1, \ldots, n$$

where $\sqrt{nP}$ is the matrix square root of nP such that $\left(\sqrt{nP}\right)^T \left(\sqrt{nP}\right) = nP$, $\left(\sqrt{nP}\right)_i$ is the ith row of $\sqrt{nP}$, $\bar{x}$ is mean and P is the error covariance of the state vector x. These sigma points when propagated through the nonlinear equation capture the true mean and covariance of the random variable. The mathematical details of UKF algorithm have been intentionally skipped and readers are referred to [5–8] for more details.

4.1   The UKF State Model

The state model for the UKF is formulated based on the error states, i.e., the difference between the encoder and gyroscope measurement is used as state variable for the UKF. The 3-element state vector consists of position error states e_x, e_y, e_θ. This can be obtained by calculating the position of the robot based on gyroscope and encoders separately and then taking the difference of the two values. The results obtained from the UKF (error states) are then added to the encoder based position values to exactly find the position of the robot. The state vector x_k therefore serves as an error compensator for encoder data. The state model of UKF can be mathematically put as:

$$x_k = \begin{bmatrix} e_{x,k} \\ e_{y,k} \\ e_{\theta,k} \end{bmatrix} = \begin{bmatrix} e_{x,k-1} + v_{k-1}(\cos\theta_{g,k-1} - \cos\theta_{e,k-1})\Delta t \\ e_{y,k-1} + v_{k-1}(\sin\theta_{g,k-1} - \sin\theta_{e,k-1})\Delta t \\ e_{\theta,k-1} \quad + \quad (w_{g,k-1} - w_{e,k-1})\Delta t \end{bmatrix} + w_{k-1} \tag{5}$$

The subscript 'g' and 'e' refer to the data from gyro and encoder, and w is the zero mean process noise with covariance Q_k. The error state '$e_{x/y/\theta}$' is defined as the difference in gyro and encoder measurement. This can be mathematically put as:

$$x_k = \begin{pmatrix} e_{x,k} \\ e_{y,k} \\ e_{\theta,k} \end{pmatrix} = \begin{pmatrix} p_{x,gyro} - p_{x,enc} \\ p_{y,gyro} - p_{y,enc} \\ \theta_{gyro} - \theta_{enc} \end{pmatrix} \tag{6}$$

In the case of slip detection, the encoder data is no more reliable, so a different formulation of the above state vector is used. When slip is detected, we replace the encoder data with the accelerometer data in the above equation.

4.2   The UKF Measurement Model

The measurement model is based on the difference of the orientation of the robot measured by the gyroscope and encoders. This model can be mathematically written as follows:

$$z_k = e_{\theta,k} + v_k = \begin{bmatrix} 0 & 0 & 1 \end{bmatrix} \begin{bmatrix} e_{x,k} \\ e_{y,k} \\ e_{\theta,k} \end{bmatrix} + v_k \tag{7}$$

$$e_{\theta,k} = \begin{bmatrix} \theta_{gyro} - \theta_{enc} \end{bmatrix}$$

Where v_k is the zero mean measurement noise with covariance R_k. As in the case of state model, the measurement model is also changed when slip occurs. Here again, the encoder data is replaced with the accelerometer data in case of slip.

Finding the UKF noise parameters (Q, R) is a tedious job and requires repeated experiments with the model and sensors. The diagonal elements of the measurement noise covariance matrix represent the square of the standard deviation of the error in corresponding parameters. So the measurement noise covariance matrix can be found by experimenting several times with the robot encoders, gyroscope, and accelerometer to calculate the square of the standard deviation of the error. Finding the process noise covariance is more subtle. Often times the best method of estimating Q is by tuning of filter i.e. adjusting the value of Q to obtain the optimal state [12].

5   The VOR Modeling

After the execution of the first block, i.e. the robot motion information block, we now have the robot position information with respect to the previous position. Figure 2 shows the coordinate system used to calculate the camera rotation angle.

The camera rotation angle can be calculated from the following equation

$$\theta_{rotate} = \tan^{-1}\left(\frac{r_{y,k}}{r_{x,k}}\right) = \tan^{-1}\left(\frac{-(X - p_{x,k})\sin\theta_{heading} + (Y - p_{y,k})\cos\theta_{heading}}{(X - p_{x,k})\cos\theta_{heading} + (Y - p_{y,k})\sin\theta_{heading}}\right) \tag{8}$$

The DC motor is used to rotate the camera towards the target based on the voltage generated corresponding to the θ_{rotate}.

6   The Slip Detector

Our system has two different models for slip and no-slip cases. The slip therefore must be detected beforehand. The data from accelerometer and encoders is first compared to ascertain the occurrence of slip. The occurrence of slip is confirmed if the difference

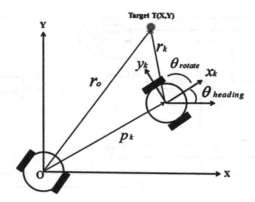

Fig. 2. Coordinate frame for calculating camera rotation angle

of encoder and accelerometer position is greater than a threshold value k. The threshold value can be adjusted as desired and should also accommodate the accelerometer errors. Based on the output of this slip detector, either of the two UKF models is used (Fig. 3).

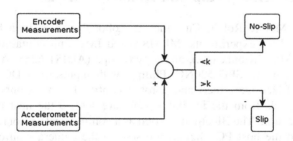

Fig. 3. The slip detector

7 The Complete System

The complete system consists of two UKF models incorporating data from three sensors i.e. gyroscope, accelerometer and encoders. The two UKF models are labeled as the slip UKF model and no-slip UKF model. Based on the output of this slip detector, either of the two UKF models is used. The slip-UKF model integrates data from inertial sensors only while the no-slip UKF model integrates data from encoders and gyroscope. The transition between the UKF models will be automatic based on whether slip occurs or not. The overall system is shown in the block diagram below (Fig. 4).

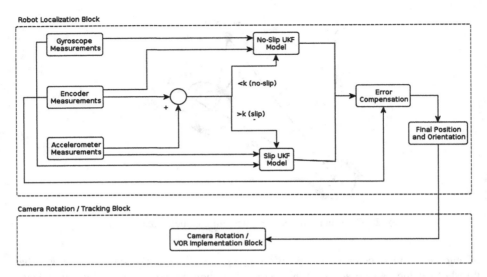

Fig. 4. The complete system including the slip detector. The constant value, k, is chosen to accommodate the error in accelerometer data.

8 The Experimental Setup and Results Analysis

A mobile robot (Mobile Robot, Customer & Robot Co., Ltd) with build-in wheel encoders was used for experiments. MEMS based IMU unit containing an accelerometer (KXPS5-3157, Kionix Inc.), and a gyroscope (ADIS1 6255, Analog Devices Inc.), a rotating camera (SPC 520NC, Philips) with inputs from DC motor (Series 2619, MicroMo Electronics Inc.) and a robot control PC were onboard the robot platform. The signals from the inertial sensors are sent to the host control PC via RS232 communication. The Bluetooth is used as a communication source between the robot control and the host PC. Signals are sent to the camera control motor using RS485 communication protocol. The program to get the sensor data from encoders and inertial sensors is written in Microsoft Visual C++ 6.0 (Fig. 5).

Fig. 5. System configuration

8.1   The Localization Experiments

Experiments are first conducted to test the robot motion information block (block1). The results have been tested by comparing position output with an accurate distance laser sensor (DLS-BH 30, Dimitix). Intentional slip is generated by rolling a paper sheet under the robot wheels. It would be interesting to compare block1 results in slip case. During the slip experiment, using only encoder data gives 238 mm position error (for a 1 m linear motion) while our UKF system gives only 96 mm error. The comparison of our system block1 performance with encoders and laser sensor is given in Fig. 6.

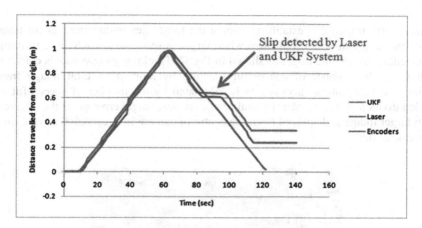

Fig. 6. Distance from the origin as measured by our algorithm, encoders and accurate laser sensor during simple translational motion (slip case).

8.2   The Tracking System Experiments

Experiments were also conducted to test the accuracy of the vision tracking system. The robot camera is initially fixed at the target. The robot is then moved in various trajectories to test the efficiency of vision tracking system. The SURF (Speeded Up Robust Features) algorithm is used to detect the target inside the camera image. The following three types of experiments were conducted:

1. Simple Translational Motion: Robot moved 1 m forward, and then moved 1 m backward. The initial distance between the robot camera and the target was kept at 1.5 m and the velocity of the robot is 0.2 m/s.
2. Translational Rotational Combined Motion: Robot moved 1m forward, rotated 90° counterclockwise and then moved 1 m forward. The initial distance between the robot camera and the target is once again kept at 1.5 m and the translational and rotational velocities of robot are 0.2 m/s and 30°/s respectively.
3. The Square Motion: The robot followed the square path described in Fig. 6. The initial distance, translational and rotational velocities of robot remain same.

The results obtained are summarized in the table below (Table 1).

Table 1. Summary of the experimental results

Experiment	Tracking success rate (%)	Pixel error (RMS) (pixel)	Angle error (RMS)	Recognition success rate (%)
Experiment 1 Linear motion	100	21.62	1.96°	100
Experiment 2 Rotation + translation	99	32	2.83°	92
Experiment 3 Square motion	96	33	2.59°	92

Successful tracking is established when the target lies within the camera image. Successful recognition is established when target is recognized by SURF, forming a red boundary around the target as shown in Fig. 7. Tracking success rate is calculated by dividing the number of successful tracking by number of total image frame. Similary the recognition success rate is calculated by the number of successful recognition divided by the number of total image frames. Angle error gives the degree by which target image is displaced from the centre of camera image. SURF is used for all those calculations.

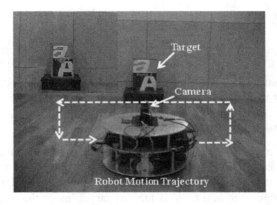

Fig. 7. Experiment 3: square motion

9 Conclusion

We have presented the implementation of a vestibulo-ocular-reflex (VOR) based vision tracking system for mobile robots. Since the cost of inertial sensors is very low and most of the mobile robots have built-in wheel encoders, this system can be implemented with a very low cost. The system used an advanced filtering technique called the unscented Kalman filter (UKF) for sensor data fusion.

The system is also designed to work in the case of slip in the robot wheels. The system detects the slip by comparing data from encoders and accelerometer and accordingly switches to one of the sensor fusion model built for slip and no-slip cases. The no-slip sensor fusion model integrates data from encoders and gyroscope while the slip case sensor fusion model integrates data from gyroscope and accelerometer. The system implemented in this work has been experimentally tested for accuracy and gives satisfactory results both in slip and no-slip cases. However, the performance of the system is not up to the expected level in the case of robot wheel slip. This is due to the fact that the slip sensor fusion model uses data from accelerometer, which is very noisy. A more careful and robust modeling of the accelerometer noises is required to achieve better performance in the case of slip.

References

1. Jaehong, P., Wonsang, H., Hyun-il, K., Jong-hyeon, K., Chang-hun, L., Anjum, M.L., Kwang-soo, K., Cho, D.: High performance vision tracking system for mobile robot using sensor data fusion via Kalman filters. In: Proceedings of IEEE/RSJ International Conference on Intelligent Robots and Systems (IROS 2010), Taipei, Taiwan
2. Shim, E.S., Hwang, W., Anjum, M.L., Kim, H.S., Park, K.S., Kim, K., Cho, D.: Stable vision system for indoor moving robot using encoder information. In: Proceedings of 9th IFAC Symposium on Robot Control (SYROCO) (2009)
3. Barshan, B., Durrant-Whyte, H.F.: Inertial navigation systems for mobile robots. IEEE Trans. Robot. Autom. **11**(3), 328–342 (1995)
4. Dieterich, M., Brandt, T.: Vestibulo-ocular reflex. Curr. Opin. Neurol. **8**(1), 83–88 (1995)
5. Merwe, R.V., Wan, E.A., Julier, S.I.: Sigma-point Kalman filter for nonlinear estimation and sensor-fusion, application to integrated navigation. In: Proceedings of the American Institute of Aeronautics and Astronautics Guidance Navigation and Control Conference (2004)
6. Julier, S., Ulhmann, J., Durrant-Whyte, H.F.: A new method for the nonlinear transformation of means and covariances in filters and estimators. IEEE Trans. Autom. Control **45**(3), 477–482 (2000)
7. Van der Merwe, R.: Sigma-point Kalman filter for probabilistic inference in dynamic state space models. Ph.D. Thesis, OGI School of Science and Engineering at Oregon Health & Science University, Portland, OR (2004)
8. Simon, D.: Optimal State Estimation: Kalman, H-Infinity, and Nonlinear Approaches, 1st edn. Wiley, Hoboken (2006)
9. Hwang, W., Park, J., Kwon, H., Anjum, M.L., Kim, J., Lee, C., Kim, K., Cho, D.D.: Vision tracking system for mobile robots using two Kalman filters and a slip detector. In: Proceedings of International Conference of Control, Automation and Systems (ICCAS), October 2010
10. Kwon, H., Park, J., Hwang, W., Kim, J., Lee, C., Anjum, M.L., Kim, K., Cho, D.: Sensor data fusion using fuzzy control for VOR-based vision tracking system. In: Proceedings of IEEE/RSJ International Conference on Intelligent Robots and Systems, (IROS 2010), Taipei, Taiwan (2010)
11. Jia, Z., Balasuriyaa, A., Challa, S.: Sensor fusion-based visual target tracking for autonomous vehicles with the out-of-sequence measurements solution. Robot. Auton. Syst. **56**(2), 157–176 (2008)
12. Anjum, M.L., Jaehong, P., Wonsang, H., Hyun-il, K., Jong-hyeon, K., Changhun, L., Kwang-soo, K., Cho, D.: Sensor data fusion using unscented Kalman filter for accurate localization of mobile robots. In: Proceedings of International Conference of Control, Automation and Systems (ICCAS), October 2010

Visual Homing of an Upper Torso Humanoid Robot Using a Depth Camera

Alan Broun[1]([✉]), Chris Beck[2], Tony Pipe[1], Majid Mirmehdi[2],
and Chris Melhuish[1]

[1] Bristol Robotics Laboratory, Bristol, UK
{alan.broun,tony.pipe,chris.melhuish}@brl.ac.uk
[2] Visual Information Laboratory, University of Bristol, Bristol, UK
{csxcb,majid}@compsci.bristol.ac.uk

Abstract. We present a system which can automatically home an upper torso humanoid robot so that its true joint angles are known using only information from a depth camera, coupled with incremental encoders. This means that extra components, such as absolute encoders, resolvers, homing switches etc., are not needed. This in turn means that the cost of the system may be reduced and reliability improved as potential component failures are eliminated. The system uses exploratory moves to locate the robot's end effector and to measure the pose of the robot's wrist frame in camera space. Multiple measurements of the wrist frame pose are combined together in a kinematic calibration step to obtain the true joint angles and thus home the robot. We conduct experiments to explore the accuracy and reliability of our homing system, both in simulation and on a physical robotic platform.

1 Introduction

In order to make use of a kinematic model, a robot needs to know the current angle of its joints relative to a starting or *home* position. During operation, the joint angles relative to the home position can be tracked with incremental joint encoders. However, if power is lost, or if the system shuts down, the robot may lose track of its joint angles, because the joints may move before power is restored. For example, incremental encoders give an angular reading relative to the position of the joint at power on and so movements using these angles will be inconsistent, unless the robot is powered on at exactly the same starting position each time it is used.

Modern robotic systems employ a variety of methods for coping with a lack of knowledge about the home position. For robots having only incremental encoders, the simplest solution is to always start the robot from a known home position. However, it can be tedious to manually position the robot and it may be hard to do this with a high level of accuracy. Another method is to install homing switches on the robot, or to have hard mechanical end-stops that robot joints can be driven against, so that a control program can automatically seek

A. Natraj et al. (Eds.): TAROS 2013, LNAI 8069, pp. 114–126, 2014.
DOI: 10.1007/978-3-662-43645-5_13, © Springer-Verlag Berlin Heidelberg 2014

and find the home position. Mechanical end-stops may not always be available however, i.e. on joints designed for continuous rotation, and homing switches increase system complexity and may fail.

Absolute encoders or resolvers solve the problem of losing the position relative to a home position, but absolute encoders can often be both more expensive and have lower resolution than incremental encoders. This is because incremental encoders can be implemented cheaply with hall sensors, and get a resolution increase from any reduction gearing which may be present in the joint.

In this paper, we present an alternative approach to homing which is to have the robot autonomously find its home position using visual information from a depth camera, such as a Microsoft Kinect. We envisage that this system would find use on service robots in the home where the robot may only have incremental encoders in order to keep costs low, but at the same time is likely to have a depth camera available as a cheap and flexible sensor modality.

The system seeks to find the home position for a known kinematic model. The pan and tilt joints in the neck of the robot are first homed by locating an Augmented Reality (AR) marker board, fixed to the base of the robot. After this, the hand of the robot is located in the robot's vision system by using exploratory motions. By tracking the movement of the hand, using depth data from the Kinect, it is possible to measure the position and orientation of the robot's wrist frame in Kinect camera space. At this point the system then uses inverse kinematics to identify possible joint configurations which would match the observation.

As the inverse kinematics of the robot may give rise to multiple possible joint configurations for a given pose of the wrist frame in camera space, we need a way of choosing between possible joint configurations to identify the correct one. This is done by having the robot move its hand to different points in its field of view in order to take more measurements of its wrist frame. These extra measurements are combined with the initial wrist pose in a kinematic calibration step, which seeks to find the set of joint angles which best explains the observed wrist frame measurements. Finally, the estimated joint angles can be compared to joint angles measured by the robot's incremental encoders, in order to find the robot's offset from its true position. This effectively homes the robot.

Visual homing of a humanoid robot's joints does not seem to have been the subject of much research so far. Jin and Xie [7] describe a system that performs visual homing by detecting and tracking a number of feature points on a robot's arm. They do not give a quantitative evaluation of the accuracy of their system however, and so it is not clear how robust their system would be when viewing the feature points from oblique angles or in the presence of varying illumination. Santos et al. [9] give an elegant method that uses homography estimation to home the panning joints in the pan-tilt joints of an iCub's head. However, as this method seems to be restricted to pan-tilt joints that move a camera, it would not be suitable for homing joints in a robot's arms.

The main contribution of this paper is to present a novel visual homing system that uses motion, and the data from a depth camera, in order to achieve reliable homing of an upper torso humanoid robot. Also, another contribution we make is to explicitly consider how the joint limits of the robot and uncertainty about the starting position of the robot restrict the areas of joint space in which the robot can move whilst homing itself.

An important point to make is that, whilst we describe the use of our system on an upper torso humanoid robot, there is no reason why the method could not be applied to a variety of other robotic systems.

The rest of the paper proceeds as follows. Section 2 describes the robotic platform used for our work, Sect. 3 presents our method for performing visual homing, and Sect. 4 describes the experiments we carried out to validate our system and to determine its accuracy. Finally, Sect. 5 discusses the system performance and ideas for future work.

2   Robotic Platform

The work presented in this paper was carried out using a Bristol and Elumotion Robotic Torso (BERT) robot (Fig. 1a). This is an upper torso humanoid robot which is composed of harmonic drives and brushless DC motors. It also has a Microsoft Kinect mounted on its neck, which acts as its vision system.

Figure 1b shows the kinematic model of BERT that we are trying to home. The model is expressed with Denavit-Hartenberg parameters [4], and in homing BERT, we are essentially trying to find the joint angle offsets that will bring the model into line with the true position of the robot. In this work, we focus on homing the robot's neck and its left arm comprising of 9 degrees of freedom in total. The techniques presented are equally applicable to the right arm.

3   Visual Homing

Our method for Visual Homing is outlined in Fig. 2. The robot when turned on, first homes its neck motors by locating an AR marker board fixed to its

(a)   (b)

Fig. 1. (a) BERT, the robotic torso used in our work. (b) Kinematic model of BERT with his left arm raised, showing the camera and wrist coordinate frames.

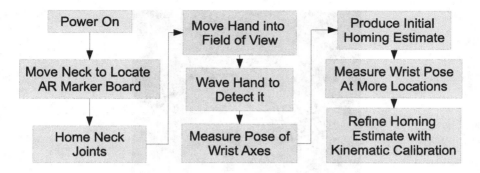

Fig. 2. An overview of the visual homing process.

base. It then proceeds to move its arm, to bring its hand into the field of view of its camera. The hand is identified through the use of exploratory waving motions, which allows the robot to distinguish its hand from the background. Once the hand has been identified, we are able to use the Iterative Closest Point (ICP) algorithm [1] to track small movements of the hand in 3D camera space. This ability means that the current position and orientation of the robot's wrist axes can be measured in camera space. Being able to measure the position and orientation of the wrist axes exposes the current pose of the wrist frame. At this point, inverse kinematics is used to find a feasible pose for the robot which could give rise to the observed wrist frame.

An alternative to measuring the pose of the wrist frame would be to measure the pose of the robot's hand, as the centre of the robot's hand is related to the wrist frame by a fixed 3D transformation. The measurement could be performed using 3D object pose estimation techniques, such as those presented by Hinterstoisser et al. [5]. However, we found that we were more reliably able to measure the pose of the wrist frame using motion.

If there was just one possible set of joint angles, which could give rise to an observed pose for the wrist frame in camera space, then at this point the robot would be homed. We could simply compare the joint angles we read from the incremental encoders, with the joint angles from the inverse kinematics, and this would give us the necessary angle offsets needed to home the robot. Unfortunately, with a redundant kinematic chain such as our robot's arm, there may be infinitely many poses which could give rise to a particular wrist frame measurement in camera space. Therefore, more measurements of the pose of the wrist frame are taken, and the estimate of the joint angles is refined using kinematic calibration.

3.1 The Exploratory Joint Space

Making use of visual homing carries with it some level of risk, in that if a joint is powered on too close to its joint limit, then the exploratory motions needed to perform visual homing may take the joint beyond its range of motion. Depending upon the design of the robot, this may cause physical damage. For example, on

Fig. 3. The AR marker board used to home the neck joints.

our robot there are some joints where movement beyond safe limits will cause cables to break. We mitigate this risk by assuming that each joint starts within a certain distance α of its true zero position. This is reasonable because it is not an onerous task to make sure that the robot is roughly in a pose close to the true zero position before start up. The greater challenge is ensuring that the robot is finely homed, so that it can do useful work, and that problem is addressed by this paper.

Assuming that a joint is subject to limits that confine it to the angular range $[q_{low}, q_{high}]$ where, without loss of generality, $q_{low} \leq 0$ and $q_{high} \geq 0$, knowing that the initial zero angle of the encoder is within α degrees of the real zero of the encoder means that the joint limits are reduced to $[\min(q_{low} + \alpha, 0), \max(q_{high} - \alpha, 0)]$.

We describe the initial homing error to which the robot is subject to, as the homing *offset*. This emphasises the fact that in attempting to home the robot, we are looking for the distance that the robot's joints were from their home positions when the robot was powered on. An offset $\mathbf{o}$ is a vector in $\mathbb{R}^N$ where $\mathbf{o} = \{\alpha_1, \alpha_2, ..., \alpha_N\}$ and N is the number of joints to be homed.

The initial distances α_i, may be different for each joint. The only requirement is that the resulting limits should provide enough movement to locate the AR marker board used for homing the neck and enough movement so that the hand can be brought within the field of view of the robot's camera. Once the robot has been homed, the joint limits can be restored to their unrestricted form, and the possible range of motion of the robot is extended.

3.2   Homing the Neck Joints

The pan and tilt joints in the neck of the robot are homed by locating an AR marker board in the robot's field of view. An AR marker board is a collection of AR markers arranged in a grid. We use the ArUco library[1] to locate and

---

[1] ArUco: http://www.uco.es/investiga/grupos/ava/node/26

track AR markers. By looking for a collection of AR markers, the detection and localisation process is made more robust. Figure 3 shows the AR marker board we use. We assume that for a pair of pan and tilt angles, q_{pan} and q_{tilt}, we know what the image coordinates of the centre of the AR marker board (x_b, y_b) should be.

Upon powering on, the robot performs a search from left to right and top to bottom in the exploratory joint space of its two neck joints, in order to locate the AR marker board. When the AR marker board is found, the robot adjusts its joint angles in order to bring the marker board to the recorded position (x_b, y_b) in the image space of the Kinect. When the marker board has reached this position, we know the angles of the neck joints, and therefore, the neck joints of the robot are successfully homed.

3.3   Finding the Hand

Once the neck joints are homed, we move the robot to a preprogrammed pose which has a high likelihood of putting the robot's hand somewhere in its field of view. We do not know where the hand is in the field of view, so we use a waving motion in order to locate the hand. This technique is drawn from a forthcoming publication [2], and presented here briefly, so that its place in the visual homing system can be understood.

To generate a waving motion, we pass a sine wave through the wrist actuator, which means that the angle of the wrist actuator γ as a function of time t is

$$\gamma(t) = A\sin(2\pi ft), \tag{1}$$

where A is the amplitude of the wave and f is the frequency of the wave. The instantaneous linear velocity v of a point on the hand at distance r from the actuator axis is

$$v(t) = r\,\omega(t), \tag{2}$$

where ω, the angular velocity of the actuator, i.e. the derivative of the wrist actuator angle, is

$$\omega(t) = \frac{d\gamma}{dt} = 2\pi Af\sin(\frac{\pi}{2} - 2\pi ft). \tag{3}$$

The proportion of the linear velocity that will be seen on the image plane v_{image} when the moving hand is viewed by the Kinect is

$$v_{image}(t) = \sin\beta v(t), \tag{4}$$

where β is the angle between the Kinect camera axis and instantaneous velocity vector at the observed point. Combining (2), (3), and (4), we see that the sine wave we put into the actuator will be phase shifted and have a very different amplitude by the time it is observed on the image plane. However, it will still be recognisable as a sine wave and it will still have the same frequency.

In order to determine the velocity of points in the Kinect image, we use a block matching algorithm [6]. This algorithm works by dividing up the image

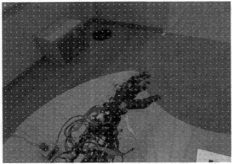

Fig. 4. An example of the optical flow (right) obtained from the block matching algorithm, in response to an exploratory waving motion (left).

with a grid of blocks (we use 8×8 blocks). Then, for each block, the algorithm estimates the optical flow to the next frame by searching in a small window about the centre of the block for the best match in the next frame in a Sum of Absolute Differences sense. An example of the optical flow obtained from this algorithm is shown in Fig 4.

To detect its hand, the robot starts a *detection episode* by briefly holding its hand still, and then passing a repeated sine wave through its wrist actuator (the exploratory motion). Whilst the sine wave plays out of the wrist actuator, the Kinect records images of the hand, and optical flow is calculated for each recorded image. A small delay follows the end of the sine wave to allow all of the motion to be recorded by the camera, and then the detection episode is concluded. The optical flow from all of the recorded images is then concatenated to give a time dependent optical flow series for each block.

In order to identify the input sine wave in the output optical flow signal, we use Normalised Cross Correlation (NCC). This involves treating the input sine wave and the output optical flow as two random processes X and Y, and then calculating the correlation coefficient $\rho(X, Y)$, defined as

$$\rho(X, Y) = \frac{cov(X, Y)}{\sqrt{var(X)var(Y)}}, \qquad (5)$$

where $var(.)$ and $cov(.,.)$ are variance and covariance, respectively. The correlation coefficient will be a value in the range $[-1, 1]$, with 1 signifying a perfectly matching pair of signals, -1 signifying that X is an inverted version of Y, and 0 signifying that the two signals are completely uncorrelated.

In order to detect the hand at a particular point in the Kinect image, we therefore calculate (5) between the optical flow, and repeatedly delayed versions of the input signal. If the correlation coefficient goes above a given threshold λ, for both the x and y components of the optical flow, then we identify the block as having motion that matches the exploratory wave. Figure 5a illustrates a typical application of NCC to the optical flow of a block containing the hand. Once NCC

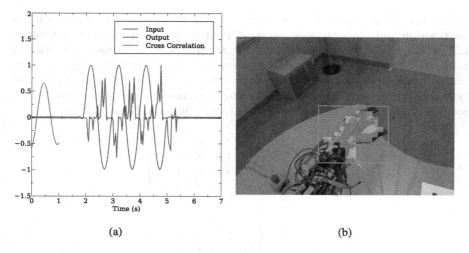

(a)   (b)

Fig. 5. (a) An example of performing NCC on the input signal and on optical flow from an area of the screen containing the hand. (b) Typical output from the wave detection process. The regions in yellow represent blocks where the NCC response was above a preset threshold λ.

has been run on all blocks, the hand is declared to be the largest connected set of positively identified blocks and a bounding box gives the expected extents of the hand. An example of this is shown in Fig. 5b.

3.4 Locating Joint Axes

The previous step tells us where the hand is in camera space and gives us a rough bounding box which allows us to filter the point cloud from the depth camera, leaving only points from the hand. Using this knowledge of which points make up the hand, we are able to track the movement of the hand in camera space by using the ICP algorithm. The ICP algorithm, introduced in works by Chen and Medioni [3] and Besl and McKay [1] is a well known algorithm for aligning 3D point clouds, which gives good results, providing that a reasonably accurate, initial estimate of the 3D transformation is available. In our case, the fact that we can move the robot's actuators in small steps means that we only need to use the identity transform as the initial transformation when estimating the 3D transformation of the hand between frames.

Briefly, given two point clouds $P = \{\mathbf{p}_1, \mathbf{p}_2, ..., \mathbf{p}_N\}$ and $Q = \{\mathbf{q}_1, \mathbf{q}_2, ..., \mathbf{q}_N\}$, a single step of the ICP algorithm involves finding the optimal transformation $\mathbf{T}_{opt}$ such that

$$\mathbf{T}_{opt} = \arg\min_{\mathbf{T}} \sum_{i=1}^{M} \left\| \mathbf{T}\mathbf{p}_i - \mathbf{q}_{match(i)} \right\|^2 \tag{6}$$

where $match(i)$ gives the index of a point in Q which is closest to a given point p_i. This step is iterated to convergence, thus giving the ICP algorithm its name.

We use this ability to track the movement of the hand, and in turn, to estimate the position and orientation of the wrist joint axes. This is done by moving an axis actuator by a small angle ϕ and measuring the transformation $\mathbf{T}_\phi$ that is induced by the movement. *Chasles' theorem* [10], states that "Any rigid transformation can be represented by a rotation about a fixed axis, followed or preceded by a translation along that axis". Therefore, we can use the screw-axis decomposition [10] to obtain the axis from the transformation $\mathbf{T}_\phi$ as follows

$$\mathbf{T}_\phi = \begin{pmatrix} \mathbf{R} & \mathbf{t} \\ \mathbf{0} & 1 \end{pmatrix} = \begin{pmatrix} r_{11} & r_{12} & r_{13} & t_x \\ r_{21} & r_{22} & r_{23} & t_y \\ r_{31} & r_{32} & r_{33} & t_z \\ 0 & 0 & 0 & 1 \end{pmatrix} \tag{7}$$

$$\mathbf{l} = \begin{pmatrix} r_{32} - r_{23} \\ r_{13} - r_{31} \\ r_{21} - r_{12} \end{pmatrix} \tag{8}$$

$$\hat{\phi} = sign(\mathbf{l}^\top \mathbf{t}) \left| \cos^{-1}\left(\frac{r_{11} + r_{22} + r_{33} - 1}{2} \right) \right| \tag{9}$$

$$\mathbf{m} = \frac{(\mathbf{1}_{3\times3} - \mathbf{R}^\top)\mathbf{t}}{2(1 - \cos\hat{\phi})} \tag{10}$$

$$\mathbf{s} = \frac{\mathbf{l}}{2\sin\hat{\phi}} \tag{11}$$

where $\mathbf{1}_{3\times3}$ is the 3×3 identity matrix, $\mathbf{m}$ is a point on the axis, $\mathbf{s}$ is the direction of the axis, and $\hat{\phi}$ is the measured angle of rotation. Ideally, $\hat{\phi}$ should match ϕ. It may be that $\hat{\phi} = -\phi$, as a rotation around an axis is equivalent to the negative rotation around an axis pointing in the opposite direction. In this case, the direction of $\mathbf{s}$ is reversed. If however, the absolute value of $\hat{\phi}$ differs too much from that of ϕ, then we assume that the measurement of the axis has failed and retake the measurement.

We use this technique to measure two of the robot's wrist axes. These axes are sequential in the kinematic model, they intersect, and they are practically orthogonal. We can therefore use the cross product to find a mutually orthogonal axis and so measure the position and orientation of the wrist frame controlled by the last wrist axis.

3.5   Kinematic Calibration

The final step in the visual homing process is a kinematic calibration step. This uses the measured pose of the wrist frame to estimate the true joint angles of the robot. Using the methods given in previous sections of the paper, the robot's hand is located, and K measurements are taken, with the hand at different positions in the robot's field of view. For each measurement, we have a measured vector of joint angles $\mathbf{p}_k$ in exploratory joint space, along with the measured pose

$\mathbf{x}_k$ of the robot's wrist in camera space. We now seek to find the optimal offset $\mathbf{o}_{opt}$ such that

$$\mathbf{o}_{opt} = \arg\min_{\mathbf{o}} \sum_{k=1}^{K} error(\mathbf{p}_k + \mathbf{o}, \mathbf{x}_k) \qquad (12)$$

where $error(\mathbf{p}_k + \mathbf{o}, \mathbf{x}_k)$ gives the error between the wrist pose predicted by a given offset and the actual measured wrist pose. A variety of functions could be used as the error function. We chose to use the square of the Euclidean distance between the predicted wrist frame position and orientation, and the measured position and orientation. Equation 12 is non-linear as it makes use of the forward kinematics of the robot, which are non-linear. We minimise this equation using the Broyden-Fletcher-Goldfarb-Shanno (BFGS) method [8], coupled with random restarts to avoid local minima.

4   Experiments

We conducted a number of experiments to explore the performance of the visual homing system, both in simulation and on the BERT robot. In simulation, we were able to test how well the system performed for a large number of randomly generated joint offsets. This gave us some idea of the convergence properties of the algorithm throughout the joint space. Two aspects of the algorithm which we explored in simulation were (a) the effect of varying the number of extra wrist poses that were gathered when trying to refine the initial guess, and (b) the effect of noise when measuring the position and orientation of the wrist frame in camera space. When exploring a particular set of parameters for the system, we generated 100 random offsets for the robot and then recorded the proportion of the offsets which were correctly identified to within $1°$ for all joints. Figures 6a and 6b show the results that were obtained from the simulation. The graphs show both the result of varying levels of noise when measuring the pose of the wrist frame, and varying the number of wrist measurements which were taken for use in the kinematic calibration step. Figure 6a also provides justification for using a separate homing step for the neck joints. Initially, we tried to home the entire kinematic model by just observing the wrist frame in camera space. However, we found that as more noise was introduced to the wrist frame measurements, the number of different wrist poses that needed to be measured increased to an impractical level. Figure 6b shows that if the neck joints have already been homed, then the kinematic calibration step requires much fewer wrist pose measurements in order to converge to the correct joint angles.

The system was also implemented and evaluated on our BERT robot. To evaluate system performance, the robot was manually homed by carefully positioning its joints prior to start up. This meant that the true joint angles were available to us as they were tracked by BERT's incremental encoders. A number of homing errors were then simulated by manually generating joint offsets, allowing us to both observe how well the system was able to identify these offsets and to evaluate the performance of the system. In a trade off between the accuracy

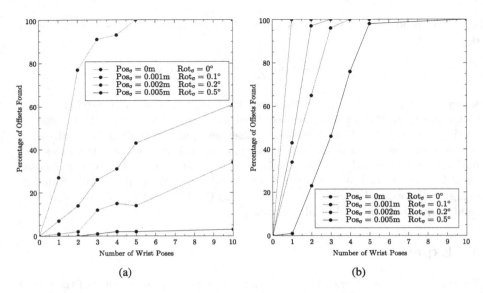

Fig. 6. The effect of varying the number of measured wrist poses in simulation. Different lines reflect the results that were obtained for different levels of zero mean Gaussian noise, added to the simulated measurements of the wrist frames position and orientation. (a) Results of homing the entire skeleton (including neck joints) by locating the hand. (b) Much better results obtained after the neck joints have already been homed.

Table 1. Grouping of 100 random offsets in terms of the accuracy to which they were recovered by our system.

Maximum joint error (°)	Number of offsets
0.5	23
1.0	74
2.5	92
5.0	100

of the result and the time required for the homing method, we set the system to gather 5 measurements of the wrist pose for use in the kinematic calibration step.

Table 1 shows the results of testing the system with 100 randomly generated offsets and classifies each in terms of the accuracy that was achieved. This shows that for 92 % of the tested offsets, our visual homing system was able to recover the offset with a maximum absolute error of not more than 2.5°, and all of the test cases were homed successfully to within 5°.

5   Conclusion

We have presented a system which allows a robot to home itself using visual information from a depth camera coupled with exploratory motions that allow it to measure the pose of its wrist frame in camera space. The homing process can be carried out autonomously in a relatively short period of time provided that the robot has been roughly homed before being powered on. In our implementation on the BERT robot, joints can start up to 20 to 30° away from the home position, depending upon the joint.

We are unable to home the robot from a completely arbitrary position. As, with no knowledge about a joint's position, the robot may accidentally move a joint beyond its range of motion and thus damage itself. However, by explicitly defining and considering an exploratory joint space for the robot, we are able to ensure that the robot does not damage itself whilst performing the exploratory moves needed to home itself.

It would be preferable if the neck joints of the robot did not have to be homed in a separate step, prior to homing the rest of the robot. However, tests in simulation showed that without this step, it was not possible to determine the robot's pose unambiguously in the presence of noise.

Tests on a real robot show that our method is reliable and gives reasonably accurate results. In future work, we will look for ways to simplify and speed up the homing process, whilst maintaining or increasing the accuracy of the system.

Acknowledgements. This work was supported by funding from the Leverhulme Trust, project number F/00182/CI.

References

1. Besl, P.J., McKay, N.D.: A method for registration of 3-D shapes. IEEE Trans. Pattern Anal. Mach. Intell. **14**, 239–256 (1992)
2. Broun, A., Beck, C., Pipe, T., Mirmehdi, M., Melhuish, C.: Bootstrapping a robot's kinematic model. Robotics and Autonomous Systems (2013) (To be published)
3. Chen, Y., Medioni, G.: Object modeling by registration of multiple range images. In: IEEE International Conference on Robotics and Automation, pp. 2724–2729 (1991)
4. Craig, J.J.: Introduction to Robotics: Mechanics and Control, 3rd edn. Prentice Hall, Englewood Cliffs (2004)
5. Hinterstoisser, S., Cagniart, C., Ilic, S., Sturm, P., Navab, N., Fua, P., Lepetit, V.: Gradient response maps for real-time detection of textureless objects. IEEE Trans. Pattern Anal. Mach. Intell. **34**, 876–888 (2012)
6. Jain, J., Jain, A.: Displacement measurement and its application in interframe image coding. IEEE Trans. Commun. **29**(12), 1799–1808 (1981)
7. Jin, Y., Xie, M.: Vision guided homing for humanoid service robot. In: Proceedings 15th International Conference on Pattern Recognition. ICPR-2000, vol. 4, pp. 511–514 (2000)
8. Nocedal, J., Wright, S.: Numerical Optimization, 2nd edn. Springer, New York (2006)

9. Santos, J., Bernardino, A., Santos-Victor, J.: Sensor-based self-calibration of the iCub's head. In: 2010 IEEE/RSJ International Conference on Intelligent Robots and Systems, pp. 5666–5672, October 2010
10. Siciliano, B., Khatib, O.: Springer Handbook Of Robotics. Springer, Heidelberg (2008)

Efficient Construction of SIFT Multi-scale Image Pyramids for Embedded Robot Vision

Peter Andreas Entschev[✉] and Hugo Vieira Neto

Graduate School of Electrical Engineering and Applied Computer Science,
Federal University of Technology – Paraná, Curitiba, Brazil
peter@entschev.com, hvieir@utfpr.edu.br
http://www.cpgei.ct.utfpr.edu.br

Abstract. Multi-scale interest point detectors such as the one used in the SIFT object recognition framework have been of interest for robot vision applications for a long time. However, the computationally intensive algorithms used for the construction of multi-scale image pyramids make real-time operation very difficult to be achieved, especially when a low-power embedded system is considered as platform for implementation. In this work an efficient method for SIFT image pyramid construction is presented, aiming at near real-time operation in embedded systems. For that purpose, separable binomial kernels for image pyramid construction, rather than conventional Gaussian kernels, are used. Also, conveniently fixed input image sizes of $2^N + 1$ pixels in each dimension are used, in order to obtain fast and accurate resampling of image pyramid levels. Experiments comparing the construction time of both the conventional SIFT pyramid building scheme and the method suggested here show that the latter is almost four times faster than the former when running in the ARM Cortex-A8 core of a BeagleBoard-xM system.

Keywords: Multi-scale image pyramid · Binomial filtering kernel · Embedded robot vision

1 Introduction

The design of autonomous mobile robots will benefit immensely from the use of physically small, low-power embedded systems that have recently become available, such as the BeagleBoard-xM [1] and the Raspberry Pi [2] boards. These platforms are based on ARM processors that are able to run the Linux operating system and the OpenCV library [3], which makes them attractive for the implementation of embedded robot vision applications.

As robot vision applications are usually computationally intensive and demand relatively large amounts of memory, there are challenges for real-time operation with the limited processing resources of an embedded system. The BeagleBoard-xM is particularly interesting in this sense because it is based on a mid-range single-core ARM Cortex-A8 processor running at 1 GHz with 512 MB

A. Natraj et al. (Eds.): TAROS 2013, LNAI 8069, pp. 127–134, 2014.
DOI: 10.1007/978-3-662-43645-5_14, © Springer-Verlag Berlin Heidelberg 2014

of DDR RAM, and a fixed-point Texas Instruments C64x+ family DSP running at 800 MHz. Regarding energetic autonomy, the BeagleBoard-xM consumes as little as 5 W at full load, against several dozens of watts consumed by a conventional personal computer.

Moreover, the BeagleBoard-xM is equipped with a dedicated camera bus, in which a CMOS camera can be connected directly to the main processor, virtually eliminating image acquisition overheads that are normally present in traditional types of camera interface, which use USB or FireWire connections. Fully programmable CMOS cameras of up to 5 MP are available, whose image size may be conveniently configured to reduce acquisition bandwidth and the resources needed for resampling processes that are often present in multi-scale robot vision algorithms.

There are many powerful object recognition methods available in the literature that support multi-scale feature extraction – for example, SIFT [4,5] and GLOH [6] – which are particularly interesting for robot vision applications. These methods are relatively expensive to compute as their core algorithms involve the construction of multi-scale image pyramids. In this work our intent is to investigate efficient implementations of the image pyramid construction scheme used in multi-scale object recognition algorithms, in order to allow near real-time execution in an embedded platform such as the BeagleBoard-xM.

2   Related Work

Objects can be detected in images by matching some of their unique visual features, usually edges and corners. In [4], Lowe presented his seminal work on the Scale Invariant Feature Transform (SIFT), demonstrating that it is possible to extract distinctive local features from an object in multiple scales, and match these features successfully afterwards, independently of scale, rotation, affine transformations or occlusions. This technique was later improved in [5].

Stable distinctive features that describe an object can be detected by computing a multi-scale Laplacian pyramid, as originally proposed in [7] and later made more efficient in [8]. In practice, the Laplacian pyramid is obtained by the differences between adjacent levels of a Gaussian pyramid built from successive low-pass filtering and down-sampling of the original input image.

After the computation of the difference of Gaussians, the location of distinctive local features (keypoints) can be found by detecting extrema (maxima and minima) among adjacent levels of the Laplacian pyramid, a function called Laplacian jet. Keypoints that are stable both in scale and space usually correspond to corners of objects.

Differences of Gaussians are less expensive to compute than computing the Laplacian directly, but even so, building the difference of Gaussians pyramid is one of the most computationally expensive processes that are executed in order to extract object features using the SIFT framework. For this reason, in this work we investigate techniques which are less computationally expensive but maintain the main property of detecting stable keypoints to describe objects.

In [9], it is demonstrated that a binomial difference of Gaussians can be used to approximate a conventional difference of Gaussians in a less computationally expensive way. The main difference is that scales are approximated by convolving the input image with a binomial kernel instead of a Gaussian kernel.

There are several other methods available in the literature that use the same principles to extract object features that can be used for scale invariant recognition [6]. Another well-known method called SURF (Speeded-Up Robust Features) builds the scale-space using a rather different approach, which involves the concept of integral images [10]. However, here we concentrate our efforts in methods that use standard convolution techniques.

3   Image Pyramid Construction

The original SIFT algorithm proposed in [4] uses an image pyramid, in which each scale consists of the previous scale convolved with a Gaussian kernel. Successive convolutions with a Gaussian kernel are applied in order to obtain different scales.

The pyramid construction method we use in this work is based on [9], in which instead of using Gaussian kernels in order to obtain different scales of the input image, a binomial kernel is used. The main advantage of using a binomial kernel instead of a Gaussian kernel is the reduced computational cost for the convolution process. For example, if the construction of a Gaussian pyramid with scales separated by a factor of $\sigma = \sqrt{2}$ is desired, it is necessary to convolve the input image in both horizontal and vertical directions with a separable 1D Gaussian kernel with a minimum length of seven elements; in order to achieve the same scale separation using successive convolutions with a binomial kernel, a length of only three elements is needed.

Building Gaussian pyramids using 2D kernels is also possible, but this is often avoided because it is more computationally expensive and has the exact same result of using separable 1D kernels. The binomial kernels studied here also present the property of separability, which is used throughout this work because our aim is specifically to reduce the processing time needed for image pyramid construction. For the work described here, two separable binomial kernels are especially relevant – one is the three-element kernel given by $\frac{1}{4} \times [1\ 2\ 1]$ and the other is the auto-convolution of the first, which is the five-element kernel given by $\frac{1}{16} \times [1\ 4\ 6\ 4\ 1]$.

3.1   Binomial Filtering

The kernel $\frac{1}{16} \times [1\ 4\ 6\ 4\ 1]$ approximates a Gaussian kernel with $\sigma = 1$, i.e. in order to obtain an approximation of a Gaussian blur of $\sigma = 1$, two consecutive convolutions with the three-element kernel $\frac{1}{4} \times [1\ 2\ 1]$ are needed.

In terms of complexity and if only separable 1D kernels are used, the practical meaning of an image convolution with three elements is that three multiplications and two additions per pixel per dimension (horizontal and vertical) are needed.

For a kernel with seven elements, it is necessary to perform seven multiplications and six additions per pixel per dimension.

The advantage of using separable binomial kernels with fixed-point coefficients is that in either one convolution with the kernel $\frac{1}{16} \times [1\ 4\ 6\ 4\ 1]$ or two consecutive convolutions with the kernel $\frac{1}{4} \times [1\ 2\ 1]$, two of the multiplications involved are multiplications by a factor of 1, making them unnecessary. The number of operations per pixel per direction is then reduced to a total of four multiplications and four additions per pixel per dimension for each scale. The total number of operations per scale of the pyramid is then $N = 8 \times R \times C \times 2$, where R is the number of rows and C is the number of columns in the pyramid level.

Yet another implicit advantage of using separable binomial kernels to perform image convolutions is that they can easily be used in implementations for fixed-point DSP cores. For instance, the built-in Texas Instruments C64x+ family DSP available in the BeagleBoard-xM supports fixed-point arithmetic and could be used in future implementations.

As shown in [9], when the image is convolved multiple times with the separable kernel $\frac{1}{16} \times [1\ 4\ 6\ 4\ 1]$, if the images at every three convolutions are stored, the resulting pyramid is separated by scale steps of $\sigma = \sqrt{2}$.

In order to improve efficiency by reducing the amount of data to be processed, instead of blurring the input image multiple times at the same octave, i.e. maintaining its original dimensions, the input image is down-sampled to half its size in each dimension every time that the scale reaches $\sigma = 2^N$ [9].

As described in [5], in order to be able to detect SIFT keypoints, at least four different scales are necessary for each octave of the Gaussian pyramid, but performing multiple convolutions of the image with a binomial kernel results in only three scales per octave – the original and two blurred ones, with $\sigma = \sqrt{2}$ and $\sigma = 2$ with respect to the original scale, respectively.

As can be seen in Fig. 1, the third image of the current octave is down-sampled in order to result in the initial scale of the next octave. Because this third image already has twice the size of the desired image in each dimension, it is possible to down-sample it using a nearest-neighbour approach with minimal loss of information. In this case, only every other pixel of each column and each row is kept, which is computationally inexpensive.

3.2   Image Acquisition and Resampling

For the resampling process, there is a great advantage provided by the built-in camera port of the BeagleBoard-xM. It is possible to keep the original image borders in all resampled scales if the acquired image has $2^N + 1$ pixels in each dimension – e.g. 129×257 pixels or 513×1025 pixels. With a fully programmable CMOS camera, images can be acquired with conveniently configured dimensions, which is a capability not always available in conventional USB or FireWire cameras for personal computers, for example.

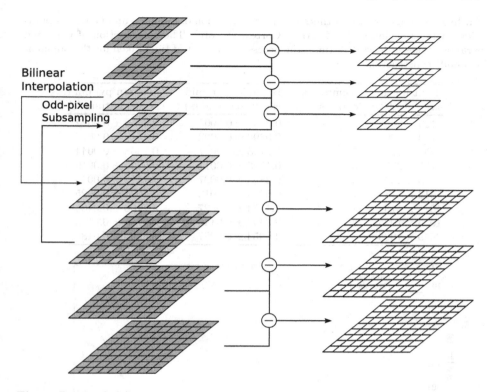

Fig. 1. Binomial difference of Gaussians pyramid construction. In adjacent octaves, the first scale is obtained by nearest-neighbour down-sampling and the fourth scale by bilinear interpolation (up-sampling).

The down-sampling process of images with $2^N + 1$ pixels in each dimension is straightforward and can be done using a nearest-neighbour approach, in which every other pixel in each dimension is kept, including the pixels of the image borders.

In the next octave of the pyramid, we can continue blurring the image with the binomial kernel with a size of five elements and get new scale levels, also separated by steps of $\sigma = \sqrt{2}$. However, the previous pyramid octave still has only three different scales, and at least a fourth one is needed to find SIFT keypoints. This problem can be solved by doubling each dimension of the second level of the next level using bilinear interpolation (see Fig. 1).

4   Experiments and Results

In this section, we present experiments and results obtained while executing the algorithms in the ARM Cortex-A8 core of a BeagleBoard-xM system running the Linux operating system. Both image pyramid construction schemes, based on binomial and Gaussian kernels, were implemented using the OpenCV library.

Table 1. Average execution times for the construction of binomial and Gaussian pyramids on the BeagleBoard-xM (ARM Cortex-A8 core). The construction of Gaussian pyramids takes at least 3.72 times the necessary amount of time taken for the binomial pyramids construction.

Input image size (pixels)	Number of octaves	Binomial pyramid execution time (s)	Gaussian pyramid execution time (s)
129 × 65	4	0.0207 ± 0.0001	0.0770 ± 0.0022
129 × 129	5	0.0398 ± 0.0002	0.1512 ± 0.0027
257 × 129	5	0.0768 ± 0.0007	0.2968 ± 0.0044
257 × 257	6	0.1508 ± 0.0011	0.5725 ± 0.0019
513 × 257	6	0.2893 ± 0.0007	1.1712 ± 0.0049
513 × 513	7	0.5914 ± 0.0012	2.2671 ± 0.0025
1025 × 513	7	1.1764 ± 0.0027	4.7103 ± 0.0108
1025 × 1025	8	2.3447 ± 0.0149	9.3758 ± 0.0103
2049 × 1025	8	4.6954 ± 0.0287	18.8793 ± 0.0183

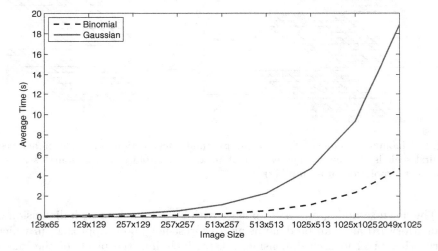

Fig. 2. Average execution times for the construction of binomial and Gaussian pyramids for multiple image sizes on the BeagleBoard-xM (ARM Cortex-A8 core). The black dashed line corresponds to execution times for the construction of binomial pyramids whereas the gray solid line corresponds to execution times for the construction of Gaussian pyramids.

For these experiments, we executed the construction of binomial and Gaussian pyramids for several different image sizes ranging from 129 × 65 to 2049 × 1025 pixels, with total sizes doubling at each step. Each instance was executed 200 times – the average execution times obtained, along with their standard deviations, are shown in Table 1.

Inspecting the average execution times in Table 1, it is clearly visible that the construction of the binomial pyramid is faster than the Gaussian pyramid. Calculating the ratio between the average time to build the Gaussian pyramid

and the binomial pyramid, a minimum of 3.72 times is observed for images of
129×65 pixels in size, raising up to 4.02 for images of 2049×1025 pixels in size.

In order to obtain the results in Table 1, equivalent binomial and Gaussian
pyramids were computed. We kept down-sampling the image to half its size while
both dimensions were still greater than eight pixels. In other words, for the image
of 129×65 pixels, four octaves were computed, for the image of 129×129, five
octaves, and so on.

In Fig. 2, it can be seen graphically how the execution time of both pyramid
construction schemes is affected by the size of the input image.

5   Conclusions

Autonomous mobile robots that aim to use vision as perceptual input will benefit
greatly from designs using physically small, low-power embedded systems which
can run the Linux operating system and the OpenCV library. However, efficient
implementation of robot vision algorithms are necessary in order to allow near
real-time operation.

This work has presented a less computationally expensive method for the
construction of multi-scale SIFT [4,5] pyramids. For this, we have focused on the
work of Crowley and Riff [9], which describes the advantages of using separable
binomial kernels over Gaussian kernels in order to build multi-scale pyramids.

Results of experiments conducted in a real embedded platform based on
an ARM Cortex-A8 processor have shown that the binomial pyramid building
scheme discussed in this work takes about one fourth of the time needed for building the conventional Gaussian pyramid. The approximation method described
here reduces the overall time necessary for extracting SIFT features, making it
more suitable for near real-time processing, especially on embedded platforms,
in which limited computational resources are available.

Future improvements in the technique detailed here include using the fixed-
point DSP core available in the BeagleBoard-xM. As the Texas Instruments
C64x+ family DSP shares a limited portion of the available DDR RAM with
the ARM Cortex-A8 processor, a hybrid approach using the parallel processing
power of both cores to compute the image pyramids is possible. For near real-
time continuous image feature extraction, a pipeline technique can be used to
share the execution of processes between the DSP and the ARM processor.

Experiments and discussions about the stability of the extracted keypoints
using both pyramid construction schemes are the subject of future work – this
deserves special attention, given that it is necessary to assess the best parameters
for selection of the most stable SIFT features.

References

1. Coley, G: Beagleboard-xM System Reference Manual – Revision A2. http://
   beagleboard.org (2010)
2. Upton, E., Halfacree, G.: Raspberry Pi User Guide. Wiley, New York (2012)

3. Bradski, G., Kaehler, A.: Learning OpenCV: Computer Vision with the OpenCV Library. O'Reilly Media, Sebastopol (2008)
4. Lowe, D.G.: Object recognition from local scale-invariant features. In: Proceedings of the 7th IEEE International Conference on Computer Vision, vol. 2, pp. 1150–1157. IEEE (1999)
5. Lowe, D.G.: Distinctive image features from scale-invariant keypoints. Int. J. Comput. Vis. **60**(2), 90–110 (2004)
6. Mikolajczyk, K., Schmid, C.: A performance evaluation of local descriptors. IEEE Trans. Pattern Anal. Mach. Intell. **27**(10), 1615–1630 (2005)
7. Burt, P., Adelson, E.: The Laplacian pyramid as a compact image code. IEEE Trans. Commun. **31**(4), 532–540 (1983)
8. Crowley, J.L., Stern, R.M.: Fast computation of the difference of low-pass transform. IEEE Trans. Pattern Anal. Mach. Intell. **2**, 212–222 (1984)
9. Crowley, J.L., Riff, O.: Fast computation of scale normalised Gaussian receptive fields. In: Griffin, L.D., Lillholm, M. (eds.) Scale Space 2003. LNCS, vol. 2695, pp. 584–598. Springer, Heidelberg (2003)
10. Bay, H., Ess, A., Tuytelaars, T., Van Gool, L.: Speeded-up robust features (SURF). Comput. Vis. Image Underst. **110**(3), 346–359 (2008)

An Evaluation of Image-Based Robot Orientation Estimation

Juan Cao$^{(\boxtimes)}$, Frédéric Labrosse, and Hannah Dee

Department of Computer Science, Aberystwyth University,
Aberystwyth SY23 3DB, UK
{juc3,ffl,hmd1}@aber.ac.uk

Abstract. This paper describes a novel image-based method for robot orientation estimation based on a single omnidirectional camera. The estimation of orientation is computed by finding the best pixel-wise match between images as a function of the rotation of the second image. This is done either using the first image as the reference image or with a moving reference image. Three datasets were collected in different scenarios along a "Gummy Bear" path in outdoor environments. This carefully designed path has the appearance of a gummy bear in profile, and provides many curves and sets of image pairs that are challenging for visual robot localisation. We compare our method to a feature-based method using SIFT and another appearance-based visual compass. Experimental results demonstrate that the appearance-based methods perform well and more consistently than the feature based method, especially when the compared images were grabbed at positions far apart.

Keywords: Robot orientation · Quadtree · SIFT · Visual compass

1 Introduction

When mobile robots move, one of the basic problems which needs to be solved is for the robot to know its orientation as accurately as possible. A range of sensors, such as laser, digital compass, wheel encoders, gyroscope and GPS can be used to perform this measurement. However, since digital cameras have become more affordable, more research has been devoted to supplement traditional systems using visual cues. Indeed, visual methods do not suffer from problems linked to proprioceptive sensors such as GPS availability or wheel slippage.

Various solutions to the problem have been proposed. We can categorize these solutions into two main groups: feature-based and appearance-based. Feature-based methods try to detect distinctive and robust points or regions between consecutive images, while appearance-based methods concentrate their efforts on the information extracted from the pixel intensity, the whole image being represented by a single descriptor, without local feature extraction. The change in orientation between frames is then computed by aligning the features or images, using a calibration of the projection onto the image plane. This process can

A. Natraj et al. (Eds.): TAROS 2013, LNAI 8069, pp. 135–147, 2014.
DOI: 10.1007/978-3-662-43645-5_15, © Springer-Verlag Berlin Heidelberg 2014

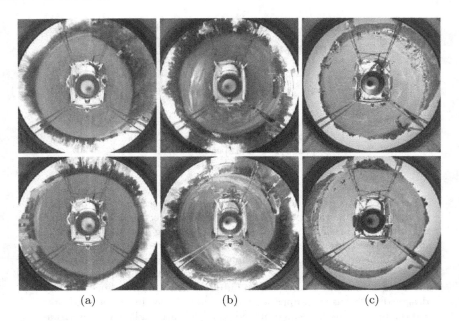

(a)   (b)   (c)

Fig. 1. Example images of the (a) FIELD, (b) CARPARK, and (c) TENERIFE datasets

be achieved incrementally from frame to frame as the robot moves, which can have the drawback that it generally becomes less and less accurate as integration introduces additive errors at each step. Alternatively, a fixed reference image can be used and all orientations calculated from it. However, as the frames become more different, the orientation estimation becomes less reliable.

In this paper we aim to address the question "What image-based techniques are best for orientation estimation?". We do this by comparing appearance-based methods such as the visual compass [5], image-based techniques which use quadtree methods to reduce noise [1] and feature matching techniques based on descriptors such as SIFT (Scale-Invariant Feature Transform) [7]. In order to thoroughly compare these methods, we measure their performance on a number of datasets captured in outdoor dynamic environments: FIELD, CARPARK, and TENERIFE. Some images from these datasets are shown in Fig. 1 while details can be found in Sect. 4.1. These datasets consist of sequences of images captured along a "Gummy Bear" path (see Fig. 2) by our four wheel drive, four wheel steering, electric vehicle *Idris*. This carefully designed path has the appearance of a gummy bear in profile, and provides many curves and sets of image pairs that are challenging for visual robot localisation (for example the ear region contains a sequence of images on a tight curve).

The rest of this paper is organised as follows: Sect. 2 reviews related work. Section 3 outlines the three different methods that we compare. Section 4 details the experiments undertaken, and reports results. Finally, Sect. 5 concludes the paper and outlines possible future improvements.

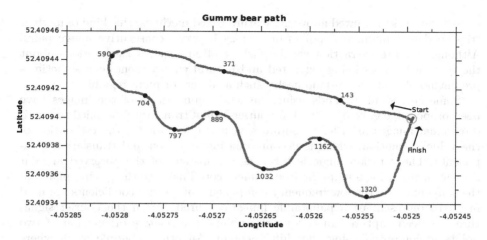

Fig. 2. RTK-GPS track from the "Gummy Bear" path for the FIELD dataset. Image numbers of key positions are marked in blue. This path is designed so that there are pinch points (at the "neck" and "knees") where the robot is quite close to where it has been before, but is clearly not in the same place (e.g. images 143 and 1162 might be expected to be similar). The path finishes at the start point but with Idris rotated through 90° (Color figure online).

2   Related Work

Vision-based motion estimation has a long history in robotics. Methods have been proposed using monocular [11,14] and stereo cameras [8,10–12]. Methods can be divided into two categories: feature-based and image-based methods. Here, we review some of these works; a more extensive survey can be found in [13].

The earliest work on estimating a vehicle's motion from visual imagery alone is [10], where the basic algorithm identifies corner features in each camera frame and estimates the depth of each feature using stereo. Then potential matches are found by normalized cross correlation. Finally, motion is computed by estimating the rigid body transformation that best aligns the features at two consecutive robot positions. However, this kind of system suffers from poor accuracy and is unstable, partly because it relies on scalar models of measurement error in triangulation. Based upon this work, [8] uses 3D Gaussian distributions to model triangulation error and incorporates the error covariance matrix of the triangulated features into the motion estimation between successive stereo pairs. The motion estimation in this work was pure translation, without considering orientation. This is extended in [12] by incorporating an absolute orientation sensor such as a compass, a sun sensor or a panoramic camera providing periodic orientation updates, with the Förstner corner detector used as feature detector. The results indicated that the error growth can be reduced to a linear function of the distance travelled and outperform previous visual odometry results. A Harris corner detector was used in [11].

All the works reviewed above are feature-based methods; this kind of method tries to detect distinctive points or regions between consecutive image pairs. Although feature extraction can be fast, it often requires assumptions about the type of features being extracted and natural environments can sometimes present no obvious visual landmarks, such as desert or plain regions.

Some successful methods using the whole appearance of the images have been proposed, e.g. [2–5,9]. In [4] a Fourier-Mellin transform is applied to omni-directional images in order to obtain a visual descriptor for the estimation of the vehicle's motion, which is decomposed into rotation and translation components. The rotation angle is taken as the median of the observed angular displacements using a mapping from camera coordinates to the ground plane. In the same manner, the low frequency components of Fourier coefficients are used in [2] to represent each panoramic image captured. When the Fourier signature has been captured in two nearby points, the relative orientation of two points is computed using the shift theorem. An other example is [9] where the colour images captured from a perspective camera are first converted to grey images, then each pixel column is summed and normalized to form a one-dimensional array. The resulting arrays are used to extract the rotation information.

More recently, both appearance-based and feature-based methods were presented in [3] to compute incrementally the motion transformation between two consecutive images. The phase information of the Fourier signature is used to compute the robot orientation, and SURF features are used to detect the interest points for image comparison.

3   Computing Robot Orientation

For the purpose of this paper, we compare the performance of our proposed method with a feature-based method and the visual compass presented in [5] using real images in different environments. Orientation is estimated from panoramic images created by unwrapping omni-directional images (see Fig. 1).

3.1   Feature-Based Method: SIFT

In this method, SIFT features [7] are used to align the images. Features are extracted from the two considered frames and are matched using a distance ratio of 0.6 (two features are matched if their distance in feature space is less than 0.6 times the distance to second closest feature). Once features have been matched, the two images are aligned by computing the average horizontal displacement of the features, taking care to wrap around positions in the panoramic images. In order to obtain a reliable solution in the presence of outliers, we use a Gaussian distribution to model the error and cut off the matches that are one standard deviation away from the mean.

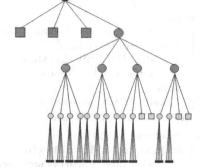

(a) Recursively decomposed image pair using quadtree

(b) Corresponding quadtree representation of image pair comparison

Fig. 3. Quadtree decomposition. There has been camera motion between the two images but the only difference detected is the car, showing robustness to small changes.

3.2 Visual Compass

In the visual compass proposed in [5], the relative rotation between pairs of panoramic images is obtained by finding the best match between images based on their column shift and using the Euclidean distance in image space. The orientation estimation is done from a moving reference image, the decision on when to change it being made using a measure of difference between images. This offers a compromise between accumulating error and comparing similar images to get a better estimation of the change in orientation. Because the sides of the panoramic images contain rotational and translational information, only the parts of the images that correspond to the front and back of the moving robot are used in the matching process.

3.3 Quadtree Method

The core technique we use for image pair comparison is quadtree decomposition combined with a number of standard image distance measures. This technique enables robustness of perceptual aliasing (the images we used are mostly made of repetitive features) and can cope with the appearance of new objects in the environment without prior information. Figure 3(a) is a visualisation of recursive image comparison and Fig. 3(b) the corresponding tree-based visualisation. Figure 3(a) shows that the decomposition into sub-regions provides us not only with robustness to noise, but also with an indication of the locations of visual change between image pairs. In Fig. 3(b), the root of the tree corresponds to the comparison of the two original images. Circles represent internal nodes of the tree, and leaf nodes correspond to quadrants that are either similar or too small.

Our method recursively compares quadrants of the two images to be compared using a metric (discussed below) until either the two quadrants are judged

```
// Base case
begin
    // Calculate the distance between I_new and I_ref
    dist = distance(I_new, I_ref);
    if dist > THRESHOLD then
        | BuildQuadTree(rootNode);
    end
    Quadtree building stopped;
end
// Quadtree building
BuildQuadTree(Node *n) begin
    // Calculate the distance between I_new and I_ref for pixels of the
        node
    dist = distance(I_new, I_ref, n → pixels);
    if dist > THRESHOLD and n → size > MIN then
        // Break image or patch into smaller patches
        for i = 0 to n → nbChildren do
            | nodeIn = BuildNode(n → child[i]);
            | BuildQuadTree(nodeIn));
        end
    end
end
```

Algorithm 1. Pseudocode of our image comparison algorithm

similar (distance below a threshold) or too small. When two quadrants are not similar enough they are split into four quadrants and the process is repeated. This is described in Algorithm 1. For panoramic images, the second level is made of six squares to allow finer decomposition. The similarity measure of the two images is then given as the area of the two images that are similar. The value of the threshold largely influences the similarity between images. We experimentally determined thresholds based on the metric and the colour space used.

To estimate the change in orientation between two images, one image is column-wise shifted (the shift corresponding to a rotation of the camera) and the similarity computed for each rotation. The maximum similarity gives the rotation between the two images.

Image metrics are used to quantitatively evaluate the similarity between two images or image regions. There are a range of metrics; we have tried Euclidean distance, χ^2 distance and Pearson's correlation coefficient. We discuss details of implementation, including the choice of image distance metric to use within the quadtree algorithm, the choice of the threshold and the smallest region size.

To determine which distance metric is appropriate for our application, we compute the similarity between one image at the middle of a sequence of images and all other images of the same sequence, using our quadtree algorithm with the three metrics mentioned. The sequence of images was captured in our laboratory using the panoramic camera mounted on a robot moving on a straight line, an

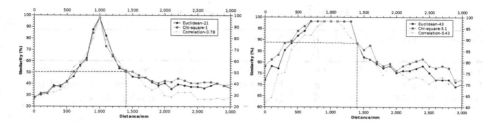

Fig. 4. Comparison of different metrics used in our quadtree similarity measurement; The left plot has a threshold set low, the right high, iso-similarity point occurs for a displacement of approximately 1,400 mm.

image being captured every 10 cm. We seek a similarity measure that is (a) smooth and (b) not too "steep" around the reference image. Figure 4 shows the similarity values for the three metrics and different values of the threshold used in the quadtree construction.

In order to produce a fair comparison between the three different metrics (with thresholds on different scales) we define an iso-similarity point for each test; this sets the threshold for quadtree decomposition so that the three metrics produce identical similarity measures. It can be seen that for low thresholds, our quadtree measure is sensitive to small displacements but that with higher thresholds we are able to determine similarity between images on a broader scale.

Briefly summarising our tests, we can see that all three metrics behave in much the same way when we find a threshold that defines an iso-similarity point. Between the two results given here, the similarity of the iso-similarity point increases from 50 % to 89 %. Pearson's correlation coefficient seems to be the most sensitive to small displacements and is also the most computationally intensive. Euclidean and χ^2 metrics are both fast and easy to compute, with the comparison results showing little difference between them. Thus, for the rest of this paper, we will present results on Euclidean distance only due to its simplicity.

In our implementation, the smallest region size (MIN) is set to 10×10 pixels, which represents a horizontal field of view of $10°$. Such a field of view implies that we can see objects 35 cm wide 2 m away. For the quadtree decomposition our experiments suggest this value is a good compromise between reducing computational load and distinguishing between objects in the robot's environment.

4   Experiments

4.1   Experimental Setup

Test images were captured from an omni-directional camera approximately one and a half meters above the ground surface as the robot moves. The CIE $L^*a^*b^*$ colour space is used in our experiments. The FIELD dataset was collected in a field-type area, with some buildings in sight but mostly trees and grass.

Table 1. Characteristics of datasets

Dataset	Frames	Length (m)	Rate (Hz)	Notes
FIELD	1525	60	6	Flat but rough surface, can see about 50 m
CARPARK	2101	60	8	Flat, can see 30 m, light changes, moving objects
TENERIFE	2156	60	8	Bumpy, can see 100 m, moving objects

Table 2. Root Mean Square Error, Mean Error and Standard Deviation Error for each dataset (F: FIELD; C: CARPARK; T: TENERIFE; VC: visual compass, subscript indicates fixed reference or moving with the corresponding skip value).

Method	RMSE			Mean			SD		
	F	C	T	F	C	T	F	C	T
VC	**11.42**	19.51	41.18	**4.63**	−1.33	−24.47	**10.43**	19.47	33.12
SIFT$_f$	24.15	40.67	**21.83**	10.61	−3.57	**−2.89**	21.70	40.52	21.63
SIFT$_5$	53.06	68.00	93.48	−35.28	−48.82	−69.55	39.68	47.38	62.52
SIFT$_{20}$	106.02	108.39	107.35	−17.01	48.63	−5.36	105.30	96.87	107.69
QT$_f$	29.93	**16.62**	23.88	16.00	**−0.76**	−11.59	25.30	**16.60**	**20.88**
QT$_5$	62.79	81.43	79.29	−38.42	−55.79	9.70	49.73	59.32	78.76
QT$_{20}$	12.51	24.06	24.43	5.90	5.66	−10.00	11.10	23.38	22.39

The CARPARK dataset was captured from a carpark with trees around, where few cars were parked, one car moved, and some parts of ground were wet from rain providing challenging reflections and shadows. The TENERIFE dataset was obtained at the El Teide National Park, Tenerife. Its flat landscape with fine textures of volcanic sand, pebbles and occasional rocky outcrops are similar to those encountered on the surface of Mars. Some tourists were walking around during the data acquisition. Characteristics of the datasets are given in Table 1.

The performance of each method is evaluated using a real-time kinematic (RTK) GPS system, which is theoretically accurate up to 1 cm in an horizontal plane. However, the accuracy and reliability depends on factors such as satellite availability, baseline length and sufficient redundancy of GPS observations. To allow comparison, absolute GPS heading is filtered using a Kalman filter and converted to relative bearing by subtracting the absolute heading of the starting point of "Gummy Bear" trajectory, and changed to a range between 0 and 360. This is used as ground truth.

4.2 Experimental Results

For both quadtree and feature based methods we estimate the orientation in two ways. The first uses the first image as a reference from which the orientation is calculated. The second uses a moving reference image and accumulates changes in orientation. In the second case we present results skipping a fixed number of images between pairs. The visual compass method from [5] uses a

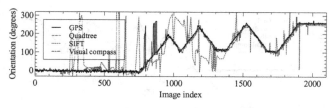

(a) Orientation estimation and groundtruth

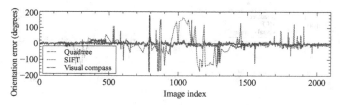

(b) Orientation error from groundtruth

Fig. 5. Experimental results for dataset CARPARK with a fixed reference image

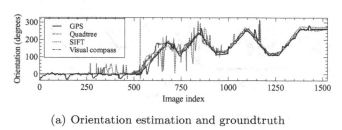

(a) Orientation estimation and groundtruth

(b) Orientation error from groundtruth

Fig. 6. Experimental results for dataset FIELD with a fixed reference image

moving reference with automatically adjusted skips. It is therefore compared to the methods using a fixed reference image. Table 2 gives quantitative results for all cases.

Figures 5, 6 and 7 show the results for the three methods with a fixed reference image for quadtree and SIFT (the visual compass uses a moving reference image but this is internal to the method and not exposed). These show that both appearance-based methods perform well and consistently for the whole path of

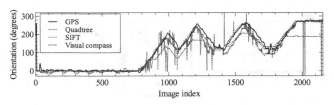

(a) Orientation estimation and groundtruth

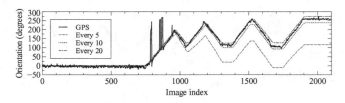

(b) Orientation error from groundtruth

Fig. 7. Experimental results for dataset TENERIFE with a fixed reference image

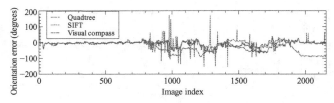

(a) Orientation from the quadtree method

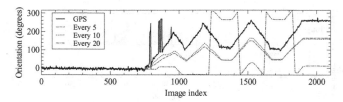

(b) Orientation from the SIFT method

Fig. 8. Experimental results for dataset CARPARK with a moving reference image

the robot. The feature-based method performs well when the images are close to the reference image but poorly when far away with many frames where no features were found to match. This is due to a lack of matched SIFT features due to the distortions introduced by the camera. Indeed, the orientation could not be calculated using the SIFT method for many frames: 36 % for CARPARK, 62 % for FIELD and 67 % for TENERIFE.

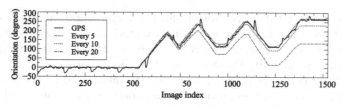

(a) Orientation from the quadtree method

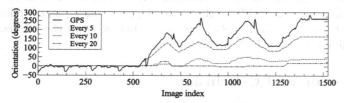

(b) Orientation from the SIFT method

Fig. 9. Experimental results for dataset FIELD with a moving reference image

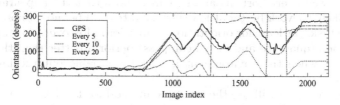

(a) Orientation from the quadtree method

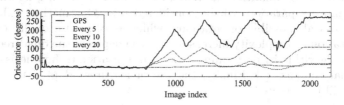

(b) Orientation from the SIFT method

Fig. 10. Experimental results for dataset TENERIFE with a moving reference image

Overall, the SIFT method performs better than both appearance-based methods on the TENERIFE dataset where enough features are detected (only 33 % of the frames for this dataset). This is because the boundary between sky and land is very strong, not visible all around the robot and slanted. Alignment of the images using pixel values will therefore tend to align the skyline introducing a bias due to the slant.

Figures 8, 9 and 10 show the results for the incremental quadtree and SIFT methods that use a moving reference. For both methods, pairs were created skipping a fixed number of images and results are given for different values of the number of images skipped. These clearly show that choosing the correct compromise between better short term rotation estimation and accumulating error is critical. In fact, none of these results are as good as that of the visual compass. This is due to the subpixel processing and the automatic, adaptive estimation of the best compromise in the visual compass. Nevertheless, the quadtree method performs similarly to the SIFT method, but skipping more images. This is in line with the fact that the SIFT method performs better when the reference image is not too different from the processed images.

5   Discussion and Conclusion

In this paper we have evaluated three methods for robot orientation estimation with panoramic images. The results gathered from the three different scenarios in two different ways show that the quadtree method performs better than the SIFT method when the distance between images becomes high, while the SIFT-based method does well over short distances. This implies that the appearance-based method is likely to work better at lower frame rates and be more appropriate for loop closure tasks, at least for orientation estimation. Moreover, the appearance-based methods (quadtree and visual compass) perform better than the feature-based method when the environment is visually variable but the images lack contrast.

In future work, we will use the quadtree method for localisation tasks where its handling of partially similar images will be of importance. In order to improve the robustness of the quadtree method with respect to varying lighting conditions, luminance separation will be considered [15]. Our method exhaustively searches across all possible orientations, which is computationally expensive, and computational complexity increases with the depth of the quadtree. We can improve our algorithm by performing a local minimisation technique as employed by the visual compass [5]. Finally, the use of features adapted to panoramic images [6] will be explored.

References

1. Cao, J., Labrosse, F., Dee, H.: A novel image similarity measure for place recognition in visual robotic navigation. In: Herrmann, G., Studley, M., Pearson, M., Conn, A., Melhuish, C., Witkowski, M., Kim, J.-H., Vadakkepat, P. (eds.) TAROS-FIRA 2012. LNCS, vol. 7429. Springer, Heidelberg (2012)
2. Fernández, L., Payá, L., Reinoso, Ó., Amorós, F.: Appearance-based visual odometry with omnidirectional images - a practical application to topological mapping. In: Proceedings of ICINCO, vol. 2, pp. 205–210 (2011)
3. García, D.V., Rojo, L.F., Aparicio, A.G., Castelló, L.P., García, O.R.: Visual odometry through appearance- and feature-based method with omnidirectional images. J. Robot. **2012**, 1–13 (2012)

4. Goecke, R., Asthana, A., Pettersson, N., Petersson, L.: Visual vehicle egomotion estimation using the Fourier-Mellin transform. In: IEEE Intelligent Vehicles Symposium, pp. 450–455 (2007)
5. Labrosse, F.: The visual compass: performance and limitations of an appearance-based method. J. Field Robot. **23**(10), 913–941 (2006)
6. Lourenco, M., Barreto, J.P., Vasconcelos, F.: sRD-SIFT: keypoint detection and matching in images with radial distortion. IEEE Trans. Robot. **28**(3), 752–760 (2012)
7. Lowe, D.: Distinctive image features from scale-invariant keypoints. Int. J. Comput. Vis. **60**(2), 91–110 (2004)
8. Matthies, L., Shafer, S.: Error modeling in stereo navigation. IEEE J. Robot. Autom. **3**(3), 239–250 (1987)
9. Milford, M.J., Wyeth, G.F.: Single camera vision-only slam on a suburban road network. In: Proceedings of ICRA, pp. 3684–3689 (2008)
10. Moravec, H.: Obstacle avoidance and navigation in the real world by a seeing robot rover. Technical report CMU-RI-TR-80-03, Robotics Institute, Carnegie Mellon University (1980)
11. Nistr, D., Naroditsky, O., Bergen, J.: Visual odometry for ground vehicle applications. J. Field Robot. **23**(1), 3–20 (2006)
12. Olson, C.F., Matthies, L.H., Schoppers, M., Maimone, M.W.: Rover navigation using stereo ego-motion. Robot. Auton. Syst. **43**(4), 215–229 (2003)
13. Scaramuzza, D., Fraundorfer, F.: Visual odometry [tutorial]. Robot. Autom. Mag. **18**(4), 80–92 (2011)
14. Tomasi, C., Shi, J.: Direction of heading from image deformations. In: Proceedings of CVPR, pp. 422–427 (1993)
15. Woodland, A., Labrosse, F.: On the separation of luminance from colour in images. In: Proceedings of VVG, pp. 29–36 (2005)

A Heuristic-Based Approach for Flattening Wrinkled Clothes

Li Sun[✉], Gerarado Aragon-Camarasa, Paul Cockshott,
Simon Rogers, and J. Paul Siebert

School of Computing Science, University of Glasgow, Glasgow,
Scotland G12 8QQ, UK
l.sun.1@research.gla.ac.uk

Abstract. In this paper, we present a heuristic-based strategy to flatten a crumpled cloth by eliminating visually detected wrinkles. In order to explore and validate visually guided clothing manipulation, we have developed a hand-eye interactive learning system that incorporates a clothing simulator to close the effector-garment-visual sensing interaction loop. We also propose a criterion by which to evaluate the various approaches used to flatten cloth. In this paper, our heuristic-based method is applied to virtual cloth in our simulator and the resulting flattening performance is compared to that obtained by manual flattening methods. These experiments demonstrate that the effectiveness and efficiency of our heuristic-based garment flattening method approaches that of manual flattening.

Keywords: Flattening clothes · Range map analysis · Robotics · Cloth simulation

1 Introduction

Interacting with clothing requires dexterous manipulation and hand-eye coordination capabilities that come naturally to humans but represent a challenge to current autonomous robotic systems. Such robot capabilities have matured to some extent as demonstrated by [2, 3]; however, several limitations still remains that must be circumvented before autonomous robotic laundry systems become viable.

Specifically, current research and development in perception and manipulation tasks for robotic laundry systems consist of isolating clothes from a heap [4, 5, 7], performing clothes classification [6] and then folding the clothes [2, 3, 7]. Researchers usually assume that garments are in a tractable state, but do not measure the state of the garment being flattened prior to applying any folding procedure. Therefore, garment or cloth flattening has not been fully considered in current robotics developments. In this paper, we devise and demonstrate a novel heuristic-based approach for detecting and eliminating wrinkles in clothes automatically. A garment can take different forms and shapes depending on the external forces in the environment. In order to mitigate the unpredictable nature of clothes, we propose to tackle the problem initially under simulation, before transferring our algorithm to a real scenario. In this case, our simulator renders a virtual piece of cloth after it has been separated from the heap and then simulates the remaining steps required to flatten it.

A. Natraj et al. (Eds.): TAROS 2013, LNAI 8069, pp. 148–160, 2014.
DOI: 10.1007/978-3-662-43645-5_16, © Springer-Verlag Berlin Heidelberg 2014

This paper is structured as follows: Sect. 2 presents a literature review of related approaches for manipulation and perception of clothes. Section 3 provides an overall description of our approach while Sect. 4 describes the cloth-flattening problem and how we approached the problem. Section 5 details our heuristic-based flattening approach and Sect. 6 presents a simulated demonstration of our approach. Finally, Sect. 7 concludes this paper and provides future directions of this work.

2 Literature Review

A variety of different approaches have been proposed to implement parts of the laundry process described in Sect. 1. Below, we provide a discussion of successful methods used for perception and manipulation tasks for clothes which have inspired our heuristic-based approach.

The most relevant paper to our work is reported by Willimon, et al. [1] who proposes an interactive perception-based cloth unfolding strategy that relies upon detecting depth discontinuity corners in range images. For each iteration, the highest depth corner on the observed crumpled cloth is grasped and pulled away from the centroid of the cloth. A limitation of this method is that it cannot handle large garments. If a cloth of considerable size is used, the strategy of grasping corners and pulling these away from each other can potentially generate wrinkles while the cloth is hanging on the robot's grippers.

Other relevant research in terms of perception and manipulation includes that reported by Ramisa, et al. [5] where he proposes a supervised learning approach to grasping highly wrinkled garments. In Ramisa's grasp point detection method, highly wrinkled regions are manually segmented in order to train logistic regression and χ^2 SVM using bag-of-words descriptors based on SIFT and Geodesic-Depth Histogram features. Both classifiers are used to select candidate grasp-points and the final candidate is then determined by means of a robust wrinkledness feature. Wrinkledness, as defined in [5], serves to enable a robot to measure the current state of the cloth in terms of range-map wrinkles. Accordingly, we employ this visual representation in order to evaluate the state of the garment as described Sect. 6.2.

Maitin-Shepard, et al. [7] proposed a robot system that folds a square towel autonomously using a two-handed robot. Their method involves iteratively finding the lowest point of a suspended towel (grasped by the previously found lowest corner and initialized with a random grasp) until the robot is grasping a corner in each hand. The robot then stretches the garment and verifies that it does not present any wrinkles before the folding procedure starts. This strategy is based on a "lowest point algorithm", a heuristic that consists of tracing the towel's border by inducing a motion field map on the robot's field of view in order to select candidate-grasping points at corners. However, their method cannot be extended to manipulating garments as it relies on a predefined control sequence tailored to manipulating one small and specific towel. This lowest-point algorithm is unable to resolve ambiguities introduced by wrinkles.

Yamakazi, et al. [8] proposed applying Gabor filtering to an intensity image in order to detect wrinkles. Intensity image based features are extremely sensitive to the intrinsic texture observed on the manipulated cloth and, in consequence, this limits the robot's perceptual ability to detect concave and/or convex regions around wrinkled areas. Therefore, in this paper, we propose to avoid intensity image based descriptions and instead employ depth (range) features that capture the surface manifold shape of the cloth robustly. In this paper, we adopt a local measure of average absolute deviation extracted from densely sampled patches in the range image. This is a computationally inexpensive feature description that avoids the need to use complex descriptions that could potentially reduce the robot's performance as discussed above.

3   The Virtual Cloth Manipulation System

Our cloth simulator enables us to determine the force's action point accurately while avoiding sources of error such as the robot's joints and noise introduced by sensing systems. Therefore, a virtual world allows us to demonstrate the performance of our approach in isolation from uncontrolled variables before it is transferred to a real scenario. The overall design of our virtual cloth manipulation system is illustrated in Fig. 1.

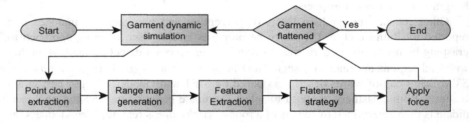

Fig. 1. The flow chart of the manipulation system.

This virtual system follows the perception-action cycle: the processing cycle starts by either initializing randomly the starting state of the cloth, or applying given forces to deform the cloth. After the cloth has reached static equilibrium, a point cloud is generated from all the particles that compose the cloth (perception). A range map is then generated from the point cloud (an example is shown in Sect. 3.2), which is passed to the flattening algorithm (action) as described in Sects. 4 and 5.

3.1   Cloth Simulation

Our simulated virtual cloth is composed of particles that are governed by: structural, shear and bending constraints [9]. The motion of these particles is modelled in terms of Newton's Laws of Motion and a Mass-Spring model, as described in [9], operating under gravity. For each of the above constraints, the interaction of each particle is further restricted by an offset that limits the range of distances between all connected particles (i.e. to prevent particles from getting too close or far away from each other).

We have also incorporated the frictional forces exerted by a virtual table acting on the cloth into our kinematic simulation by adopting Bridson's friction model (as detailed in [11]) that includes both static and sliding frictional forces. Our simulator is implemented in Virtual Studio 2010 (C++) running in Windows 7, 64-bits. OpenGL [10] is used to render the 3D surface of the cloth.

3.2   From Point Cloud to Range Map

Figure 2 illustrates the overall pipeline developed to render a range image from our virtual cloth. In order to generate a surface from the point cloud, we employ a cubic convolution interpolation algorithm [13] that allows us to generate a 2.5D range image (Fig. 2(c)). As our goal is to obtain perspective range maps that resemble those obtained by stereo-photogrammetry systems, or RGB-D sensors, points that are occluded from the camera point-of-view on the simulated world are deleted by means of a hidden point removal algorithm [12]. This step removes points that can potentially affect the surface construction and interpolation process. In this paper, we have selected the camera's point of view to be perpendicular to a ground-plane; this loosely resembles the perspective from which a robot might observe a garment lying on a table.

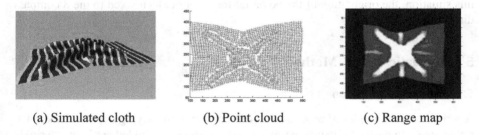

(a) Simulated cloth   (b) Point cloud   (c) Range map

Fig. 2. (a) Cloth rendered in our virtual world. (b) Generated point cloud: this point cloud is composed of 2475 particles (a square cloth of 54 by 45 particles). (c) Computed range map from the point cloud.

4   Flattening a Wrinkled Cloth

4.1   Problem Description

The input to our cloth flattening algorithm is the simulated cloth range map image (Fig. 2(c)) from which we extract image features (as described in Sect. 5.1) that are then fed to our "heuristic-based strategy" (Fig. 1). The output of our model is the predicted force and the force's point of action that minimizes the wrinkledness of the cloth. In this paper, the force is applied in a direction parallel to the table plane; therefore, we can only infer the orientation of the force in a 2D plane.

4.2 Simplifying the Problem

In order to determine the optimal force to be applied to the cloth, our heuristic-based strategy searches for the best force in all directions. This implies that our model has an infinite search space, which in practice is intractable. Therefore, we restrict our model to consider only 8 "compass" directions of applied force: 4 orientations parallel to the X-Y axes (north, south, west and east) and 4 diagonal orientations (northwest, northeast, southwest and southeast). We restrict the force's action point to act only on the closest edge for force directions parallel to the X-Y axes. The force action point is restricted to the corresponding four corners of the cloth for diagonal force orientations. For X-Y aligned forces, it is still possible that these forces can influence (i.e. disrupt) the cloth's edges. Therefore, we further constrain the X-Y aligned forces by selecting the force action points close to the detected wrinkle, as described below.

In order to maximise the force's effect, the putative force orientation is set to the angle bisector of the lines connecting the start and end points of the wrinkle (see Fig. 4 and Sect. 5.2), and we finally select the edge that is closer to the wrinkle's centre (in pixel units) as the force action point. It is noticeable that, the edges or corners will be rolled if we apply a non-orthogonal force to the edges as currently we only apply a fixed force for a fixed duration for each manipulation. In order to avoid this situation, the orientation of the potential force (bisector) is fixed to the 8 compass directions.

5 Heuristic-Based Method

5.1 Feature Extraction

A wrinkle can be modelled as a continuous structure with varying depth values in the range map. In order to detect wrinkles, we measure "wrinkledness" or "wrinkle strength" by computing the local average absolute deviation of range values within a square patch for every pixel in the range map. Specifically, for each pixel p we set a square patch with a side-length equal to 31 around the location of p (this patch size is based on empirical evidence). We then compute the average absolute deviation $Deviation_p$ of the mean depth value within the pixels in the patch,

$$Deviation_p = \frac{\sum_{N_{patch}} |depth(p_i) - d_{mean}|}{N_{patch}} \tag{1}$$

Where N_p refers to the number of pixels in this patch and p_i is the ith pixel, while $depth(p_i)$ and d_{mean} refer to the depth value of p_i and the mean depth value in this patch. A Gaussian smoothing process is then applied to create a feature map that

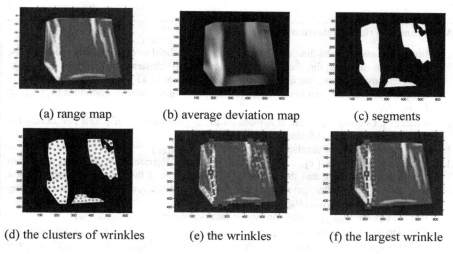

(a) range map   (b) average deviation map   (c) segments

(d) the clusters of wrinkles   (e) the wrinkles   (f) the largest wrinkle

Fig. 3. (a) The input range map of the currently observed state of the cloth in the simulator. (b) The average absolute deviation features as described in Sect. 5.1. (c) The average absolute deviation features are employed to segment putative wrinkle segments (average absolute deviation values less than 3 are no longer considered). (d) Result after applying k-means and the hierarchical clustering approach on the x-y coordinates of the resultant pixels (the cluster centres after k-means are the small circles while hierarchical groups are depicted with the same colour). (d) Result after applying the hierarchical clustering approach in order to join groups. (e) Detection of salient wrinkles. Blue lines parameterised the shape of the wrinkle while squares are the wrinkle's end points and centres. (f) The most salient wrinkle in terms of the area in the image that occupies (largest wrinkle in the current state of the cloth) (Color figure online).

preserves continuity across pixels. An example of the range map and the corresponding feature map is shown in Fig. 3(a, b), respectively.

5.2 Methodology

As discussed before, our approach is predicated in terms of the perception-action cycle (Fig. 1). For each iteration, the largest wrinkled is detected by Algorithm 1 and its parameters input to our heuristic-based strategy (as shown in Algorithm 2) in order to calculate the optimal force to removes this wrinkle from the observed cloth. This cycle is executed until all wrinkles are removed from the cloth.

Algorithm 1. Wrinkle Detection

Input: the average absolute deviation map, I, a threshold to distinguish wrinkle pixels, σ_1, and a distance threshold for the hierarchical clustering algorithm, σ_2 (from empirical investigation, we found that a threshold equal to 35 works well in practice)

Output: the largest wrinkle $w_{largest}$(its length l and centre c, endpoints $(x,y)_{start}$, $(x,y)_{end}$).

1. k-means clustering algorithm is applied on the x-y coordinates of pixels that satisfy I > σ_1, in order to obtain 100 clusters $\{c_1,\ldots,c_{100}\}$.
2. Record the number of pixels in each cluster $\{n_1,\ldots,n_{100}\}$.
3. Cluster the k clusters $\{c_1,\ldots,c_{100}\}$ in step 1 using a Hierarchical Clustering algorithm (from bottom to top), and the clustering will terminate if the distance is larger than a threshold σ_2, thereafter, get new clusters(wrinkles) $\{w_1,\ldots,w_n\}$ and the corresponding number of pixels $\{N_1,\ldots,N_n\}$
4. Get the cluster $w_{largest}$ with the largest number of pixels in $\{N_1,\ldots,N_n\}$
5. The length $1 = \max\ (|(x_{max},y)-(x_{min},y)|,|(x,y_{ymax})-(x,y_{ymin})|)$ and the endpoints are those which have the lager distance, (x_{max},y), (x_{min},y) are x-y coordinates in w_i that has the maximum and minimum coordinate in the x axis, while (x,y_{ymax}), (x,y_{ymin}) are the maximum and minimum for the y axis. $c = mean_{p \in w_{largest}}((x_p,y_p))$

Algorithm 2. Inferring the optimal force and orientation

Input: the wrinkle's start, end point and centre: $(x,y)_{start}$, $(x,y)_{end}$ and c respectively. The threshold σ to classify long and short wrinkles

Output: the force orientation d and point of action p_{act}

1. Compute the wrinkle's angle bisector V_b, and then match it with the 8 "compass" directions described in Section 4.2 and get the estimated angle bisector V_b'.
2. if V_b' is on x-y orientation and $1 > \sigma$
    p_{act} is the cross point between the nearest edge and V_b'.
    $d \in (V_b', -V_b')$ is the one away from wrinkle's centre c
    elseif V_b' is on diagonal orientation
    p_{act} is on the nearest corner .
    $d \in (V_b', -V_b')$ is the one away from wrinkle's centre c
3. Return d, p_{act}

In Algorithm 1, we apply *k-means* clustering algorithm initially on the x-y coordinates of those pixels labelled as highly wrinkled in order to form small clusters of "wrinkle" pixels. Thereafter, we join these clusters using a bottom-up hierarchical clustering algorithm in order to group them into significant salient wrinkles. The two end points forming a wrinkle, and also the wrinkle centre, can be computed from the final clustering (as shown in Algorithm 1). An example is demonstrated on Fig. 3.

Once the algorithm has detected the largest wrinkle, the optimal force orientation and its action point can then be predicted (Algorithm 2). More specifically, the wrinkle's angle bisector is first compared with the 8 "compass" directions (Sect. 4.2) in order to find the closest orientation. If the force orientation is on a diagonal axis, the force is applied to the nearest corner with respect to the wrinkle; otherwise, if the force direction is parallel to the X or Y axis, the force is applied to the closest edge with respect to the wrinkle's centre.

In order to avoid generating new wrinkles after applying a force, we determine that if the length of the wrinkle (the length between two endpoints (squares in Fig. 3(e, f)) is less than 100 pixels, it will be treated as a short wrinkle. Therefore, our strategy adds a force on the nearest corner of the cloth to pull it away from the centre of the wrinkle. Figure 4 illustrates the above cases.

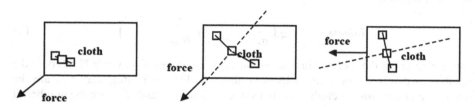

Fig. 4. The three types of wrinkles that can potentially emerge and the corresponding predicted force. (a) The wrinkle is small, thus we first find the nearest corner (the lower-left corner in this case) and, consequently, we can infer that the force's orientation in this corner can only be south-west. (b) The wrinkle is large (diagonal orientation case), therefore, the closest orientation to the angle bisector is southwest and northeast, and the possible action points are the lower-left and upper-right corners. In this case, the nearest corner (lower-left) is selected and the corresponding force orientation is determined to be southwest. (c) The wrinkle is large (X-Y aligned orientation case), therefore, the closest "compass" directions to the angle bisector is west and east; thus, the possible action point is either on the left or right edge. The nearest point to the left edge is selected as the force action point because it is nearer to the centre than the right edge.

6 Experiments

In order to demonstrate our heuristic-based flattening strategy approach, we prepared 8 different flattening experiments where the piece of cloth is heavily wrinkled. The initial state of the cloth in our simulator was generated randomly by grasping and then dropping the cloth on a virtual table. These 8 experiments (as shown in Fig. 5) are employed throughout the evaluation process.

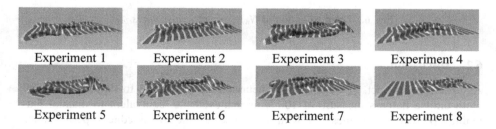

Fig. 5. Wrinkled experiments

6.1   Halting Criterion

In order to evaluate the performance of our approach, we measure the flatness of the cloth by computing two measures of wrinkledness: the cloth wrinkledness (the mean wrinkledness measure [5] of cloth range map) and the number of wrinkled pixels, which are named $wrinkledness_{cloth}$ and $N_{wrinkled}$, respectively. Both measures capture the quality and quantity, respectively, of the cloth's flatness. In that respect, we propose a halting criterion that is based on the above measures in order to measure the flatness percentage as follows

$$flatness = 1 - \frac{1}{2}\left(\frac{\sigma - \sigma_{min}}{\sigma_{max} - \sigma_{min}} + \frac{N - N_{min}}{N_{max} - N_{min}}\right) \tag{2}$$

where σ refers to the mean wrinkledness value of the image size, while N refers to the number of pixels whose wrinkledness is above 3 (Fig. 3(c)). σ_{min} and σ_{max} are the maximum and minimum cloth wrinkled values, while N_{min} and N_{max} are the number of pixels of the detected wrinkles.

To obtain ground truth we flattened in all 8 experiments by manually specifying force inputs by means of a GUI and we recorded the actions undertaken. The $wrinkledness_{cloth}$ and $N_{wrinkled}$ values were then computed on the 8 wrinkled examples and 8-flattened examples. We then calculated the median values of $wrinkledness_{cloth}$ and $N_{wrinkled}$ of the initial wrinkled examples and flattened examples to provide reference values for the wrinkled cloth and the flattened cloth. For completeness, the values of $\sigma_{min}, \sigma_{max}, N_{min}$, and N_{max} are the following median values (1.1595, 2.9865, 39020, and 317440, respectively). The median values are shown as green triangles in Fig. 6 and in Table 1.

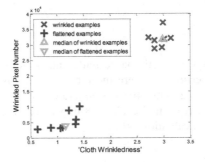

Fig. 6. The distribution of $N_{wrinkled}$ and $wrinkledness_{cloth}$.

6.2   Discussion

As described in Sect. 5, we allow our heuristic-based method to flatten the virtual cloth until a flatness measure of 100 % is achieved, i.e. equivalent to manual flattening. (Percentage flatness is with reference to the wrinkledness performance obtained above by manual flattening, and therefore 100 % flat does not actually mean

Table 1. The *wrinkledness_cloth* and $N_{wrinkled}$ computed in wrinkled and flattened examples

Wrinkled examples	Flattened examples
2.9496/354120	1.3434/44190
2.7239/322740	0.8790/33150
2.9755/291370	1.3527/58790
2.7408/285140	1.4237/101450
3.1411/320960	0.6300/28270
3.0151/318200	1.0605/32060
2.9975/370830	1.0998/33850
2.8379/313630	1.2192/88700
2.9865/317440	**1.1595/39020**

that the cloth is 100 % planar.) We then compare the number of cycles with respect to the manually flattened examples. Figure 7 shows a complete flattening procedure of experiment 5.

The required number of iterations (RNI) for each experiment and the difference between two methods are summarized in Table 2. As observed, the manual approach

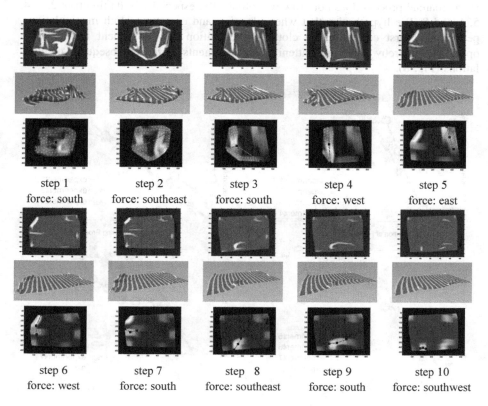

step 1	step 2	step 3	step 4	step 5
force: south	force: southeast	force: south	force: west	force: east

step 6	step 7	step 8	step 9	step 10
force: west	force: south	force: southeast	force: south	force: southwest

Fig. 7. The whole flattening process of experiment 5

Table 2. The statistic of required number of iteration. The third row shows the difference between two methods.

Experiment No.	1	2	3	4	5	6	7	8	Average
RNI of our method	13	11	9	14	10	13	9	12	11.375
RNI of manual method	11	10	11	11	13	11	9	10	9.75
Difference	−2	−1	2	−3	3	−2	0	−2	−1.625

is marginally better than our heuristic-based approach in 6 of the 8 experiments. For our heuristic-based approach, the average number of iterations required to flatten a wrinkled cloth is 11.125 iterations. This is 16.6 % (1.625/9.75) higher than the average manual method (9.75).

We also recorded the flatness percentages; the results of our heuristic-based strategy are shown on the left in Fig. 7 while the results for the manual approach are on the right. Figure 7 shows that our heuristic-based strategy is able to approach 70 % of cloth flatness rapidly, as opposed to the manual approach; however our approach requires more iterations to stabilise. By inspecting Fig. 7, we found that the heuristic-based method often makes the low-level wrinkling worse as a result of applying excessive force. However, when compared with heuristic-based method, the flatness in the manual process does not grow monotonically, especially in flatting task 2, 3, 4, 5, 6 and 8. We hypothesise that when planning and executing cloth manipulations, people may first configure the cloth in preparation for subsequent manipulation operations, thereby deferring flattening improvements to these subsequent manipulations (Fig. 8).

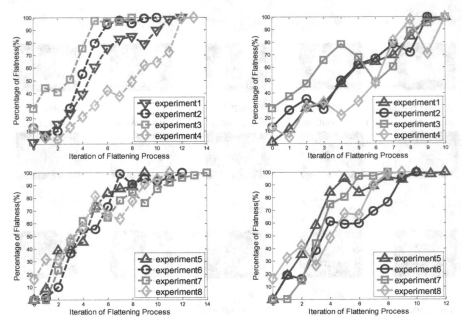

Fig. 8. The variation of flatness in every iteration of flattening process

7 Conclusion

In this paper we present a heuristic-based strategy for flattening a cloth that advances the current state-of-the-art in autonomous robotics for laundry processes, as described in Sect. 2. Our approach detects wrinkles in the cloth and then predicts the required force orientation and action point in order to gradually flatten the cloth. We demonstrate that our approach is able to flatten the cloth without constraining the initial configuration of the cloth. Moreover, we believe that our method can be extended to flatten different garments shapes by simply using the same constraints described in Sect. 4.

In order to apply actuation forces more accurately (in magnitude and orientation) and to improve the flattening achieved, we plan to adjust the duration over which an actuation force is applied by tracking the magnitude of individual wrinkles from cycle-to-cycle. We also propose to investigate the use of a greater number of force action points in order to improve the performance our method. Similarly, we propose to investigate whether reinforcement learning can be employed to optimise the magnitude and direction of the force. Finally, we intend to apply the approach presented in this paper within a robotic test bed used to investigate autonomous perception and manipulation of textiles and garments (European FP7 Strategic Research Project, CloPeMa; www.clopema.eu). The CloPeMa robot is equipped with two robotic arms, three RGB-D sensors for wide-angle vision and an active binocular robot head for foveated vision. Our robotic head is able to maintain camera convergence [4], explore its visual field for multiple same-class instance recognition [15], visually inspect garments and compute highly detailed range maps images (16 MP) in less than 2 s [14].

References

1. Willimon, B., Birchfield, S., Walker, I.: Model for unfolding laundry using interactive perception. In: Proceedings of the IEEE/RSJ International Conference on Intelligent Robots and Systems (IROS), pp. 4871–4876 (2011)
2. Van-Den-Berg, J.., Miller, S., Goldberg, K., Abbeel, P: Gravity-based robotic cloth folding. In: Hsu, D., Isler, V., Latombe, J.-C., Lin, M.C. (eds.) Algorithmic Foundations of Robotics IX, pp. 409–424. Springer, Berlin Heidelberg (2011)
3. Miller, S., Van-Den-Berg, J., Fritz, M., Darrell, T., Goldberg, K., Abbeel, P.: A geometric approach to robotic laundry folding. Int. J. Robot. Res. **31**(2), 249–267 (2012)
4. Aragon-Camarasa, G., Fattah, H., Siebert, J.P.: Towards a unified visual framework in a binocular active robot vision system. Robot. Auton. Syst. **58**(3), 276–286 (2010)
5. Ramisa, A., Alenya, G., Moreno-Noguer, F., Torras, C.: Using depth and appearance features for informed robot grasping of highly wrinkled clothes. In: IEEE International Conference on Robotics and Automation, pp. 1703–1708 (2012)
6. Willimon, B., Birchfield, S., Walker, I.: Classification of clothing using interactive perception. In: IEEE International Conference on Robotics and Automation, pp. 1862–1868 (2011)
7. Maitin-Shepard, J., Cusumano-Towner, M., Lei, J., Abbeel, P.: Cloth grasp point detection based on multiple-view geometric cues with application to robotic towel folding. In: IEEE International Conference on Robotics and Automation pp. 2308–2315 (2010)

8. Yamakazi, K., Inaba, M.: A cloth detection method based on image wrinkle feature for daily assistive robots. In: IAPR Conference on Machine Vision Applications, pp. 366–369 (2009)

9. Lander, J.: Devil in the blue-faceted dress: real-time cloth animation. Game Dev. Mag. p. 21 (1999)

10. Shreiner, D.: OpenGL Programming Guide: The Official Guide to Learning OpenGL, Versions 3.0 and 3.1. Addison-Wesley Professional, Upper Saddle River (2009)

11. Bridson, R., Fedkiw, R., Anderson, J.: Robust treatment of collisions, contact and friction for cloth animation. ACM Trans. Graph. **21**(3), 594–603 (2002)

12. Katz, S., Tal, A., Basri, R.: Direct visibility of point sets. ACM Trans. Graph. **26**(3), 24 (2007)

13. Keys, R.: Cubic convolution interpolation for digital image processing. IEEE Trans. Acoust. Speech Signal Process. **29**(6), 1153–1160 (1981)

14. Cockshott, W.P., Oehler, S., Aragon Camarasa, G., Siebert, J., and Xu, T.: A parallel stereo vision algorithm. In: Many-core Applications Research Community Symposium (2012)

15. Aragon-Camarasa, G., Siebert, J.P.: Unsupervised clustering in Hough space for recognition of multiple instances of the same object in a cluttered scene. Pattern Recogn. Lett. **31**(11), 1274–1284 (2010)

Vision-Based Cooperative Localization
for Small Networked Robot Teams

James Milligan[1], M. Ani Hsieh[1](✉), and Luiz Chaimowicz[2]

[1] Scalable Autonomous Systems Lab, Drexel University,
Philadelphia, PA 19104, USA
{milligan.james,mhsieh1}@drexel.edu
http://drexelsaslab.appspot.com
[2] VeRLab Computer Science Department,
Federal University of Minas Gerais, Belo Horizonte, MG, Brazil
chaimo@dcc.ufmg.br
http://homepages.dcc.ufmg.br/~chaimo/

Abstract. We describe the development of a vision-based cooperative localization and tracking framework for a team of small autonomous ground robots operating in an indoor environment. The objective is to enable a team of small mobile ground robots with limited sensing to explore, monitor, and search for objects of interest in regions in the workspace that may be inaccessible to larger mobile ground robots. In this work, we describe a vision-based cooperative localization and tracking framework for a team of small networked ground robots. We assume each robot is equipped with an LED-based identifier/marker, a color camera, wheel encoders, and wireless communication capabilities. Cooperative localization and tracking is achieved using the on-board cameras and local inter-agent communication. We describe our approach and present experimental and simulation results where a team of small ground vehicles cooperatively track other ground robots as they move around the workspace.

Keywords: Networked robots · Localization · Applications development

1 Introduction

We are interested in the development of heterogeneous teams of air and ground vehicles for exploration and mapping of three dimensional (3-D) spaces where control and coordination strategies exploit the complementary actuation and sensing capabilities of aerial and ground vehicles. In particular, we envision the deployment of aerial vehicles cooperating with mobile ground robots equipped with range finders and cameras operating in an indoor environment and operating with smaller less capable robots. The aerial vehicles, with their ability to move in 3-D spaces, can provide information from vantage points inaccessible to the ground robots. The larger ground vehicles, with their larger payload capacity, can be equipped with high fidelity range sensors and cameras. Aerial and larger

A. Natraj et al. (Eds.): TAROS 2013, LNAI 8069, pp. 161–172, 2014.
DOI: 10.1007/978-3-662-43645-5_17, © Springer-Verlag Berlin Heidelberg 2014

ground robots can collaboratively build 3-D maps of the environment, help localize the team, and help deploy smaller ground robots. Once deployed, the smaller ground robots can be used to explore, search, and/or provide information from areas inaccessible to the larger robots.

We are interested in exploiting the capabilities of small ground robots to access regions in the workspace that are inaccessible to aerial vehicles and other larger ground robots. Examples include deploying a team of small ground robots to search for suspicious packages concealed under tables and benches, or provide information from vantage points inaccessible to the other robots in the team, e.g., views through an air vent. Since these robots must be able to navigate in small confined spaces, their payload, and therefore their sensing capabilities, must be limited. We describe a cooperative localization and tracking strategy for teams of ground robots. The objective is to enable localization of a target located within the workspace that is not visible to the aerial vehicles and larger ground vehicles. We describe a strategy where a team of small ground robots cooperatively establish a local coordinate frame using their on-board cameras and local inter-agent communication. Cameras on the ground vehicles are used to observe identifying LED markers on each ground vehicle. Mutual observations define a visibility graph that is then used to triangulate the location of the ground vehicles in the local coordinate frame similar to [15]. Finally, assuming an aerial vehicle or larger ground robot can visually localize at least one element of each connected component of the small ground vehicle visibility graph, the local coordinate frame of the team of small ground robots can then be transformed into the global coordinate frame maintained by the aerial or larger ground vehicles.

In this work, smaller ground robots first establish a local coordinate frame using their on-board cameras and odometry. Information from an aerial vehicle is simulated using a static camera in the workspace and the estimate of the global state for one of the ground robots is communicated via the 802.11 wireless network. This information is then used to determine all the global poses for the team of ground robots within the workspace. Once localization has been achieved, one or a subset of the ground vehicles can then be tasked to drive to some region of interest to locate and localize a target of interest in the environment. The described approach can be scaled for teams of micro aerial vehicles (MAVs) cooperating with a team of larger unmanned ground vehicles (UGVs) and enables aerial vehicles to not only deploy teams of ground vehicles, but also to serve as localization anchors that can support a multitude of small teams of ground vehicles that can bring sensors to bear in regions inaccessible to MAVs and larger ground robots. The ability for the smaller UGVs to localize within a global coordinate frame enables more complex tasks such as autonomous exploration and searching and tracking of persons or items of interest hidden in the environment.

Existing work in cooperative localization [3,6] has been shown using a wide selection of sensor combinations including range-only [17], bearing-only [14], and a combination of the two [16,18]. Extensions of the general cooperative localization to that of a distributed system has also been addressed in [12]. In our work, we rely on on-board cameras to extract relative orientation information and thus the localization problem considered constitutes a bearings-only localization problem. The majority of bearings-only vision-based localization in existing literature rely on known landmarks within the environment [1,2,7,8,10,13]. Similar to [15], we rely on LED markers on each robot to enable a team of small ground vehicles to establish a local coordinate frame. State estimation and tracking is then achieved by fusing the information from the team using an Extended Kalman Filter (EKF) similar to [2,4,5,7,9,11,13].

The remainder of this paper is organized as follows: Sect. 2 briefly describes the scenario and the experimental testbed. Our Methodology is described in Sect. 3. We conclude with a brief summary and discussion of our experimental and simulation results in Sect. 4 and conclude in Sect. 5.

2   Experimental Testbed

As stated in the previous section, our objective is to develop a collaborative/ cooperative localization and tracking strategy for a team of small, sensing limited team of ground robots for applications like exploration, target search, and providing situational awareness from vantage points inaccessible to aerial vehicles and larger ground robots. Since the objective is to use small ground robots, our robots are less capable than larger ground robots, i.e., an iRobot Create platform. As such, we assume that our robots are only equipped with a single camera, odometry, and wireless communication capabilities.

Specifically, we consider a team of mSRV-1 robots (see Fig. 1). The mSRV-1 is a differential-drive robot equipped with a 600 MHz Blackfin embedded processor, 802.11b wireless communication, a color camera, and wheel encoders. The robot has a footprint of approximately $10\,\mathrm{cm} \times 9\,\mathrm{cm}$. The state of each vehicle is given by $X_i = [x_i\, y_i\, \theta]^T$ and summarized in Fig. 2. Additionally, each robot on the team is equipped with an LED marker. The LED markers can be programmed with unique blinking patterns to enable unique identification of the individual robots similar to those described in [15].

3   Methodology

In this section we describe the development of our cooperative localization and tracking framework which consists of a state estimation step for estimating the state of the robots as they move within the workspace and an image processing step for identifying and localizing the robots within a local coordinate frame.

3.1   Extended Kalman Filtering

In our implementation, we use an Extended Kalman Filter with two independent update steps that incorporate information from the image processing step and from the robot odometry. As mentioned before the robot state is defined

Fig. 1. mSRV-1 robot

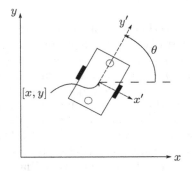

Fig. 2. mSRV-1 state

as $[x, y, \theta]$. In the implementation of the EKF it is necessary to predict the system measurements in order to determine some residual value for filtering. In the image processing step, he primary output is a list of relative orientations to any LED markers in the field of view of any one of the robot cameras. Included with that are markers to denote which camera the information was taken from as well as which robot's LED markers are seen though this information is not directly used in the EKF.

Predicted outputs for the image processing step depend on both the field of view of the cameras as well as the state estimate for each of the robots. Since the state of a robot is defined with respect to a central point, there is some offset from this which defines the position of the camera. Figure 3 illustrates this idea. Here $[x', y', \phi']$ represent the position and relative orientation with respect to the camera. It is this ϕ' value that is being provided as a measurement. Since the EKF is based on the measurement residual between expected and measured, it is not necessary to bring this ϕ' value into the robot frame but rather the estimated measurement value could be carried out with respect to the camera

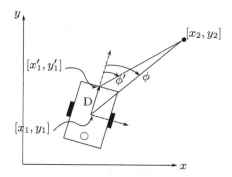

Fig. 3. Camera coordinate transformation

position. The measurement prediction results in the following relationship:

$$h_\phi(\hat{x}_k) = \arctan\left(\frac{y_i - y_j - D\sin(\theta_j)}{x_i - x_j - D\cos(\theta_j)}\right) - \theta_j \tag{1}$$

where D represents camera offset distance from the robot center, the i components refer to the state information of the given LED and the j components, that of the camera.

Each robot is also equipped with wheel encoders and thus the output of the odometry is given as a column for the total distanced traveled at each time step.

3.2   Image Processing

At each time step, each robot is tasked with acquiring an image from its camera. In our system, we only consider the intensity of the robot LED markers in the images rather than the marker colors, however, our approach can be easily extended to include color as another distinguishing feature. Given two images obtained by robot i at time steps k and $k+1$ denoted by $I_i(k)$ and $I_i(k+1)$, let $\Delta I_i = I_i(k+1) - I_i(k)$ denote the change in intensity in the image. As a robot moves through an image, the area where the robot was previously will have a decreased intensity while the intensity of the new location will increase as a result of the active LED markers. Thus, the region in ΔI_i, denoted as R_i with intensity values greater than some threshold $\delta_I > 0$ should give the new location of the LED marker in view and thus the robot.

Given that the robot dimensions are known, to improve the identification of the location of the LED marker in ΔI_i and filter out false positives, we take into account the distance of R_i from the ground plane and the size of R_i in relation to the known size of the robot. We then perform an opening operation followed by a closing operation to eliminate any smaller artifacts that may have shown up in ΔI_i. While both of these techniques improve the quality of the information extracted from the images, there is still no guarantee that a point of interest is in

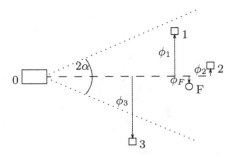

Fig. 4. Example case for robot 0

fact a robot. As such, the centroids of all R_i in ΔI_i whose intensities are greater than δ_I and whose areas are greater than $\delta_A > 0$ are then determined. Given the field of view of the camera, the centroid of each R_i in ΔI_i is then converted into a relative orientation of the point with respect to the camera as follows:

$$\phi_k = \alpha \left(\frac{w - 2p}{w} \right) \tag{2}$$

where ϕ_k represents the relative orientation of the LED marker with respect to the camera according to the right hand convention at time step k, α is the field of view of the camera, expressed as a $\pm$ value with respect to the camera normal, w is the width of the image, and p is the location within that image that is determined to be the centroid of the region of interest.

As mentioned before there are no guarantees that a region of interest in an image actually corresponds to a robot. Furthermore there is as of yet no information regarding which robot a given region of interest may correspond to. To address these issues, a vector, ρ, is created representing the robots that are in view for any given camera. Figure 4 illustrates one possible scenario for a robot.

Based on the estimated position of each of the robots at the previous time step, an estimated list of which robots are in view of Robot 0 can be created. In our scenario, each robot can potentially obtain an initial estimate of its position from an aerial vehicle or from another ground robot. In this instance, it is known that Robots 1 and 2 are within the field of view of Robot 0 denoted by the dotted lines, while Robot 3 is outside of Robot 0's field of view. The point F in Fig. 4 represents a false positive in the image for which an orientation was provided though it does not correspond to a known robot.

This information is combined with the relative orientation computed using each of the estimated robot poses which we denote by $\hat{\phi}_j(k)$ to create a vector ψ_i, such that ψ_i for the example given in Fig. 4 is given as: $\psi_0 = [Null, \hat{\phi}_1, \hat{\phi}_2, Null]^T$ where both the orientation to the robot itself and any orientations to robots outside the field of view are given by $Null$. The $Null$ values insure that no information from these values is propagated through the system.

At this point, each ϕ given by Eq. 2 is compared with each of the entries in ψ_i and matched with the entry that provides the smallest difference in orientations. This is essentially computing the residual values of the relative orientation measured and the ones estimated between all possible robots in view and provides an initial approximation for the image processing output. For any given robot there may be multiple orientations provided. For instance, in the example shown in Fig. 4, ϕ_F would be attributed to Robot 2 as well as ϕ_2 since that is the closest match. Each robot is checked to ensure only one match is found in a given image. If there are multiple matches, the match with the least distance to the predicted orientation, based on the estimate pose, is taken to be the correct match. Since the size of the intensity blobs in the difference image is roughly the size of the robot itself, small errors in the position estimates at this point are not likely to drastically change the values returned by the imaging system, i.e., attributing an incorrect match to a robot.

With one match assigned per robot, another check is performed to make sure that all matches are within a given threshold. For the example shown in Fig. 4, if the LED marker for Robot 2 did not appear in the image and yet the F did, ϕ_F would be returned as the relative orientation to Robot 2 since it was the best and only match for that robot. However, by ignoring differences in the relative orientation greater than that some appropriately chosen threshold, our system returns no information for Robot 2 instead of incorrect information at point F.

The vector ψ now represents a column vector in which each row corresponds to a robot in the team and the entries represent relative orientations as reported by the cameras. Since the EKF requires both the orientation as well as which robot camera the information came from, ψ is expanded to the matrix O for which each row entry reflects the following information: $[\phi, camID, ledID]$. In our EKF implementation, $camID$ and $ledID$ values are used to determine the locations for the Jacobian blocks of the observation matrix. For the scenario provided the values for ψ, O, and H would be as follows:

$$\psi_0 = \begin{bmatrix} Null \\ \phi_1 \\ \phi_2 \\ Null \end{bmatrix} \tag{3}$$

$$O = \begin{bmatrix} \phi_1 & 0 & 1 \\ \phi_2 & 0 & 2 \end{bmatrix} \tag{4}$$

$$H = \begin{bmatrix} H_C & H_L & 0 & 0 \\ H_C & 0 & H_L & 0 \end{bmatrix} \tag{5}$$

where H_C represents the derivative of Eq. 1, as presented in Sect. 3.1, with respect to the camera states, and H_L the derivative with respect to the LED states. Each entry shown in Eq. 5 consists of a 1×3 block so that each row of H corresponds with an image system output and each column a state of the system.

In the full system, values for O as provided by each camera are concatenated and a single H matrix is created to include all entries provided by each camera. For demonstration purposes the above values are assuming that Robot 0 is the only available camera in the team.

In summary, consider the case where the team is deployed in an environment where one or a subset of the robots is tasked to drive from their initial locations to some goal positions in the workspace. Our collaborative localization and tracking strategy for each robot can then be summarized as follows:

```
if the robot has not reached the goal location
    compute desired velocities (linear and angular)
    drive the robot
    predict current robot state based on control inputs (EKF predict)
    collect new measurements (images & odometry)
    extract relative orientation information
    update the state and covariance estimates (EKF update)
    check if the robot has arrived
end
```

4   Results

In our experiments, we consider a team of 5 mSRV-1 robots where one robot was tasked to drive from an initial position to a final goal position using the estimated state provided by the team. Since our robots are quite resource constrained, all image processing, state estimation, and computation was implemented on a centralized computer to ensure the experiments can be completed in a reasonable amount of time and robots can move at a reasonable speed. The general implementation of the calculations and image processing, while done on a central computer, was carried out as if it were on individual processors, simulating a decentralized implementation. With demonstrated success in this manner experimentally, we believe our approach can be easily decentralized once the image processing has been successfully implemented on each robot's embedded processor. Furthermore, an initial state estimation for at least two of the robots is provided by an overhead localization system. The overhead localization system provides information that can be provided by an aerial vehicle or larger ground robot with view of at least one of the ground robots.

In order to validate the results of our collaborative localization and state estimation framework, several experiments were conducted where the estimated states are compared to ground truth measurements provided by our overhead localization system. The accuracy of our overhead localization system is approximately ± 20 mm. Figure 5 shows three trial runs and the predicted error ellipses.

The error at each time step was calculated as the 2-norm between the ground truth and the estimated position. The mean error over the full run as well as the standard deviation of the error is provided for each trial. A total of 12 trials were completed with varying start and end positions comprising a total of 278 pose estimates. The average error in position over all 278 estimates was 132 mm with

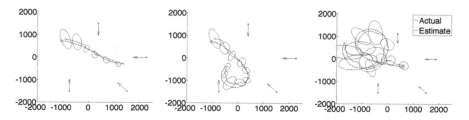

Fig. 5. Three example trials. Start and end positions are shown as ○ and × respectively and cameras are denoted with squares, oriented as shown. Axes shown in mm.

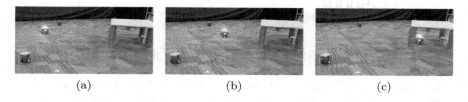

(a)   (b)   (c)

Fig. 6. Surveillance of suspicious object under table.

a standard deviation of 67 mm. Despite the lack of high fidelity odometry sensors on our robots, the accuracy of our strategy is about a robot's bodylength. Based on the covariance matrices of the EKF it is also possible to see the assumed error in the measurements from the robot's perspective. The ellipses shown on Fig. 5 represent the location of the robot with 80 % confidence based on noise in the measurements. As can be seen the number of robots that can see the traveling robot as well as the direction from which the robot is seen has a large affect on the error in the system.

In addition to evaluating the position error of the system, we considered an example scenario where the smaller ground vehicle cannot be easily localized by an aerial vehicle or larger ground robot. This is shown in Fig. 6 where a robot was tasked with capturing an image of a suspicious object underneath a low table. The view of the object is obstructed from the point of view of any aerial vehicles. Furthermore, the low clearance means the space is inaccessible to larger ground vehicles. Using our cooperative localization and tracking strategy the mSRV-1 is able to infiltrate the space to localize the target.

Finally, the show the scalability of the proposed strategy, we simulated a scenario where a team of five robots must simultaneously track three moving robots. In our simulation, the error on relative orientation measurements was assume to follow a Gaussian distribution based on errors obtained in experiments. Figure 7 shows one such trial in which three robots were tasked to drive 10 m towards three different goal locations. Using the methodology presented, the two static robots were successful in providing the information necessary to localize and track the three others.

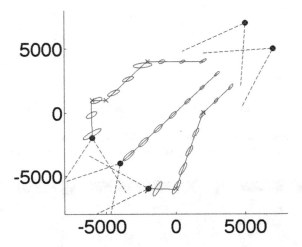

Fig. 7. Multiple driver simulation. Dashed lines represent field of view of the robots, solid lines show the paths traveled, interim goals are shown as ×, axes are in mm.

5   Conclusions

In this work, we presented a cooperative localization and tracking framework specifically designed to enable small ground vehicles to operate in regions of the workspace that is inaccessible to aerial vehicles and/or larger ground robots. Specifically, we address the localization and navigation of small ground vehicles in such confined spaces. An EKF-based position update scheme was presented in which bearing information received from on-board cameras and odometry information results in pose estimates for the team that are accurate to approximately one robot body length. One direction for future work is full distributed implementation of the strategy on a team of robots capable of on-board image processing. A second direction for future work includes the integration of the approach with a team of aerial vehicles capable of providing the initial global pose estimates for the team.

For all of the scenarios presented, the location of the static robots responsible for tracking the moving robots were predefined. The positioning of these tracking robots has a large effect on the accuracy of the system. As illustrated in Fig. 8, a system consisting of two robots with parallel views is much more susceptible

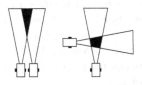

Fig. 8. Error envelopes for different camera poses.

to large errors due to measurement noise than a system of two robots with perpendicular views.

In Fig. 8, lines are shown to represent an error envelope on the relative orientations as provided by the cameras. This error is then extended into the $[x, y]$ position measurements as represented by the shaded region. The position error in the parallel view scenario is unbounded and highly undesirable. This idea of *best sensing locations* is examined in [16], in which observers are directed to move in order to reach the optimal imaging location given a target's position. We are interested in extending our approach to include such formation control strategies and eliminate the need for predefining the positions of the tracking robots.

References

1. Bishop, A.N., Anderson, B.D.O., Fidan, B., Pathirana, P.N., Mao, G.: Bearing-only localization using geometrically constrained optimization. IEEE Trans. Aerosp. Electron. Syst. **45**(1), 308–320 (2009)
2. Bonnifait, P., Garcia, G.: A multisensor localization algorithm for mobile robots and its real-time experimental validation. In: IEEE International Conference on Robotics and Automation, pp. 1395–1400, April 1996
3. Howard, A., Matarić, M.J., Sukhatme, G.: Putting the 'I' in 'team': an ego-centric approach to cooperative localization (2003)
4. Huang, G.P., Trawny, N., Mourikis, A.I., Roumeliotis, S.I.: On the consistency of multi-robot cooperative localization. In: Proceedings of Robotics Science and Systems, June 2009
5. Jetto, L., Longhi, S., Venturini, G.: Development and experimental validation of an adaptive extended Kalman filter for the localization of mobile robots. IEEE Trans. Robot. Autom. **15**(2), 219–229 (1999)
6. Kurazume, R., Nagata, S., Hirose, S.: Cooperative positioning with multiple robots (1994)
7. Leonard, J., Durrant-Whyte, H.: Mobile robot localization by tracking geometric beacons. IEEE Trans. Robot. Autom. **7**(3), 376–382 (1991)
8. Loizou, S.G., Kumar, V.: Biologically inspired bearing-only navigation and tracking. In: 2007 46th IEEE Conference on Decision and Control, pp. 1386–1391 (2007)
9. Mourikis, A., Roumeliotis, S.: Performance analysis of multirobot cooperative localization. IEEE Trans. Rob. **22**(4), 666–681 (2006)
10. Reza, D.S.H., Mutijarsa, K., Adiprawita, W.: Mobile robot localization using augmented reality landmark and fuzzy inference system. In: 2011 International Conference on Electrical Engineering and Informatics (ICEEI), pp. 1–6 (2011)
11. Roumeliotis, S.I., Bekey, G.A.: Collective localization: a distributed Kalman filter approach to localization of groups of mobile robots (2000)
12. Roumeliotis, S., Bekey, G.: Distributed multi-robot localization. IEEE Trans. Robot. Autom. **15**(5), 781–795 (2002)
13. Se, S., Lowe, D., Little, J.: Local and global localization for mobile robots using visual landmarks. In: Proceedings of the 2001 IEEE/RSJ International Conference on Intelligent Robots and Systems, vol. 1, pp. 414–420 (2001)
14. Sharma, R., Quebe, S., Beard, R., Taylor, C.: Bearing-only cooperative localization. J. Intell. Rob. Syst., 1–12 (2013), http://dx.doi.org/10.1007/s10846-012-9809-z

15. Shirmohammadi, B., Taylor, C.J.: Self localizing smart camera networks. ACM Trans. Sens. Netw. **8**(2), 11:1–11:24 (2012)
16. Zhou, K., Roumeliotis, S.I.: Multi-robot active target tracking with combinations of relative observations. IEEE Trans. Rob. **27**(4), 678–695 (2011)
17. Zhou, X., Roumeliotis, S.: Robot-to-robot relative pose estimation from range measurements. IEEE Trans. Rob. **24**(6), 1379–1393 (2008)
18. Zhou, X.S., Roumeliotis, S.I.: Determining the robot-to-robot 3D relative pose using combinations of range and bearing measurements (Part II). In: 2011 IEEE International Conference on Robotics and Automation (Part II), pp. 4736–4743, May 2011

Active Vision Speed Estimation from Optical Flow

Sotirios Ch. Diamantas$^{(\boxtimes)}$ and Prithviraj Dasgupta

C-MANTIC Lab, Department of Computer Science,
University of Nebraska, Omaha, Omaha, NE 68182, USA
{sdiamantas,pdasgupta}@unomaha.edu
http://cmantic.unomhaha.edu

Abstract. In this research, we address an important problem in mobile robotics - how to estimate the speed of a moving robot or vehicle using optical flow obtained from a series of images of the moving robot captured by a camera. Our method generalizes several restrictions and assumptions that have been used previously to solve this problem - we use an uncalibrated camera, we do not use any reference points on the ground or on the image, and we do not make any assumptions on the height of the moving target. The only known parameter is the camera distance from the ground. In our method we exploit the optical flow patterns generated by varying the camera focal length in order to pinpoint the principal point on the image plane and project the camera height on the image plane. This height is then used to estimate the speed of the target.

Keywords: Speed estimation · Principal point estimation · Optical flow · Active vision

1 Introduction

In this paper we describe our research on estimating a moving vehicle's speed using optical flow vectors. A few decades ago and up to date most of the effort on speed estimation has been focused on using Time-of-Flight (TOF) sensors. Since the advent of visual perception sensors a significant number of research has been devoted on estimating speed using cameras. One main reason is that cameras provide an efficient yet inexpensive means for speed estimation. Furthermore, the use of computational techniques has advanced the field by providing satisfactory estimations. Moreover, cameras are also used, for example, in road traffic for purposes beyond speed estimation such as monitoring, plate recognition, and vehicle tracking.

In this research we make two contributions. The first contribution is that we estimate the principal point of an uncalibrated camera by exploiting the

This research has been supported by the U.S. Office of Naval Research grant no. N000140911174 as part of the COMRADES project.

A. Natraj et al. (Eds.): TAROS 2013, LNAI 8069, pp. 173–184, 2014.
DOI: 10.1007/978-3-662-43645-5_18, © Springer-Verlag Berlin Heidelberg 2014

inherent properties of today's modern digital cameras. In particular, we make use of active vision by varying the focal length of the camera, that is the optical zoom, with the view to estimate the point in the image plane that is invariant to rotation and lens translation. In every lens there is an imaginary optical axis that is coming out of its center. The projection of the optical axis onto the image plane is the principal point. However, the optical axis of the lens is almost never aligned with the center of the image sensor. For this reason, in order to estimate parameters such as the principal point, most of the methods employed rely upon camera calibration techniques. In our approach we have used an uncalibrated camera to infer the principal point by varying the focal length of the camera; a technique that can be used in real-time on most robotic systems.

The second contribution is that we estimate the speed of a moving vehicle, namely a *Corobot* mobile robot with minimal assumptions. We make no use of artificial marks on the ground nor any other assumptions such as the distance between two points or a reference object in the image plane. In addition, the distance between the camera and the moving object need not be known. The only known parameter that we employ as a reference point for estimating the speed of the vehicle is the height of the visual sensor from the ground which is available or can be calculated in most vision-based systems. Yet, the visual sensor in our method is placed parallel to the ground; a case which holds for most robotic and other systems. The proposed method can easily be applied to several robotic tasks such as navigation, localization, and height estimation.

The following section provides a literature review on vehicle speed estimation as well as the mathematics that underlie optical flow. In Sect. 3 the methodology of applying optical flow for principal point and speed estimation is described. In Sect. 4 the results of our method are presented and finally Sect. 5 epitomizes the paper with a section on the conclusions drawn from this research as well as the prospects for future work.

2   Background Work

Optical flow, that is the rate of change of image motion in the retina or a visual sensor, is extracted from the apparent motion of an agent be it natural or artificial. Although motion is an inherent property of optical flow, there is not a large number of works that employ optical flow for speed estimation. In [1] the authors present an optical flow method for speed estimation using a calibrated camera. A method of speed estimation that does not rely on a calibrated camera but instead makes assumptions drawn from a distribution about the mean height, width, and length of the vehicle is presented in [2].

In [3] the authors present a method similar to the one followed in our research for speed estimation. However, they know in advance the height of the moving vehicle. In particular, they use a measurement tape to infer the heights of the vehicle. An optical flow method for velocity estimation for Micro Aerial Vehicles (MAVs) is presented in [4]. In that work, the authors use an FPGA platform in order to extract the optical flow field. Their research is suitable for embedded systems since most of the calculations take place on an FPGA. A method for speed estimation using a known ground distance along the road axis is presented in [5]. A velocity estimation for a blimp using a non-optical but air flow sensor appears in [6] where particle filtering and a probabilistic model are employed. In [7] the authors present an embedded system for vehicle speed estimation. An RF-based method for speed estimation appears in [8]. In [9] the authors make use of magnetic signatures induced by vehicles to estimate the speed of vehicles in highways.

The problem of estimating the heights of objects by means of the focus of expansion (FOE) is addressed in [10]. In that work the authors use the cross ratio of a known object's height in the image plane in order to estimate other object's heights in the same image. Reference objects are also used in the seminal work of [11]. A method for height estimation using a single image from an uncalibrated camera and a vanishing point is described in [12]. An active vision technique for zoom tracking by varying the focal length of the camera is presented in [13]. Our research in contrast to the afore-mentioned works can estimate the speed of a vehicle with an uncalibrated camera and with no reference objects in the image plane. Furthermore, our research, similarly to the work of [10], makes use of feature points. However, we exploit the focal length of the visual sensor to achieve near (pure) translation instead of the robot movement.

2.1   Mathematics of Optical Flow

This section describes the mathematics of optical flow algorithms, and in particular, the Lucas-Kanade (LK) algorithm [14] which has been employed in this research. Following are the equations needed for a 5×5 pixels window that results in a system of 25 linear equations that need to be solved (1). However, if the window is too small the *aperture problem* may be encountered where only one dimension of the motion of a pixel can be detected and not the two-dimensional. On the other hand, if the window is too large then the spatial coherence criterion may not be met.

$$\underbrace{\begin{bmatrix} I_x(p1) & I_y(p1) \\ I_x(p2) & I_y(p2) \\ \vdots \\ I_x(p25) & I_y(p25) \end{bmatrix}}_{A=25\times 2} \underbrace{\begin{bmatrix} u \\ v \end{bmatrix}}_{u=2\times 1} = - \underbrace{\begin{bmatrix} I_t(p1) \\ I_t(p2) \\ \vdots \\ I_t(p25) \end{bmatrix}}_{b=25\times 1} \tag{1}$$

The goal on the above system of linear equations is to minimize $||Au - b||^2$ where $Au = b$ is solved by employing least-squares minimization as in (2),

$$(A^T A)u = A^T b \tag{2}$$

where $A^T A, u$, and $A^T b$ are equal to (3),

$$\underbrace{\begin{bmatrix} \sum I_x^2 & \sum I_x I_y \\ \sum I_x I_y & \sum I_y^2 \end{bmatrix}}_{A^T A} \underbrace{\begin{bmatrix} u \\ v \end{bmatrix}}_{u} = - \underbrace{\begin{bmatrix} \sum I_x I_t \\ \sum I_y I_t \end{bmatrix}}_{A^T b} \tag{3}$$

and the solution to the equation is given by (4)

$$\boldsymbol{u} = \begin{bmatrix} u \\ v \end{bmatrix} = (A^T A)^{-1} A^T b. \tag{4}$$

If $A^T A$ is invertible, i.e., no zero eigenvalues, it means it has full rank 2 and two large eigenvectors. This occurs in images where there is high texture in at least two directions. If the area that is tracked is an edge, then $A^T A$ becomes singular, that is (5),

$$\begin{bmatrix} \sum I_x^2 & \sum I_x I_y \\ \sum I_x I_y & \sum I_y^2 \end{bmatrix} \begin{bmatrix} -I_y \\ I_x \end{bmatrix} = \begin{bmatrix} 0 \\ 0 \end{bmatrix} \tag{5}$$

where $-I_y, I_x$ is an eigenvector with eigenvalue 0. If the area of interest is homogeneous then $A^T A \approx 0$, implying 0 eigenvalues. The pyramidal approach of the LK algorithm overcomes the local information problem at the top layer by tracking over large spatial scales and then as it proceeds downwards to the lower layers the speed criteria are refined until it arrives at the raw image pixels.

3   Methodology

In this section we describe the methodology followed for estimating a moving robot's speed relative to a fixed camera through principal point estimation and optical flow. For this purpose the focal length of the camera, f, is varied in order to exploit the apparent motion of the objects in the image plane. Figure 1 provides a pictorial representation of the optical axis and the principal point as well as the various focal lengths a zoom lens can have. In this figure, it can also be seen how the focal length, f, of the camera affects the field-of-view (FOV). This relationship is expressed by (6),

$$FOV = tan^{-1} \frac{d}{2f} \tag{6}$$

where d is the sensor's width. In general, in our method we attempt to keep the number of assumptions to as few as possible while at the same time develop a system that is flexible and efficient. In particular, the visual sensor we have employed is a SONY TX-9 digital camera with 4x optical zoom. The camera is uncalibrated and is placed parallel to the ground and orthogonal to the direction of motion of the mobile robot, hence, the optical axis of the camera is perpendicular to the direction of motion of the vehicle. The geometrical parameters of the camera as well as the distance between the camera and the moving vehicle are unknown. In addition, the distance that the moving vehicle has covered from time frame F_{t_i} to time frame F_{t_i+1} which is needed to perform the optical flow algorithm is unknown. The only known parameter in this method is the height of the camera from the ground.

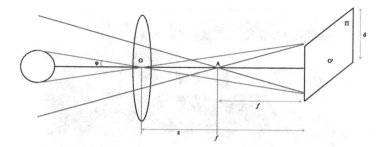

Fig. 1. Optical axis, O, of a lens and its projection onto the image plane, Π, that is, principal point, O'. In this figure a zoom lens is depicted with varying focal lengths between points OA. Zoom lens range is represented with z and lies within the points OA. Sensor's width is represented with d and ϕ is the angle between an object (e.g., ball) and the center of the lens.

3.1 Principal Point Estimation

First, we estimate the principal point using optical flow. The LK algorithm is applied to two frames taken from the same position and orientation. One frame is taken at a focal length of 4 mm and the other one at 5 mm. We then perform a combination of the *jackknife* sampling method along with the minimization of the set of linear equations[1]. Every optical flow vector represents a linear equation and the minimization of the system yields the position, P, at which the optical flow vectors converge. Thus, a set Ω_i is formed for every optical flow vector. Equations (7) and (8) show an example of two vectors,

$$P \in \Omega_1 = \{h \in \Re^2 | \underbrace{(v_1 - r_1)^T}_{\alpha_1} h = \underbrace{v_1^T \cdot r_1 - ||r_1||^2}_{\beta_1}\} \tag{7}$$

$$P \in \Omega_2 = \{h \in \Re^2 | \underbrace{(v_2 - r_2)^T}_{\alpha_2} h = \underbrace{v_2^T \cdot r_2 - ||r_2||^2}_{\beta_2}\} \tag{8}$$

where r is the position of the vector, and v is a point on a line that is perpendicular to the optical flow vector. The following Eqs. (9) and (10), show the process for $n-1$ of optical flow vectors. Noise in the system is represented by variable ϵ_i.

$$h\alpha_1 + \epsilon_1 = \beta_1$$
$$h\alpha_2 + \epsilon_2 = \beta_2$$
$$\vdots \tag{9}$$
$$h\alpha_{n-1} + \epsilon_{n-1} = \beta_{n-1}$$

$$h \in argmin \sum_{i=1}^{n-1} (h\alpha_i - \beta_i + \epsilon_i)^2 \tag{10}$$

---

[1] The non-parametric method we have developed for outlier removal aims at avoiding the known risks entailing the use of parametric methods, such as RANSAC.

Fig. 2. Principal point estimation using optical flow. The outliers (red color) have been removed using the *jackknife* sampling method. The green color vectors denote inliers. The center of blue circle is the principal point that lies along the horizon line. Some profound outliers that have accurately been detected appear on the poster and on the window; x, y point is in Cartesian coordinates (Color figure online).

$$\underbrace{\left(\sum_{i=1}^{n-1} \alpha_i \alpha_i^T + \epsilon_i\right)}_{C} h = \underbrace{\left(\sum_{i=1}^{n-1} \alpha_i \beta_i\right)}_{\gamma} \tag{11}$$

$$h = C^{-1}\gamma \tag{12}$$

In (11) C is a 2×2 matrix and $\gamma \in \Re^2$. The convergence point, P, is thus given by h. The *jackknife* sampling works by excluding one optical flow vector from the sample while estimating the point P using the $n - 1$ vectors of the data set. The Euclidean distance, $d(p_j, P_i)$, is then calculated between point p_j of the excluded vector and the convergence point, P_i. The same process is repeated n times, i.e., once for each optical flow vector within the given data set. The result is a set of n distances. The optical flow vectors whose distance is beyond the 90th percentile are considered as outliers and are disregarded. After outlier removal using *jackknife*, Eqs. (9)–(12) are used to find the convergence point. Figure 2 depicts an image with the optical flow vectors used to estimate the principal point.

3.2   Speed Estimation

After having estimated the principal point the speed of the vehicle is calculated; this is achieved by means of optical flow, too. The principal point is used as a reference point in the image in order to map heights to distances. The height is derived from the principal point on one end and the ground floor point, extracted from the motion of the robot, on the other end. The average time for the camera to acquire two continuous time frames has been estimated at $0.15\,$s. For this purpose we have used the *burst* mode of the digital camera. The well-known formula that describes the instantaneous speed, v, at time t is given by (13),

$$v(t) = \frac{\Delta s}{\Delta t} = \frac{ds}{dt} \tag{13}$$

where Δs is the distance covered in time interval Δt. Given the fact that our camera is placed parallel to the ground and the projection of the camera height onto the image plane is known, we estimate the distance traveled between the two time frames, D_r, using the cross ratio between the height of the camera from the ground in the image plane (in pixels), h_p, the real height of the camera from the ground (in cm), H_r, and the magnitude of optical flow vectors (in pixels), D_p, from the displacement of the vehicle within two time continuous frames. The following Eq. (14) describes this relationship

$$\frac{h_p}{H_r} = \frac{D_p}{D_r}. \tag{14}$$

From the previous equation it can be concluded that the distance Δs the moving vehicle has traveled in Δt is given by the variable D_r. Upon implementation of the LK algorithm on two continuous time frames, a set of optical flow vectors in generated with varying magnitudes and directions. Due to noise in sensor perception as well as due to factors such as occlusion, the optical flow vectors that appear on the moving object are not of equal magnitude. The vectors whose magnitude is below a given threshold are not considered as well as the vectors that fall outside of the tracked region, i.e., the region extracted from the difference of two images. Yet, after repeating a number of trials we have observed a number of vectors with small magnitudes. This is treated as noise due to the fact that robot motion yields optical flow vectors with significantly larger magnitudes. In order to calculate the instantaneous speed, the median value is selected from the distribution of the magnitudes of the vectors.

The magnitude of optical flow vectors in the x-direction is mapped to the height of the object in the image plane in order to convert from pixels to meters. This holds true since the height of the object does not change between the two frames. Moreover, the orientation of the optical axis of the camera being perpendicular to the direction of motion of the vehicle does not affect the height of the camera in the image plane due to perspective projection. In addition, estimating robot's speed on a ramp (Fig. 3) is as accurate as on a flat surface. This is due to the fact that optical flow vectors are parallel to the image plane.

In order to calculate the height of the vehicle in the image plane, we extract the object from its background using the difference between the time frames. We then threshold the image for the purpose of converting it into a binary one. The contour of the moving vehicle is extracted by applying a morphological

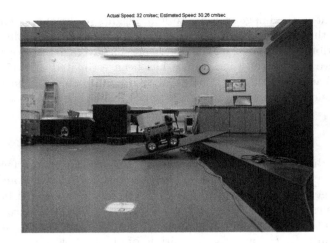

Fig. 3. Optical flow vectors at a ground truth speed of 32 cm/s and $\simeq$15° of angle. The estimated speed is 30.26 cm/s.

operation to the image. In order to remove small objects from the binary image, all connected components that have fewer than δ pixels are removed. In our images the value of δ was set to 60 pixels. This was a result of trial and error. Figure 4 shows the result of the operation of absolute differences between two frames and the implementation of the morphological operation on the same image. The contour of the moving vehicle is required in order to find the height of the camera, that is, the distance between the principal point and the ground. Figure 5 depicts the contour of the robot in two time continuous images (green color), the principal point in blue circle, and the projection of camera height onto the image plane (red color). We have tested our algorithm with varying speeds as well as multiple times with the same speed.

4   Results

In this section we present the results from the methods we have employed to estimate the principal point and hence the speed of the mobile robot. One advantage of our approach for determining the speed of the robot is that it requires only two time continuous snapshots to be taken as we do not estimate the mean speed but rather the instantaneous speed of the vehicle. Table 1 summarizes the results between actual and estimated speed as well as the absolute and the relative error between actual and estimated speed. The actual speed of the robot was calculated by averaging the speed of 5 trials between two known points on the ground while keeping constant the motor command values for speed during each set of trial. The same process was repeated for all 8 different speed estimates. Interpreting the results of Table 1 the estimated speed is underestimated

Fig. 4. Result of absolute differences between the two images used for speed estimation. Small objects have been removed using morphological operation on the image.

Table 1. Speed estimation results

Actual speed (cm/s)	Est speed	Abs error	Rel error (%)
14.5	13.18	1.32	9.1
21.92	19.84	2.08	9.49
30.38	27.74	2.64	8.69
35.89	33.76	2.13	5.93
41.91	37.49	4.42	10.55
48.62	42.88	5.74	11.81
53.52	50.04	3.48	6.5
58.27	52.83	5.44	9.34
32.0 ($\simeq 15°$ angle)	30.26	1.74	5.44

against the actual speed. The mechanism to correct this bias is through the use of linear regression.

Figure 6 shows the optical flow vectors at different robot speeds. Our results reveal that there is, on average, an error of approximately 3.4 cm/s which is considered normal since parameters such as the extraction of the contour of the robot, the measured time between frames, or a small time drifts in measuring the actual speed can influence the estimated speed.

For principal point estimation the LK algorithm proved to be a rational approach. This is due to the fact that indoor environments provide an abundance of geometrical objects and hence corners and edges. The *jackknife* sampling method successfully identified the outliers in the images although discarding optical flow vectors whose distance is beyond the 90th percentile may lead to some inliers being treated as outliers. Nevertheless, this seems not to affect the final outcome. The parameter that plays a significant role in estimating the principal point is the length of the focal points between the two successive images. We experimented with various focal lengths and we concluded that a large

Contour of robot between two frames (green square). Principal point (center of blue circle). Projection of camera height (red vertical line).

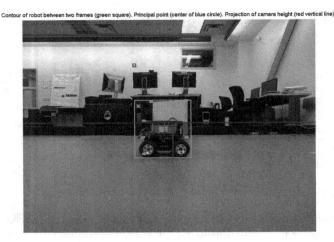

Fig. 5. Layout of the moving robot in two time continuous images (green rectangle). The blue circle depicts the principal point along the horizon line. The red vertical line is the projection of the camera height, H_r, from the ground to the principal point, onto the image plane (h_p). The magnitude of the green optical flow vectors depict the translation of robot between two images (D_p). The distance covered by the robot, D_r, is calculated from all the above (see also Eq. (14)) (Color figure online).

Actual Speed: 14.5 cm/sec; Estimated Speed: 13.18 cm/sec   Actual Speed: 53.52 cm/sec; Estimated Speed: 50.04 cm/sec

Fig. 6. Optical flow vectors at a ground truth speed of 14.5 cm/s and 53.52 cm/s (left to right). The estimated speed is 13.18 cm/s and 50.04 cm/s, respectively.

displacement between two focal lengths can influence significantly the estimation of the principal point. This has an immediate effect to the estimation of the speed of the robot.

For the estimation of the speed of the robot a critical parameter is the accurate extraction of the robot's layout, in particular with respect to the ground. This is required so as to map accurately the height of the camera on the image plane. In our method extracting the layout of the robot using morphological operators proved to be satisfactory. Nevertheless, for more complex environments a sophisticated algorithm will be needed. Yet, our method produced satisfactory results even when the snapshots of the robot were taken off the center of the image plane (Fig. 6).

5   Conclusions and Future Work

In this research we exploit the potential of optical flow to estimate both the principal point on an image as well as the speed of a mobile robot. Although speed estimation is a well-studied problem, we offer a new perspective to this problem by reducing the assumptions needed to one only known parameter, that is, the height of the camera from the ground. In spite of the fact that the grabbing of images was taken offline due to exclusively technical reasons (i.e., on one hand, we could not process the images while they were taken by the digital camera; on the other hand, plain webcameras do not provide optical zoom), we firmly believe that our method has all the potentials to be implemented in real-time. Furthermore, there is no need for reference points in the image or the environment. This is relaxed by knowing the distance between the camera and the ground, an insignificant problem. In general, we have shown that our method proves to be parsimonious yet effective irrespective of the vehicle's speed, and the distance between the camera and the moving vehicle.

Our findings show that there is a more accurate estimation of the principal point when the focal length displacement of the camera between two images is increased by the smallest possible amount. In our experiments, this amount equals to 1 mm between the focal points of two given images. The larger the displacement, the higher the error in estimating the principal point and hence the speed of the vehicle. This also reveals that estimating the FOE may be hard since near (pure) translation using a mobile robot can be a difficult thing to attain.

The results of this research can be extended further in several ways. Our current research involves the estimation of a robot's own speed with respect to stationary objects. In addition, the acceleration can be estimated using the methods described. Since the vanishing point is surfing on the horizon in the case of a camera being parallel to the ground, this method can provide the means of estimating more accurately the vanishing point and the vanishing lines in an image plane, an important parameter for detecting roads among others [15]. Future work will consider the application of this method on an Unmanned Aerial Vehicle (UAV).

References

1. Indu, S., Gupta, M., Bhattacharyya, A.: Vehicle tracking and speed estimation using optical flow method. Int. J. Eng. Sci. Technol. **3**(1), 429–434 (2011)
2. Dailey, D.J., Cathey, F.W., Pumrin, S.: An algorithm to estimate mean traffic speed using uncalibrated cameras. IEEE Trans. Intell. Transp. Syst. **1**(2), 98–107 (2000)

3. Dogan, S., Temiz, M.S., Kulur, S.: Real time speed estimation of moving vehicles from side view images from an uncalibrated video camera. Sensors **10**(5), 4805–4824 (2010)
4. Honegger, D., Greisen, P., Meier, L., Tanskanen, P., Pollefeys, M.: Real-time velocity estimation based on optical flow and disparity matching. In: Proceedings of the IEEE/RSJ International Conference on Intelligent Robots and Systems, pp. 5177–5182 (2012)
5. Grammatikopoulos, L., Karras, G., Petsa, E.: Automatic estimation of vehicle speed from uncalibrated video sequences. In: Proceedings of the International Symposium on Modern Technologies, Education and Professional Practice in Geodesy and Related Fields, pp. 332–338 (2005)
6. Muller, J., Paul, O., Burgard, W.: Probabilistic velocity estimation for autonomous miniature airships using thermal air flow sensors. In: Proceedings of the International Conference on Robotics and Automation, pp. 39–44 (2012)
7. Bauer, D., Belbachir, A.N., Donath, N., Gritsch, G., Kohn, B., Litzenberger, M., Posch, C., Schon, P., Schraml, S.: Embedded vehicle speed estimation system using an asynchronous temporal contrast vision sensor. EURASIP J. Embed. Syst. **2007**(1), 1–12 (2007)
8. Kassem, N., Kosba, A.E., Youssef, M.: RF-based vehicle detection and speed estimation. In: Proceedings of the 75th IEEE Vehicular Technology Conference, pp. 1–5 (2012)
9. Ernst, J.M., Ndoye, M., Krogmeier, J.V., Bullock, D.M.: Maximum-likelihood speed estimation using vehicle-induced magnetic signatures. In: IEEE International Conference on Intelligent Transportation Systems, pp. 1–6 (2009)
10. Chen, Z., Pears, N., Liang, B.: A method of visual metrology from uncalibrated images. Pattern Recogn. Lett. **27**(13), 1447–1456 (2006)
11. Criminisi, A., Reid, I., Zisserman, A.: Single view metrology. Int. J. Comput. Vis. **40**(2), 123–148 (2000)
12. Momeni-K., M., Diamantas, S.C., Ruggiero, F., Siciliano, B.: Height estimation from a single camera view. In: Proceedings of the International Conference on Computer Vision Theory and Applications, pp. 358–364. SciTePress (2012)
13. Fayman, J.A., Sudarsky, O., Rivlin, E., Rudzsky, M.: Zoom tracking and its applications. Mach. Vis. Appl. **13**(1), 25–37 (2001)
14. Lucas, B.D., Kanade, T.: An iterative image registration technique with an application to stereo vision. In: Proceedings of the 7th International Joint Conference on Artificial Intelligence, pp. 674–679 (1981)
15. Kong H., Audibert, J.-Y., Ponce, J.: Vanishing point detection for road detection. In: IEEE Conference on Computer Vision and Pattern Recognition, pp. 96–103 (2009)

Mechanical Weeding Using a Paddy Field Mobile Robot for Paddy Quality Improvement

Yasuhiro Yamada[1(✉)], Keisuke Iwakabe[1], Guanzuo Liu[2],
and Toshiyoshi Uejima[3]

[1] Mechanical Engineering, Graduate School of University of Fukui,
Fukui, Japan
yyamada@u-fukui.ac.jp
[2] School of Material Science and Engineering,
Nanchang HangKong University, Nanchang, China
[3] Dream On Inc., Agricultural Farm UEJIMA, Fukui, Japan

Abstract. In organic farming, eradicating weeds is a problem because great importance is attached to the ecosystem, and consequently agricultural chemicals are not used. Therefore, a great deal of heavy labor is needed for eradicating weeds. The aim of this research is to design and implement a robot arm for weeding mounted on a robot in a paddy field. It is hoped that this research will result in rice field managers being able to remove weeds remotely by means of being able to observe the rice field via a stereo camera. This task will make a big reduction of time and labor for rice farming.

Keywords: Mobile robot · Laser range finder · Weeding · Agricultural robotics

1 Introduction

In organic farming, the eradication of weeds is a problem. In organic farming does not allow the use of pesticides because more emphasis is put on the ecosystem than the amount of harvest. Therefore, a lot of physical effort is required to remove weeds from rice paddies. In this research, a laser range finder (LRF) is used for the purposes of obstacle recognition. A robotic arm that can be operated from a remote location using an autonomous mobile robot, is used to establish a way to effectively remove weeds in organic rice farming. If this purpose can be achieved, it would lead to a large reduction in the time and effort involved in farming practices.

2 Design of a Paddy Field Robot

The paddy field mobile robot has following features and sensors.

(1) Detection of obstacles when moving paddy
     The paddy field mobile robot is able to move 24 hours a day in the paddy fields. Obstacles in front of the robot are scanned by the LRF at all hours of the day and night.

A. Natraj et al. (Eds.): TAROS 2013, LNAI 8069, pp. 185–189, 2014.
DOI: 10.1007/978-3-662-43645-5_19, © Springer-Verlag Berlin Heidelberg 2014

(2) Automatic operation and autonomous mobility
     In order to carry automatic weeding and to thereby reduce the occurrence of
     weed, it is necessary to move autonomously in the paddy fields. This can be
     achieved with a microcomputer controlling the sensor information processing
     and actuators of the robot.
(3) Remote work and remote control movement
     The paddy field mobile robot movements and the remote weeding of grass in
     order to reduce the occurrence of weed can be teleoperated by an operator
     outside the paddy field. To that end, the operator uses a Head Mounted Display
     (HMD) for viewing surroundings of the robot by using a stereo camera mounted
     on the robot.

3 Autonomous Roboot Control

Figure 1 shows an autonomously derived pathway of the paddy field mobile robot. If
obstacle comes in the pathway of the robot, the robot avoid the obstacle and moving
forward by using the detected obstacle position data and the potential method.

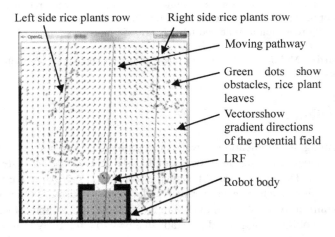

Fig. 1. Obstacle avoidance of the paddy field mobile robot

4 Robot Arm Simulator

Assuming a real paddy field environment, we create a simulator of the robot arm to
weeding the paddy weeds.

Figure 2 shows an image of a robot equipped with an articulated robot arm to
check the status of the various operations of the robot arm on the simulator. We
examined the arrangement and dimensions, such as joint angles of the robot arm.

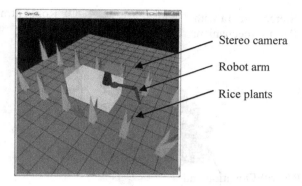

Fig. 2. Screenshot of the robot arm simulator

5 Camera Image Simulator

To create a three-dimensional model of the rice grown in paddy field, we have developed a simulator using a three-dimensional coordinate conversion formula which calculates the camera images and recognizes the environment around the robot to enable it to move around the rice paddy fields (Fig. 3).

(a) 3 days after rice   (b) 30 days after rice
    planting   planting

Fig. 3. Comparison between real rice field and simulated rice field. Upper: photograph, Lower: screenshot of the simulator

We also equipped a stereo camera on the robot for exploring the environment and for controlling the robot arm remotely.

6 Overall Structure of the Robot

Figure 4 shows the overall structure of the paddy field mobile robot, Fig. 5 shows the system configuration, Fig. 6 shows side view of the robot. The stereo camera with pan-tilt mechanism is equipped on the center of the robot. The operator view weed and the robot arm by the Head Mounted Display (HMD), and remotely operate the robot arm for weeding.

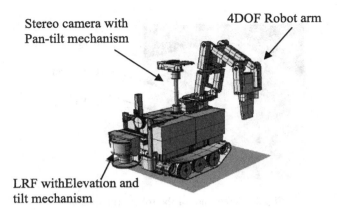

Stereo camera with
Pan-tilt mechanism

4DOF Robot arm

LRF withElevation and
tilt mechanism

Fig. 4. Overall 3D design view of the mobile robot

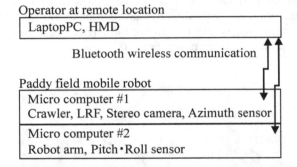

Operator at remote location
| LaptopPC, HMD |

Bluetooth wireless communication

Paddy field mobile robot
| Micro computer #1 |
| Crawler, LRF, Stereo camera, Azimuth sensor |

| Micro computer #2 |
| Robot arm, Pitch・Roll sensor |

Fig. 5. System configuration of the mobile robot

Rake
attached
at the end
of the
robot arm

Fig. 6. Side view of the completed paddy field mobile robot with a rake attached at the end of the robot arm

Rice plants

Robot's
moving
direction

Fig. 7. Snapshot of the surface of a rice field after mechanical weeding operation

7 Paddy Field Experiment

The rake, attached at the end of the robot arm, add physical stress on the soil surface of a rice field during robot movement. This mechanical weeding operation pick out fledgling weeds from the soil surface of the rice field (Fig. 7). As a result of this operation, growing weeds are decreased and the paddy quality is improved.

8 Conclusion

The rice field mobile robot is properly designed by using developed simulators: robot arm simulator and stereo camera simulator. The effectiveness of the completed rice field mobile robot has been clarified throughout the paddy field experiments. The mechanical weeding operation, adding physical stress on the soil surface, pick out fledgling weeds from the soil surface of the rice field and effectively decrease weed growth.

Acknowledgment. This work has been supported in part by the Japan Society for the Promotion of Science (JSPS) Grant-in-Aid for Challenging Exploratory Research, Grant Number 24656166.

Multi-modal People Detection
from Aerial Video Footage

Helen Flynn[✉] and Stephen Cameron

Oxford University Computing Laboratory, Wolfson Building,
Parks Road, Oxford OX1 3QD, UK
{helen.flynn,stephen.cameron}@cs.ox.ac.uk

There has been great interest in the use of small robotic helicopter vehicles over the last few years. One potential class of applications for these are in searching for people. Recently small cameras sensitive to infra-red light have become available. Although still expensive they are based on a technology whose cost could drop with mass production, and they are light enough to be added to even small helicopter unmanned aerial vehicles (UAVs). The work reported here stems from the observation that infra-red and visual light imagery should complement each other well for people detection. Live targets tend to be warmer than their surroundings, and so show up well in the infra-red, but with low discrimination; whereas visual-light imagery can be good for deciding whether a potential target is really a person. Our initial experiments have been with fixed cameras from high vantage points, but with the intention to fly the equipment this summer using an Ascending Technologies Falcon vehicle.

A handful of work to date has focussed on the problem of people detection from aerial imagery [3,5,6], but often the assumption is that the people are in standard poses or situated fairly close to the camera. In order to be able to detect people in highly articulated poses, we use the more sophisticated part-based algorithm of [2] in order to detect people. This consists of an appearance model for each body part (based on image gradients) along with a spatial model for the part locations, both learned in a latent support vector machine framework.

Our camera rig comprises a visible light camera and an infrared camera[1]. Footage was recorded at 15 fps in various cluttered urban scenes looking down to the ground from a height of approximately 20 m. We threshold the thermal image to identify the hot spots and then find the corresponding regions in the optical image. These regions are then analysed using a part-based people detection algorithm mentioned above. By tracking the detections over time and only considering those that are consistent, we are able to reduce the false positive rate dramatically. Figure 1(b) shows a precision recall curve computed from a number of different video sequences. This was computed by varying the detection score threshold. Not surprisingly, using the infrared image to narrow down the search space reduces the false positive rate dramatically.

Homography computation between successive pairs of 1280×960 frames takes on average 1.64 s on a Intel quad core 2.83 GHz CPU with an integrated graphics card. Processing the infrared image and scanning the reduced image space

---

[1] Thermoteknix Miricle Microcam 640 × 480 + 18.8 mm lens, 1/3″ sensor; Fig. 1(a).

A. Natraj et al. (Eds.): TAROS 2013, LNAI 8069, pp. 190–191, 2014.
DOI: 10.1007/978-3-662-43645-5_20, © Springer-Verlag Berlin Heidelberg 2014

(a) IR Camera

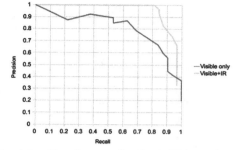

(b) Precision Recall curve for the initial results

Fig. 1. (a) IR camera (b) Precision recall curve for the initial results

with the Felzenszwalb detector takes approximately 2.4 s. This compares very favourably with the use of the same detector over the entire image, which takes around 15 s per frame, and it suggests that the algorithm should run in 'real-time' (say, under 100 ms per frame) when optimised and run on the hardware expected to be available within the next few years.

Our results show that state of the art people detection algorithms can be applied to aerial imagery with considerable success. Furthermore, the detection times suggest that this method can be employed in real-time on the sort of hardware expected to be available in the near future. Future work in this area will include: only scanning images at the most likely scales in which a person can appear; doing some image processing on the infrared image itself; and estimating the relative velocity of the person and using that to predict where to search in the next image, in order to increase confidence over multiple frames.

References

1. Dalal, N., Triggs, B.: Histograms of oriented gradients for human detection. In: Proceedings of Conference on Computer Vision and Pattern Recognition, pp. 886–893 (2005)
2. Felzenszwalb, P.F., McAllester, D., Ramanan, D.: A discriminatively trained, multi-scale, deformable part model. In: IEEE Conference on Computer Vision and Pattern Recognition (2008)
3. Gaszczak, A., Breckon, T.P., Han, J.W.: Real-time people and vehicle detection from UAV imagery. In: Proceedings of SPIE Conference Intelligent Robots and Computer Vision XXVIII: Algorithms and Techniques, Vol. 7878, No. 78780B (2011)
4. Hartley, R., Zisserman, A.: Multiple View Geometry in Computer Vision. University Press, Cambridge, ISBN: 0521540518 (2004)
5. Oreifej, O., Mehran, R., Shah, M.: Human identity recognition in aerial images. In: Proceedings of IEEE Computer Society Conference Computer Vision and Pattern Recognition, pp. 709–716 (2010)
6. Rudol, P., Doherty, P.: Human body detection and geolocalization for UAV search and rescue missions using color and thermal imagery. In: Proceedings of IEEE Aerospace Conference, pp. 1–8 (2008)

Control

Design of a Modular
Knee-Ankle-Foot-Orthosis Using Soft
Actuator for Gait Rehabilitation

S.M. Mizanoor Rahman[(⊠)]

Department of Mechanical Engineering,
Vrije Universiteit Brussel (VUB), Pleinlaan 2, 1050 Brussels, Belgium
mizansm@hotmail.com

Abstract. The design of a modular wearable knee-ankle-foot-orthosis (KAFO) using novel soft actuator for post-stroke gait rehabilitation is presented. The configuration, different modules, working principles, actuation, control concepts, novel features etc. of the KAFO are introduced. As the actuation method plays the key role for the overall performances of the KAFO, the design, configuration, working principles, kinematics, dynamics, control analyses etc. of a novel soft actuation system are presented in details. The novel actuator is a variable impedance series elastic actuator designed with one motor and two types of springs in series, which is light in weight and compact in size. The actuator model is simulated for various conditions, and the results show satisfactory dynamic performances in terms of stability, safety, force bandwidths, variable impedance, compliance, efficiency etc. Then, the fabrication of the physical KAFO and its clinical validation with stroke patients are emphasized.

Keywords: Knee-ankle-foot-orthosis · Gait · Rehabilitation · Stroke · Modular · Soft actuator · Variable impedance · Passive compliance · Adaptive shared control · HRI

1 Introduction

1.1 Severity of Stroke Diseases in the Global Perspective

According to the World Health Organization (WHO), over 5.7 million people die every year in the world due to stroke, and this number may increase to about 24 millions by 2030 [1]. Over 700,000 people per year suffer a stroke in the USA, and more than half of them survive with disability [2]. Each year, there are over 920,000 stroke cases in Europe [3]. For Australia, stroke is the second single greatest cause of death [4]. It is the leading cause of long-term disability in adults in Australia, which represents 25 % of all chronic disabilities [5]. Stroke still remains the third most common cause of death in Japan [6]. Stroke causes deaths and disabilities, it increases health care anxieties, rehabilitation and health care supports and costs [2].

A. Natraj et al. (Eds.): TAROS 2013, LNAI 8069, pp. 195–209, 2014.
DOI: 10.1007/978-3-662-43645-5_21, © Springer-Verlag Berlin Heidelberg 2014

1.2   Therapy for Stroke Rehabilitation: Manual vs. Robotic

Stroke damages neurons, which can only be repaired by repeated therapies. Stroke patients receiving frequent and repeated physical therapies have much greater chance of recovery than other forms of treatments [7, 8]. Repeated physical therapies can establish new neural pathways or can unmask the dormant pathways that control the volitional movement, and thus it can maximize the motor performances and minimize the functional deficits [9]. For manual physical therapy, one or more therapists assist and encourage the patient to follow a number of repetitive movements. However, the manual rehabilitation is not precise, it is slow and non-reproducible, and it adds burdens to the therapists. The repetitive nature of the stroke therapy makes it possible to provide the therapy by robots that may reduce the burden of the therapists or may replace the therapists and can measure the data on the patients while training them and thus can help record the patient's progress [2]. It is highly reproducible and precise, it ensures consistency in therapy [10]. This is why, robot-assisted therapies for the patients suffering from stroke, paralysis, cerebral palsy, paraplegic disease, polio, hemiplegic problems etc. have become very active areas of research [10–18].

1.3   Robot-Assisted Therapies for Stroke Rehabilitation: The State-of-the-Art

Rehabilitation robots for gait [10–16] and arm training [17, 18] have been developed. However, the gait training seems to be more important, but it is complicated due to the complex biomechanics of the human leg. Most of the current gait systems are set on treadmill [13], which is not natural. The existing lower limb systems usually do not have powered ankle joint [10] (except in few systems e.g. [12]). These systems are not so robust to adapt with changing environments. The existing systems are in generally expensive, large, heavy and do not include body-weight supports. Balance is not so good. Sizes have not been generalized to accommodate the patients of different sizes, ages, shapes etc. Again, most of the systems cannot provide patient's up-down, left-right, and front-back motions.

It is true that some existing systems are addressing the above issues. For example, a body weight support was proposed in [19]. Several gait trainers can be adjusted to the anatomic dimensions of the patients [10, 15]. Some systems allow actuated hip and knee flexion, actuated forward-backward motion, actuated abduction/adduction etc. [10]. However, proper integration of all the aforementioned required attributes in a single system has not been attempted yet.

1.4   Modular Design of Robotic Systems for Stroke Rehabilitation Therapy

Integrated robotic rehabilitation systems are usually larger than the particular requirement of the patients, and hence some of the facilities may remain unused. Again, these systems are usually heavy and costly. A large system may create local effects on the unexercised limbs or muscle groups of the patients. These problems motivated towards the modular design where different groups of muscles (e.g. spatial arm movements), limbs (e.g. wrist, fingers, legs) etc. are exercised separately [2, 9]. However, most of the systems do not provide modular design advantages [10–18].

The ankle plays a central role in the normal gait; it contributes to propulsion, shock-absorption and balance. The foot-drop disease due to stroke occurs in the ankle [9]. However, the present systems either do not include powered ankle-foot orthosis or the orthosis is powered but it runs on treadmill instead of over ground [10]. Again, a combined knee-ankle-foot system with modular operations of both ankle and knee modules as well as with all other required attributes are usually not seen.

1.5   Soft and Compliant Actuation for Rehabilitation Systems

The rehabilitation systems should have low friction, low and variable mechanical impedance, it should be back-drivable etc. to resemble its natural counterparts [20–23]. However, the present systems do not fully satisfy these requirements due to their limitations in their actuator systems that are not so soft, compliant and they do not provide variable impedance [7, 9]. As a results, most of the present systems are not so safe and human-friendly [2, 9–18, 24]. On the other hand, several soft, compliant and variable impedance actuators have been proposed [23, 25–32]. But, their applications in the design of the rehabilitation systems are not so satisfactory [10–18, 25, 33, 34].

1.6   Objectives

This paper presents the design, various modules, working methods, actuation, control concepts etc. of a novel wearable modular KAFO, and attempts to bring novelties in every sphere of the design so that it can overcome the limitations of the present gait rehabilitation systems. The novel variable impedance compliant series elastic actuator design for the KAFO is presented with detailed dynamic and control analyses. Then, the fabrication and clinical validation of the KAFO are emphasized.

2   Configuration of the Proposed Gait Rehabilitation System

The novel gait rehabilitation system is designed in its two different wearable modules: (i) the ankle module - a powered Ankle-Foot Orthosis (AFO) for rehabilitation and assistance at the ankle as shown in Fig. 1(a) [9], (ii) the Knee-Ankle-Foot-Orthosis (KAFO) module that has functionalities and motion generation capabilities at both knee and ankle simultaneously or independently for gait rehabilitation as shown in Fig. 1(b). The complete KAFO to use in the rehabilitation sessions is shown in Fig. 1(c), which includes a body-weight support and an omni-directional mobile base.

The kinematic analysis for the AFO as a representative of the whole KAFO based on Fig. 2 may help understand how the AFO will be actuated during the rehabilitation sessions [9]. In Fig. 2, a is shank length, b is foot length and x is linear displacement of the actuator output. The kinematics for the AFO can be described using Eq. (1).

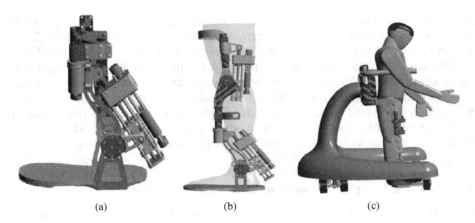

Fig. 1. (a) The AFO module, (b) the KAFO module, (c) the complete rehabilitation system.

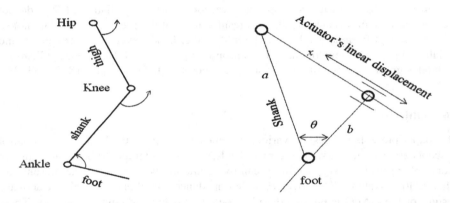

Fig. 2. Link-segment model for the kinematic analysis for the KAFO (left), and kinematic model for the AFO for the ankle-foot motion (right).

The actuator produces x, and θ changes due to the change in x. Hence, the relationship between x and θ can be established as Eq. (2). The objective is to control θ (and its derivatives) for the ankle-foot motion to assist the rehabilitation task.

The design uses carbon fiber composite material for its main structure and space grade aluminum alloy for the actuator that make it light-weight, strong and convenient [34]. A novel variable impedance compact compliant series elastic actuator (SEA) is designed for the system. The aim is to use a novel adaptive shared force control that also includes on-line learning for the adaptation to the individual patient characteristics at different stages/modules of the rehabilitation process [35]. The KAFO interface is to adapt to patient's cognitive and physical functionalities and ensure bilateral communication between them. It is also an aim to justify the design through the clinical validation of the system using the stroke patients.

$$x^2 = a^2 + b^2 - 2ab \cos \theta \tag{1}$$

$$\theta = f(x) \tag{2}$$

3 Novel Compliant Actuator Design for the Rehabilitation System

3.1 Novel Actuator Design

A novel SEA is designed to provide variable impedance to the lower limb system. The actuator design consists of (i) a servomotor, (ii) two springs, (iii) a ball screw and (iv) an output link. The design places a torsional spring in series between the servomotor and the ball screw. The nut of the ball screw is attached to the output link in series through another set of translational springs (see Fig. 3(a)). The servomotor with

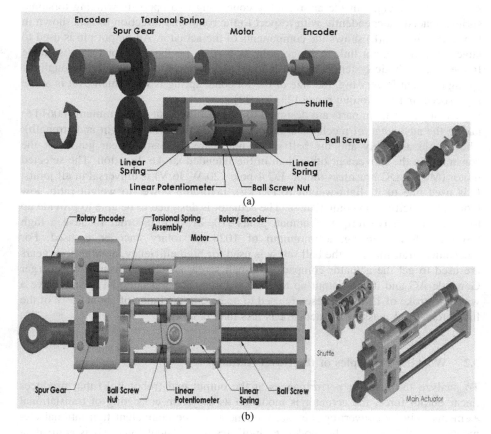

Fig. 3. (a) Mechanism of the actuator, and (b) components of the actuator with the exploded view.

an encoder is attached to the torsional spring that is attached to a spur gear and the gear is attached to another rotary encoder. The rotary encoders are used to measure the deflection of the torsional spring, and the motor torque is calculated based on the spring constant. The gear is then attached to the ball screw. The rotational speed is reduced based on the appropriate gear ratio [34, 36].

A set of translational springs is attached to the ball screw nut. If the nut moves, a deflection is created in the springs that can be measured by a linear encoder. The output force is calculated based on the spring constant and the linear deflection following the Hooke's law. The shuttle assembly is connected to the other end of the joint. The linear springs added on both ends of the ball screw nut are connected to the shuttle that can help achieve back-drivable actuation in the green-colored arrow directions. The ball screw nut can move independent of the shuttle. The key components of the force feedback are the torsional spring and two high resolution rotary encoders. The torsional spring is incorporated with a predefined spring constant and the encoders are used to measure the rotational difference between the motor (input) and the spur gear (output). To achieve the bi-directional loading of the torsional spring, it is packaged inside an assembly containing two opposite winding torsional springs loaded independently with respect to the rotational direction as it is shown in Fig. 3(a). Figure 3(b) shows the components of the actuator. An output pin is used to attach it to the output link. A hinge point is used to tie the actuator to any suitable frame of the robotic system. The translational springs are soft and small. Although the spring constant is very big, the size of the torsional spring is very small as it is at the high speed and low torque range [23, 31, 32].

The actuator main parts are designed to be machined out of aluminium (6061T6) taking the advantage of mechanical properties of the material (weight and strength). The shuttle assembly and the ball screw nut share the same linear guide for the constraint in the movement other than in the intended stroke direction. The selected motor (Maxon DC brushless motor, EC 4-pole 120 W 36 V) is universal in all joints. It is used due to its light-weight (0.175 kg), favorable power to weight ratio, low moment of inertia, and compactness. The actuator is designed to be able to provide up to 60 N m assistive torque at human joints [10, 21, 22]. In order to have a high resolution force sensing, a minimum of 1024 ppr rotary encoder is used. For mechanical transmission, the ball screw is used for high efficiency, and the spur gears are used to get the actuator compact. The ball screw is selected from Eichenberger Gewinde AG and it can output up to 1500 N force. The translational springs have a working stroke of 12 mm. These are used to operate in the range of about 25 % of the full force. Total mass of the actuator is less than 0.85 kg.

3.2   Working Principles of the Novel Actuator

To analyze the actuator performances at the output end (robot link) that produces linear output force, the actuator is modelled as a system consisting of translational elements only by converting the rotary elements to the equivalent translational elements as in Fig. 4(a). In the model, F_1 indicates motor input force, m_1 is equivalent

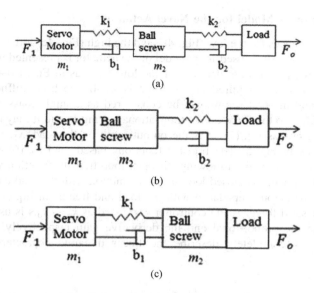

Fig. 4. (a) Working principle of the series elastic actuator for the equivalent translational motion, (b) principle for low force, (c) principle for high force case.

mass of the motor as derived in Eq. (3) where J_1 indicates moment of inertia of the motor, p indicates pitch of the ball screw, m_2 indicates equivalent mass of the ball screw as derived in Eq. (4) where J_2 indicates moment of inertia for the ball screw, k_1 indicates the equivalent translational spring constant of the torsional spring k_t as derived in Eq. (5), k_2 indicates spring constant of the translational spring, b_1 and b_2 indicate viscous damping for motor and ball screw respectively, and F_o indicates the output force. k_1 is selected to be much bigger than k_2.

At low force rehabilitation cases, the model can be reduced to as in Fig. 4(b). It means that the torsional spring is to behave as rigid and not to work as a spring, and only the translational spring is to work. Hence, the output impedance and compliance are to be produced due to only the translational spring and they are to depend on the spring constant k_2. As the allowable stroke for the translational springs is small, at high force during the rehabilitation, the translational springs are to compress and only the torsional spring is to work, thus the model is to reduce to as in Fig. 4(c). Hence, the bandwidth of the actuator is found very high at the high force cases during the rehabilitation due to big spring constant k_1. Therefore, the performances are to depend on the force range during the rehabilitation and on the difference between the spring constants k_1 and k_2. Thus, the actuator can change its output impedance and dynamic bandwidth in accordance with the force range without needing a change in the springs (hardware). The actuator thus can achieve a low output impedance and small non-linear friction because of the soft translational springs. These are also the requirements of a stroke rehabilitation robot [3, 6, 16].

3.3   The Dynamics Model for the Novel Actuator

Dynamic performances for the low (Fig. 4b) and the high (Fig. 4c) force situations are discussed as follows. The closed-loop model of the actuator is presented in Fig. 5. The dynamic motion equations are derived for the low force as in Eqs. (6)–(8) based on Figs. 4(b) and 5. If the required output force is small, the high stiffness torsional spring, motor and the ball screw may be considered as a single mass equivalent to $m_1 + m_2$. F_2 refers to the force on the translational spring k_2 and it may be calculated by the Hooke's law as Eq. (7). It is the output force on the load. x_1, x_2, x_3 are the displacements as in Fig. 5. The transfer functions based on Eqs. (6)–(8) for three conditions are derived as the following: (i) open loop transfer function with the load end fixed as in Eq. (9), (ii) closed loop transfer function with the load end fixed as in Eq. (10), and (iii) output impedance with the load end free as in Eq. (11), where F_d indicates the desired force. A PD controller expressed as $k_p + k_d.s$ is used, where k_p and k_d indicate the proportional and the derivative gain respectively. These three transfer functions completely describe the linear dynamic characteristics of the actuator [23].

$$m_1 = J_1 \left(\frac{2\pi}{p} \right)^2 \tag{3}$$

$$m_2 = J_2 \left(\frac{2\pi}{p} \right)^2 \tag{4}$$

$$k_1 = k_t \left(\frac{2\pi}{p} \right)^2 \tag{5}$$

$$F_1 = b_2(\dot{x}_2 - \dot{x}_3) + k_2(x_2 - x_3) + (m_1 + m_2)\ddot{x}_2 \tag{6}$$

$$F_2 = k_2(x_2 - x_3) \tag{7}$$

$$(F_d - F_2) \cdot (k_p + k_d s) = F_1 - F_2 \tag{8}$$

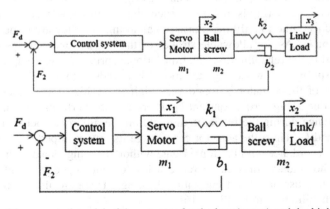

Fig. 5. Closed-loop control model of the actuator for the low (upper) and the high (lower) force situations.

$$\frac{F_2(s)}{F_1(s)} = \frac{k_2}{(m_1 + m_2)s^2 + b_2s + k_2} \tag{9}$$

$$\frac{F_2(s)}{F_d(s)} = \frac{k_dk_2s^2 + k_pk_2s}{(m_1 + m_2)s^3 + b_2s^2 + k_dk_2s^2 + k_pk_2s} \tag{10}$$

$$\frac{F_2(s)}{x_3(s)} = \frac{-(m_1k_2 + m_2k_2)s^3}{(m_1 + m_2)s^3 + b_2s^2 + k_dk_2s^2 + k_pk_2s} \tag{11}$$

If the output force is large, the low stiffness spring is to be compressed completely and the ball screw and the load are to be integrated as a single mass. So, the actuator may be activated at this condition by the torsional spring only. The dynamic equations for the high force situations are derived in accordance with Figs. 4(c) and 5 as in Eqs. (12)–(15). The transfer functions for the open loop and the closed loop force control, and the output impendence are derived as in Eqs. (16)–(18) respectively.

$$m_1\ddot{x}_1 = F_1 - k_1(x_1 - x_2) - b_1(\dot{x}_1 - \dot{x}_2) \tag{12}$$

$$m_2\ddot{x}_2 = -k_1(x_2 - x_1) - b_1(\dot{x}_2 - \dot{x}_1) \tag{13}$$

$$F_2 = k_1(x_1 - x_2) \tag{14}$$

$$F_1 - F_2 = (F_d - F_2) \cdot (k_p + k_ds) \tag{15}$$

$$\frac{F_2(s)}{F_1(s)} = \frac{k_1}{m_1s^2 + b_1s + k_1} \tag{16}$$

$$\frac{F_2(s)}{F_d(s)} = \frac{k_dk_1s + k_pk_1}{m_1s^2 + b_1s + k_dk_1s + k_pk_1} \tag{17}$$

$$\frac{F_2(s)}{x_2(s)} = \frac{-m_1k_1s^2}{m_1s^2 + b_1s + k_dk_1s + k_pk_1} \tag{18}$$

3.4 Dynamic Responses and Control Analysis for the Actuator Model

The values of m_1, m_2, b_1, b_2, k_1, and k_2 are used for MATLAB simulation for the model based on the proposed physical actuator (Table 1) for the control shown in Fig. 5. The step responses are shown in Fig. 6(a) for the small (100 N) and in Fig. 6(b) for the large (1000 N) force situations. For the small force, the PD controller's parameters are $k_p = 2.1$, $k_d = 0.03$ and the figure shows that the overshoot is 0 % and the settling time is 0.08 s. For the large force, the PD controller's parameters are $k_p = 0.16$, $k_d = 0.0003$. The overshoot is almost zero (max 2 %) and the settling time is about 0.01 s. The results thus show satisfactory dynamic performances of the design [36]. However, the control needs to be designed in such a way that a switch occurs automatically during the transformation of the system from the low to the high output force condition based on the situation or demand, and the system is also stable during the switch.

Table 1. Actuator design parameters and their values used for MATLAB simulation

Values of the actuator hardware parameters for the rotational motion	Values of the actuator hardware parameters for the equivalent translational motion
$J_1 = 8.91 \times 10^{-7}$ kg m^2	$m_1 = J_1 (2\pi/p)^2 = 8.78$ kg
$J_2 = 642 \times 10^{-9}$ kg m^2	$m_2 = J_2 (2\pi/p)^2 = 6.33$ kg
Torsional spring constant (k_t) = 1 N m/rad	$k_1 = k_t (2\pi/p)^2 = 9.86 \times 10^6$ N/m
	$k_2 = 24 \times 10^3$ N/m
$k_2 = 24 \times 10^3$ N/m	$b_1 = 800$ N s/m
Pitch of the ball screw (p) = 2×10^{-3} m	$b_2 = 1000$ N s/m

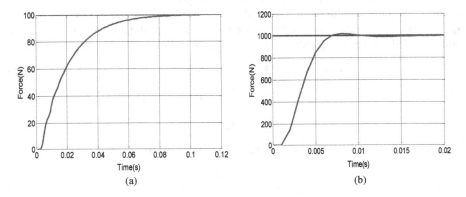

Fig. 6. Step responses for the actuator design for (a) low force (100 N), and (b) high force (1000 N) situation.

The ramp response for the actuator when both k_1 and k_2 are active is in Fig. 7. Here, the actuator performances can be checked when it transits from low to high force situation. Figure 7(a) shows almost no tracking error for both the low and the high force. The magnified graph for the point when the system enters into the high force condition (>240 N) from the low force condition (≤240 N) is shown in Fig. 7(b). The figure also shows a small tracking error (less than 5 %) when the switch occurs. The results indicate that the switch in the system control from the low to the high force is justified and there will have no tracking error, oscillations, instability etc. during the switch in the practical applications with the rehabilitation robot [36].

Bode plots for low and high force are shown in Fig. 8. For low force, the bandwidth is about 100 rad/s (16 Hz). If k_p increases the bandwidth increases, but if b_1 increases the bandwidth may decrease. There may have a possibility of resonance after 1500 rad/s (240 Hz) though the magnitude (dB) is not so high. This resonance is brought by k_1, but k_1 no longer acts as a spring for low force. In this case, the system falls within high frequency (and also high force), and hence the resonance cannot occur. For high force as in Fig. 8(b), the bandwidth is about 400 rad/s (64 Hz), and there is no possibility of resonance up to this range. This bandwidth is determined by the natural frequencies of the system as in Eq. (19) and Eq. (20), where ω_{n1} and ω_{n2}

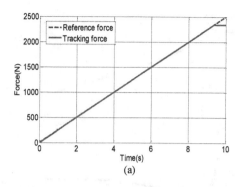
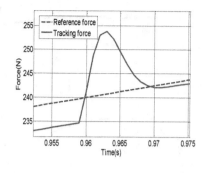

(a)

Fig. 7. Ramp responses for the actuator system. Here, (a) shows the response when both k_1 and k_2 are active, and (b) shows the magnified view when the system switches from the low to the high force situation.

are the natural frequency for the high and the low force respectively. The difference between the natural frequencies is high as the difference between k_1 and k_2 is high, which confirms that the system will not experience resonance either for low force or for high force operation. The frequencies when the magnitudes clearly deviate from zero are about 100 rad/s and 400 rad/s for the low and the high force cases respectively. Again, the frequency when the phase crosses $-180°$ for the low force is about 1500 rad/s (but, the phase does not cross $-180°$ for the high force case). It is seen that 100 rad/s is definitely very smaller than 1500 rad/s, which indicates that the system is stable. Again, the system does not show any tendency of approaching towards 0 dB and $-180°$, which also indicates the stability of the system [36].

$$\omega_{n1} = \sqrt{\frac{k_1}{m_1}} \tag{19}$$

$$\omega_{n2} = \sqrt{\frac{k_2}{m_1 + m_2}} \tag{20}$$

4 Discussion, Conclusions and Future Works

The design of a novel gait rehabilitation system for stroke patients is presented with introduction of its configuration, modules, working methods, actuation, control, novelties etc. The design, configuration, mechanisms, dynamics, control analysis etc. of a novel variable impedance compliant series elastic actuator are presented in details for the rehabilitation system. The dynamic and control characteristics of the actuator in terms of adjustment with sudden disturbances, shocks and impacts (step responses), changes in input with time (ramp responses), stability, absence of oscillations and resonance (bode plots) etc. are proven satisfactory. The actuator is compact, light-weight, it provides variable impedance, compliance, good force controllability,

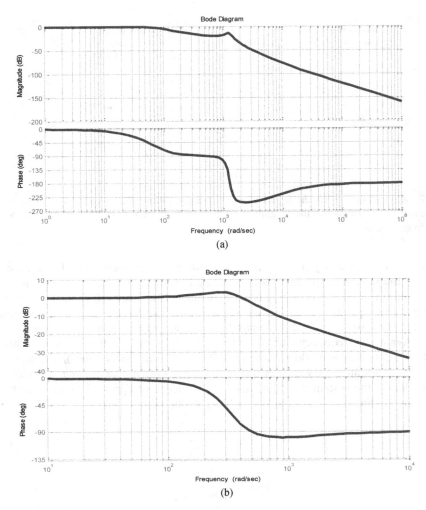

Fig. 8. Bode plot of the closed-loop control for (a) low force, and (b) high force situation.

back-drivability, large force bandwidths, high efficiency (due to ball screw) and safety, high power/mass and force/mass ratios, high assistive torque etc., and low friction, inertia, impedance etc. These criteria clearly justify the novelties and supe-riority of the design of the actuator. The rehabilitation system is designed using this novel actuator, and thus the novelties and characteristics of the actuator also benefit the rehabilitation system. The novel actuation strategy along with other novelties such as the advanced materials in construction, adaptive shared control, modularity, omnidirectional mobility, body-weight support, powered ankle, generality, over-ground application etc. clearly justify the design of the rehabilitation system.

In the future, optimization of the design for static characteristics such as power/ energy requirements, kinematic (motion) characteristics of the system etc. will be investigated and compared with human, the fabrication of the rehabilitation system (AFO, KAFO) based on the proposed design will be tried to be finished, an adaptive shared control will be implemented to control the human-robot interactions, and the system will be validated for its each module through clinical trials with stroke patients.

Acknowledgements. The author acknowledges the supports that he received from the National University of Singapore (NUS), Singapore in relation to the work presented in this paper.

References

1. http://www.myheart.org.sg/heart-facts/statistics/
2. Wheeler, J., Krebs, H., Hogan, N.: An ankle robot for a modular gait rehabilitation system. In: Proceedings of the 2004 IEEE/RSJ International Conference on Intelligent Robots and Systems, vol. 2, pp. 1680–1684
3. Brainin, M., Bornstein, N., Boysen, G., Demarin, V.: Acute neurological stroke care in Europe: results of the European stroke care inventory. Eur. J. Neurol. 7(1), 5–10 (2000)
4. Australian Institute of Health and Welfare.: Heart, stroke and vascular diseases – Australian facts 2004. AIHW Cat. No. CVD 27, Canberra, AIHW and National Heart Foundation of Australia (Cardiovascular Disease Series No. 22)
5. Australian Institute of Health and Welfare.: Secondary prevention and rehabilitation after coronary events or stroke: a review of monitoring issues. AIHW Cat. No. CVD 25, AIHW, Canberra (2003)
6. Turin, T., Kokubo, Y., Murakami, Y., Higashiyama, A., Rumana, N., Watanabe, M.: Lifetime risk of stroke in Japan. Stroke 41, 1552–1554 (2010)
7. Bharadwaj, K., Hollander, K.W., Mathis, C.A., Sugar, T.G.: Spring over muscle (SOM) actuator for rehabilitation devices. In: Proceedings of the 2004 Annual International Conference of the IEEE Engineering in Medicine and Biology Society, vol. 1, pp. 2726–2729, 1–5 Sept 2004
8. Taub, E., Uswatte, G., Pidikiti, R.: Constraint-induced movement therapy: a new family of techniques with broad application to physical rehabilitation - a clinical review. J. Rehabil. Res. Dev. 36, vii–viii (1999)
9. Rahman, S., Ikeura, R.: A novel variable impedance compact compliant ankle robot for overground gait rehabilitation and assistance. Procedia Eng. (Elsevier) 41, 522–531 (2012)
10. Veneman, J., Kruidhof, R., Hekman, E., Ekkelenkamp, R., Van Asseldonk, E.H.F., Van der Kooij, H.: Design and evaluation of the LOPES exoskeleton robot for interactive gait rehabilitation. IEEE Trans. Neural Syst. Rehabil. Eng. 15(3), 379–386 (2007)
11. Peshkin, M., Brown, D.A., Santos-Munne, J.J., Makhlin, A., Lewis, E., Colgate, J.E., Patton, J., Schwandt, D.: KineAssist: a robotic overground gait and balance training device. In: Proceedings of the 2005 9th IEEE International Conference on Rehabilitation Robotics, pp. 241–246
12. Allemand, Y., Stauffer, Y., Clavel, R., Brodard, R.: Design of a new lower extremity orthosis for overground gait training with the WalkTrainer. In: Proceedings of the 2009 IEEE International Conference on Rehabilitation Robotics, pp. 550–555
13. Zhang, X., Kong, X., Liu, G., Wang, Y.: Research on the walking gait coordinations of the lower limb rehabilitation robot. In: Proceedings of the 2010 IEEE International Conference on Robotics and Biomimetics, pp.1233–1237

14. Yoon, J., Novandy, B., Yoon, C., Park, K.: A 6-DOF gait rehabilitation robot with upper and lower limb connections that allows walking velocity updates on various terrains. IEEE/ASME Trans. Mechatron. **15**(2), 201–215 (2010)
15. Jezernik, S., Colombo, G., Keller, T., Morari, F.: Robotic orthosis lokomat: a rehabilitation and research tool. Neuromodulation **6**(2), 108–115 (2003)
16. Galvez, J.A., Reinkensmeyer, D.J.: Robotics for gait training after spinal cord injury. Technol. Strateg. Enhanc. Mobil. **11**(2), 18–33 (2005)
17. Krebs, H.I., Ferraro, M., Buerger, S.P., Newbery, M.J., Makiyama, A., Sandmann, M., Lynch, D., Volpe, B.T., Hogan, N.: Rehabilitation robotics: pilot trial of a spatial extension for MIT-Manus. J. NeuroEng. Rehabil. **1**, 5 (2004)
18. Nef, T., Mihelj, M., Colombo, G., Riener, R.: ARMin - robot for rehabilitation of the upper extremities. In: Proceedings of the 2006 IEEE International Conference on Robotics and Automation, pp. 3152–3157
19. Lee, C., Seo, K., Oh, C., Lee, J.: A system for gait rehabilitation with body weight support: mobile manipulator approach. J. HWRSERC **2**(3), 16–21 (2000)
20. Vische, D., Kathib, O.: Design and development of high performance torque-controlled joints. IEEE Trans. Robot. Autom. **11**(4), 537–544 (1995)
21. Ferris, D.P., Farley, C.T.: Interaction of leg stiffness and surface stiffness during human hopping. J. Appl. Physiol. (American Physiological Society) **82**, 15–22 (1997)
22. Yang, C., Ganesh, G., Haddadin, S., Parusel, S., Albu-Schaeffer, A., Burdet, E.: Human-like adaptation of force and impedance in stable and unstable interactions. IEEE Trans. Robot. **27**(5), 918–930 (2011)
23. Robinson, D., Pratt, J., Paluska, D., Pratt, G.: Series elastic actuator development for a biomimetic walking robot. In: Proceedings of the 1999 IEEE/ASME International Conference on Advanced Intelligent Mechatronics, pp. 561–568
24. Vallery, H., Veneman, J., Asseldonk, E., Ekkelenkamp, R., Buss, M., Kooij, H.: Compliant actuation of rehabilitation robots. IEEE Robot. Autom. Mag. **15**(3), 60–69 (2008)
25. Tsagarakis, N., Laffranchi, M., Vanderborght, B., Caldwell, D.: A compact soft actuator unit for small scale human friendly robots. In: Proceedings of the 2009 IEEE International Conference on Robotics and Automation, pp. 4356–4362
26. Hirzinger, G., Sporer, N., Albu-Schaffer, A., Hahnle, M., Krenn, R., Pascucci, A., Schedl, M.: DLR's torque-controlled light weight robot III – are we reaching the technological limits now? In: Proceedings of the IEEE International Conference on Robotics and Automation, vol. 2, pp. 1710–1716 (2002)
27. Albu-Schaffer, A., Eiberger, O., Grebenstein, M., Haddadin, S., Ott, C., Wimbock, T., Wolf, S., Hirzinger, G.: Soft robotics. IEEE Robot. Autom. Mag. **15**(3), 20–30 (2008)
28. Schiavi, R., Grioli, G., Sen, S., Bicchi, A.: VSA-II: a novel prototype of variable stiffness actuator for safe and performing robots interacting with humans. In: Proceedings of the 2008 IEEE International Conference on Robotics and Automation, pp. 2171–2176
29. Wolf, S., Hirzinger, G.: A new variable stiffness design: matching requirements of the next robot generation. In: Proceedings of the 2008 IEEE International Conference on Robotics and Automation, pp. 1741–1746
30. Tagliamonte, N.L., Sergi, F., Carpino, G., Accoto, D., Guglielmelli, E.: Design of a variable impedance differential actuator for wearable robotics applications. In: Proceedings of the 2010 IEEE/RSJ International Conference on Intelligent Robots and Systems, pp. 2639–2644
31. Pratt, G., Williamson, M.: Series elastic actuators. In: Proceedings of the IEEE/RSJ International Conference on Intelligent Robots and Systems, vol. 1, pp. 399–406 (1995)

32. Hirai, K., Hirose, M., Haikawa, Y., Takenaka, T.: The development of Honda humanoid robot. In: Proceedings of the 1998 IEEE International Conference on Robotics and Automation, pp. 1321–1326
33. Wassink, M., Carloni, R., Stramigioli, S.: Port-Hamiltonian analysis of a novel robotic finger concept for minimal actuation variable impedance grasping. In: Proceedings of the 2010 IEEE International Conference on Robotics and Automation, pp. 771–776
34. Rahman, S.: A novel variable impedance compact compliant series elastic actuator: analysis of design, dynamics, materials and manufacturing. Appl. Mech. Mater. **245**, 99–106 (2013)
35. Carlson, T., Demiris, Y.: Collaborative control for a robotic wheelchair: evaluation of performance, attention and workload. IEEE Trans. Syst. Man Cybern. B **42**(3), 876–888 (2012)
36. Rahman, S.: A novel variable impedance compact compliant series elastic actuator for human-friendly soft robotics applications. In: Proceedings of the 2012 21st IEEE International Symposium on Robot and Human Interactive Communication, pp. 19–24

Evaluation of Laser Range-Finder Mapping for Agricultural Spraying Vehicles

Francisco-Angel Moreno[1], Grzegorz Cielniak[2](✉), and Tom Duckett[2]

[1] Department of System Engineering and Automation,
University of Málaga, Málaga, Spain
famoreno@uma.es
[2] Lincoln School of Computer Science, University of Lincoln, Lincoln, UK
{gcielniak,tduckett}@lincoln.ac.uk

Abstract. In this paper, we present a new application of laser range-finder sensing to agricultural spraying vehicles. The current generation of spraying vehicles use automatic controllers to maintain the height of the sprayer booms above the crop. However, these control systems are typically based on ultrasonic sensors mounted on the booms, which limits the accuracy of the measurements and the response of the controller to changes in the terrain, resulting in a sub-optimal spraying process. To overcome these limitations, we propose to use a laser scanner, attached to the front of the sprayer's cabin, to scan the ground surface in front of the vehicle and to build a scrolling 3d map of the terrain. We evaluate the proposed solution in a series of field tests, demonstrating that the approach provides a more detailed and accurate representation of the environment than the current sonar-based solution, and which can lead to the development of more efficient boom control systems.

Keywords: Agri-robotics · 3d terrain reconstruction · Outdoor mapping

1 Introduction

Precision agriculture aims to utilise automated management and technology solutions for optimisation of various farming processes. Future agricultural systems will rely on machines performing tasks like ploughing, spraying or harvesting autonomously with minimal intervention from a human user. This work is concerned with a particular class of agricultural spraying vehicles, namely horizontal boom sprayers (see Fig. 1). The modern generation of these vehicles feature adjustable spraying booms which can be automatically controlled to maintain a constant distance from the crops. This is a critical process as the height of the boom affects the amount and distribution of the sprayed substance, which has not only financial implications, but is also becoming increasingly important in the light of tougher environmental policies on the use of fertilisers and pesticides. The current boom control systems rely on ultrasonic sensors for measuring the height and level of the booms. The ultrasonic sensors,

A. Natraj et al. (Eds.): TAROS 2013, LNAI 8069, pp. 210–221, 2014.
DOI: 10.1007/978-3-662-43645-5_22, © Springer-Verlag Berlin Heidelberg 2014

Fig. 1. A 24-meter wide sprayer model, Merlin 4000/24, by Househam Sprayers Ltd.

whilst inexpensive, are relatively slow and provide noisy information for only a small patch of the terrain immediately below the spraying boom. This results in limited effectiveness and restricts the maximum speed of the sprayer, since only a reactive control strategy is possible, resulting in a sub-optimal spraying process.

This paper investigates an alternative sensing technology based on laser range-finders (LRF) and terrain modelling that can provide predictive (i.e. with a longer "look-ahead"), higher-quality distance information. Laser range-finders, which have been widely used for over a decade in the mobile robotics community, have been applied to numerous applications such as mapping and navigation. Our proposal is to apply these well-established techniques to enhance the sensing abilities of agricultural spraying vehicles. The core component of the proposed system is a 3d map of the terrain, reconstructed from a scanning laser range-finder and additional information provided by GPS. With this approach the terrain is sensed in advance, so the controller should have more time to adjust the height of the booms. The approach not only improves the accuracy of the provided distance information but can also enable new applications such as terrain-based vehicle steering or variable-rate spraying, leading towards development of fully autonomous spraying vehicles.

2 Related Work

Robotic applications in agriculture can bring numerous economic, societal and environmental benefits (e.g. reduced production costs, more friendly working environments, reduced contamination risks, etc.) [10]. However, the future development of such systems will have to address several challenges arising from the complexity of farming processes, outdoor environments, and the mechanical complexity and physical size of agricultural machinery. To meet these challenges, the future autonomous farming vehicles will have to rely not only on GPS-based solutions but also on a detailed representation of the environment, e.g. 3d maps.

The majority of outdoor 3d mapping applications consider urban environments (e.g. [9]) where there are physical, man-made structures that assist in the registration of 3d scans, improving the quality of the resulting maps. Other examples include mapping solutions for off-road autonomous car driving [13], mining operations [5] and autonomous road inspection [8]. The latter system (RoadBot) fuses information from range scanners thanks to precise pose estimation obtained from high precision GPS (i.e. real-time-kinematics-enabled devices) with laser-aided height correction, resulting in a dense and precise map of the road. We assume a similar approach in our work. The majority of the presented mapping systems rely on laser scanners, although other sensors can be used, e.g. stereo vision [11].

Recent interest in agricultural robotics has resulted in several systems relying on 3d respresentation of the envrionment. For example, [7] presents a multi-vehicle system for automating orchard farming operations like spraying and mowing. The vehicle navigation is performed by a combination of GPS, range and vision data, enabling obstacle avoidance and row-following behaviours. BoniRob [3] is a mobile phenotyping robotic platform equipped with a range of sensors designed to automatically measure different plant properties. The system functionality includes detection of the ground and individual plants based on 3d range maps and semantic place classification. The experiments conducted with different range sensing technologies suggest that the laser-based techniques are most suitable in the farming context.

Existing autonomous spraying solutions were mostly deployed in horticulture scenarios such as orchards or vineyards. For example,[16] describes techniques for building off-line 3d models of trees and their subsequent use for precision spraying. The models are based on probabilistic techniques that separate range data from ladar sensor into ground and tree canopy. The tree height and density is later used by a variable-rate spraying controller, resulting in more precise and directed application of the chemicals. Other examples include a review of different tree area estimation methods [14], 3d fluid dynamics modelling for improved spraying [6], and range-based real-time control of the spraying [15].

Related work in technological solutions for horizontal boom sprayers considers mostly mechanical aspects of the booms, including their design [2], analysis of oscillations [4], and control based on mechanical modelling with a simple infra-red distance sensing solution [12]. In contrast, our work concentrates on the novel application of laser range-finder sensing, together with GPS information, to build a scrolling 3d model of the terrain, which could later be used for improving the control of the sprayer booms.

3   System Description

3.1   Horizontal Boom Sprayer

The main components of the current system consist of a spraying vehicle and an adjustable spraying boom which can be folded and unfolded for easier transportation and storage (see Fig. 1). The vehicle also carries a tank containing chemicals being distributed on the field from the nozzles located on the booms. The length of the booms depends on the sprayer model and ranges from 12 to

18 m on each side. The boom is automatically controlled by a custom made controller which can adjust the height of the boom platform and the boom incline on each side. The current sensory system of the sprayer is composed of a set of ultrasonic sensors regularly distributed along the booms (every 5–6 m) and pointing downwards so that they can measure the distance between the ground (or the crops) and the booms. The sonar data is provided through the internal CAN bus which also provides the information about the current configuration of the boom levelling platform including its height and incline angles.

In addition, the vehicle is equipped with a Trimble GPS receiver used for vehicle auto steering and variable-rate spraying application, providing global position measurements at a regular rate of 5 Hz. The GPS receiver operates in a differential mode, receiving positioning corrections from a nearby base station thus achieving a theoretical accuracy of a few centimetres.

3.2   Experimental Set-up

To address the limitations of the current ultrasonic sensing system, we propose to use a laser scanner, attached to the front part of the sprayer's cabin to scan the ground surface in front of the vehicle (see Fig. 2(a)) and build a scrolling 3d map of the terrain as it approaches the sprayer. We used a Hokuyo UTM-30LX outdoor laser scanner, which combines an affordable price and small size with good performance and relatively long range (~30 m). We also employed an additional set of sensors for inspection, development and evaluation of the system. Thus, we attached a short-range laser scanner (Hokuyo URG-04LX) to the left spraying boom, pointing downwards and close to one of the ultrasonic sensors, so that the laser can scan the ground just below the boom (see Fig. 2(b)). A backup consumer-grade GPS was placed on the roof of the spraying vehicle in order to provide auxiliary positioning measurements used in initial stages of the system development.

(a)   (b)

Fig. 2. (a) Overview of the proposed system. (b) Placement of the rear laser and sonar.

In order to get the information from the sprayer's telemetrics, a CAN bus interface dongle has been connected to the vehicle's panel. This connection provides access to the sonar readings and boom configuration parameters. All these components were connected to a main PC, a Dell XPSL502X laptop with an Intel CORE i7 processor running Ubuntu.

The required software was developed using the freely available Mobile Robotics Programming Toolkit [1]. The toolkit contains a set of libraries and applications covering the most common algorithms for mobile robot navigation, mapping and motion planning. It also provides a complete list of hardware and robotic sensors drivers including laser scanners, cameras and GPS. The toolkit has been used to collect data from the different sensors on the sprayer and to develop a stand alone application for off-line processing of the collected data, building the point maps from the laser scanner readings and visualisation, as well as measuring the errors between the data provided by the different sensors.

3.3 Map Building

The reconstruction of the terrain's surface was performed through the transformation of all the measurements captured with the laser scanner into one local reference frame $\mathbf{O}_E$ (see Fig. 3), which is the main reference system of the working space and whose centre is placed at the starting position of the sprayer (i.e. the first position collected by the GPS). The vehicle's reference system $\mathbf{O}_{G1}$ has its origin at the position of the sprayer's built-in GPS, with the X-axis pointing in the direction of the movement and the Z-axis pointing upwards. The reference systems $\mathbf{O}_{L1}$ and $\mathbf{O}_{L2}$ correspond to the front and the rear laser scanners, respectively, and have their X-axis pointing downwards. The $\mathbf{O}_{G2}$ reference system is placed at the position of the auxiliary GPS. Finally, $\mathbf{T}_V$ stands for the pose (i.e. position and orientation) of the vehicle's reference system within the local reference system $\mathbf{O}_E$, and $\mathbf{T}_{L1}$ and $\mathbf{T}_{L2}$ for the poses of the two laser scanners with respect to the vehicle's reference system, respectively. Note that the vehicle's pose changes at each time step as the sprayer moves, whilst the last two are fixed since they are rigidly attached to the sprayer.

The process of building the map of the terrain from the laser data is accomplished as follows. Without loss of generality, we use the readings from the front laser in this explanation. Formally, let r_i and α_i be the range and angle values measured by the front laser sensor for a certain scanned point p_i on the terrain's surface. The coordinates of that point with respect to the laser scanner's reference system $\mathbf{O}_{L1}$ are computed as $\mathbf{P}_{L1} = [r_i cos(\alpha_i), r_i sin(\alpha_i), 0]$. Such a point can be further transformed with respect to the vehicle's reference system $\mathbf{O}_{G1}$ through $\mathbf{P}_{G1} = \mathbf{T}_{L1} \oplus \mathbf{P}_{L1}$, where $\oplus$ is the point-pose composition operator.

Finally, the coordinates in the local map of the scanned point $\mathbf{P}_E$ are computed in a similar way by applying again the composition operator with the pose of the vehicle within the local reference system $\mathbf{T}_V$: $\mathbf{P}_E = \mathbf{T}_V \oplus \mathbf{P}_{G1}$. The set of points $\{\mathbf{P}_E^i\}$ referenced to the local system $\mathbf{O}_E$ computed this way constitutes the reconstruction of the terrain's surface.

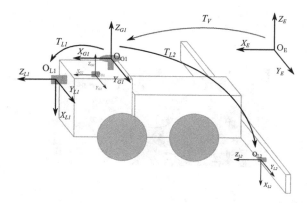

Fig. 3. All reference frames in the proposed system.

Note, that in order to build the map, it is necessary to estimate the vehicle's pose T_V at each time step, which can be achieved from the GPS readings, as well as to accurately measure T_{L1} and T_{L2}, which is addressed next.

3.4 Sensor Calibration

Since the proposed system only relies on position data provided by GPS, the sensor calibration is a critical process to assure an accurate representation of the scanned surface. The estimation of the laser scanners' poses (T_{L1} and T_{L2}) with respect to the vehicle's reference system O_{G1} has been performed as follows. First, an initial estimation of their positions was obtained by means of a standard measuring tape. Then the vehicle was driven through a field where three boxes of known size were placed on the soil (see Fig. 4). The sprayer vehicle was driven so that the three boxes were scanned twice (one in each direction) with both lasers, and one of the boxes was scanned one more time in a perpendicular direction. These data have been used afterwards to refine the initial rough estimates by aligning the range scans corresponding to the boxes appearing in the reconstructed map. Figure 4(b) shows the effects of the initial calibration errors on the quality of the map reconstruction, while Fig. 4(c) shows the final map after full calibration.

4 Experiments

4.1 Datasets

To evaluate the proposed system, we collected two data sets from different locations with different vehicle types and weather conditions (see Table 1). The estimated pose of the sensors for each data set is shown as a vector $[x, y, z, yaw, pitch, roll]$ where the translational part is expressed in meters and the rotational part in degrees. Data set #1 was recorded while the vehicle was driven through a

(a)

(b)   (c)

Fig. 4. The boxes used to perform sensor calibration (a). Detail of the map at the boxes (b) before and (c) after refining the calibration.

straight, hardened driveway while the spraying booms were suspended on an uncultivated field. Figure 5(a) shows the trajectory followed by the vehicle. This data set was used for finding the most convenient location of the sensors on the sprayer, verification of the initial set-up and assessing the laser's performance in outdoor conditions and therefore the vehicle's telemetry data was not recorded. Data set #2 was collected while the vehicle was driven on a field with short stubble left from a previous crop after harvesting. The main purpose of this data set was to assess the quality of the reconstructed map, but also for sensor calibration. Therefore three cardboard boxes were placed in known locations as depicted on Fig. 4. This time the sprayer was driven following the path presented in Fig. 5(b) while simultaneously recording data from the GPS receivers, the laser scanners and the sonar readings.

4.2   Results

Data set #1 was collected mainly for development and testing of the system and therefore we only present qualitative results in the form of a reconstructed 3d map from the front laser readings (see Fig. 6). One visible aspect in this map is a changing height pattern on both sides of the road corresponding to the rolling movement of the vehicle on uneven surface.

Table 1. Summary of the data sets.

Parameter	Data set #1	Data set #2
Date	29th of January, 2013	12th of February, 2013
Weather	Cold, wet and cloudy	Cold, dry and cloudy
Vehicle	Merlin	Air Ride
Estimated pose:		
Front laser	$[0.99, -5.80, -3.20, 0, 90, 0]$	$[1.75, 0, -0.51, 0, 90, 0]$
Rear laser	$[-5.59, 8.90, -2.19, -90, 84, -90]$	$[-5.03, 5.89, -1.93, 0, 90, 0]$
Total distance	370.7 m	290.4 m
Total points	5.7 M	4.5 M

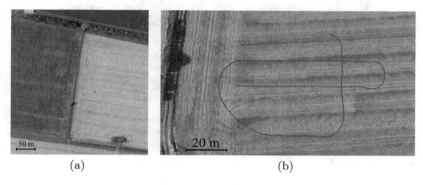

(a)   (b)

Fig. 5. A satellite view of the traversed path for (a) Data set #1 (b) Data set #2.

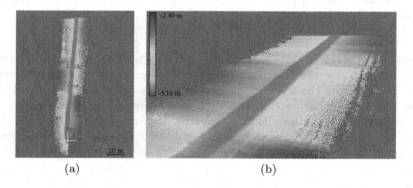

(a)   (b)

Fig. 6. The reconstructed map from Data set #1 (a) a top view, (b) map detail.

Data set #2 was used to reconstruct and assess the quality of the 3d map with all the sensors in operation. Figure 7(a) shows the distribution of readings provided by the sonar sensors. Each green dot in the figure represents a location of a single sonar reading projected within the local reference system through the process described in Sect. 3.3. The points measured from the same sonar have been joined by a distinctively coloured line. As can be seen, the number of sonar

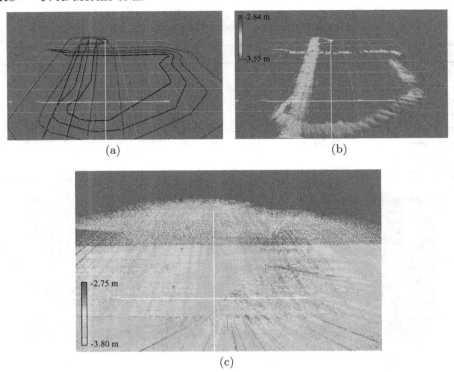

(a)   (b)

(c)

Fig. 7. The reconstructed map from Data set #2: (a) distribution of sonar readings, (b) rear laser, (c) front laser.

measurements is scarce since the readings are only updated upon their change, which demonstrates the relatively low resolution of the sonar measurements. Figure 7(b) presents the map built with the rear laser, which was placed close to the left-centre sonar so that there exists an overlapping zone between the both sensors. It can be seen that a relatively narrow section of the terrain is reconstructed due to the low positioning and short range of the rear laser. In contrast, the map reconstructed using the front laser readings (see Fig. 7(c)) results in a dense cloud of points covering the terrain which forms an accurate model of the terrain. One visible artefact in the final 3d map is a set of diagonal ridges caused by inaccuracies in pose estimation (mainly the pitch angle) due to the vehicle driving on uneven terrain.

A highlighted zone of the final map is presented in Fig. 8(a) where the difference between the tracks of the vehicle and the areas with stubble can be noticed. The same area, represented as a 2D profile, is shown in Fig. 8(b) along with the median value in height, clearly showing the difference in profile between the track and the stubble (approx. 10 cm).

In order to quantitatively assess the quality of the reconstructed surface, we have compared the scanned representation of one of the calibration boxes with its

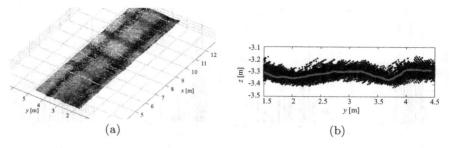

Fig. 8. (a) Detail of the map where the vehicle's tracks can be seen. (b) A 2D profile of the same area.

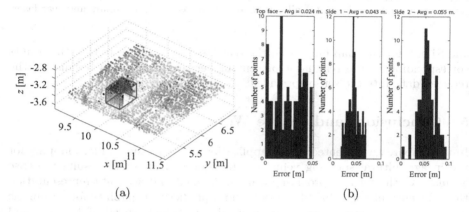

Fig. 9. The wireframe model of the test box superimposed onto the reconstructed point cloud (a). Histogram of errors between the readings and the actual box size in (b) the top face and (c,d) the lateral faces.

real, physical size. This can be seen in Fig. 9(a), where a wireframe model of the real box has been superimposed onto the point cloud. We then have measured the distance from three of the sides of the scanned box to the real sides and plotted these errors as histograms (see Fig. 9(b)). The average errors between the scanned model and the real box are 24 mm in height, and 43 and 55 mm in the lateral directions.

Finally, for comparison of different sensor types, we have computed the distance differences between the points measured by the sonar sensors and the maps reconstructed by each of the laser scanners. For that purpose, we have calculated the distance between the closest pair of readings from each sensor. Figure 10(a) and (b) show the histograms of distances between the readings from the left-centre sonar only and the maps built from the front and the rear laser scanner, presenting mean distances of 44 mm and 56 mm, respectively. The histograms of errors between the readings from all the sonars and the maps are presented in Fig. 10(c) and (d), showing a mean distance of 56 mm for the front laser's map

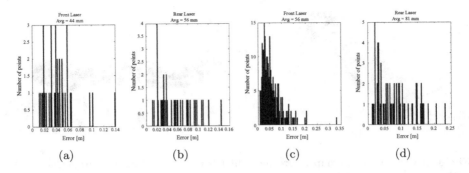

Fig. 10. Histograms of errors between (a) left-centre sonar and front laser, (b) left-centre sonar and rear laser, (c) all sonars and front laser, (d) all sonars and rear laser.

and 81 mm for the map built from the rear laser readings. Note that the results corresponding to the rear laser present higher mean distances because of the reduced data rate the sensor provides.

5   Conclusions and Future Work

In this paper, we presented a new application of laser range-finder mapping for use in agricultural spraying vehicles. The proposed approach results in dense 3d maps of the terrain, providing a more detailed and accurate representation than the current sonar-based solution and a predictive capability for automatic boom control. Based on an assumed maximum vehicle speed of 15 km/h and a distance between the front laser and the booms of 5 m, we would therefore estimate that the proposed solution can provide an additional 1.2 s (minimum) of look-ahead time for the boom controller. In addition, only one sensor is needed to cover the full width of the sprayer with the booms unfolded in comparison with the current ultrasonic set-up consisting of five sensors. The results show a mean distance between the estimated position of the ground computed from sonar and laser data of approximately 5 cm.

The major problem affecting the accuracy of the reconstructed map is the estimation of the vehicle's pose in uneven terrain. The vehicle's turning angle can only be estimated from the GPS measurements at the moment, which leads to unreliable results on curves. Moreover, when the terrain presents potholes and irregularities under the sprayer's wheels, the shaking and vibrating movements introduce significant inaccuracies in the reconstructed maps. The use of an inertial measurement unit should address this issue as the full pose of the sprayer could then be estimated more accurately. To fully assess the characteristics of the proposed system, evaluation under different types of fields, crops and weather conditions (including strong sunlight) also has to be addressed in future.

This work represents a first step toward more autonomous solutions for agricultural spraying vehicles. Future work will consider a full closed-loop boom controller based on the proposed sensing principle. In addition, the proposed mapping system can lead to other applications including automatic detection of irregularities of the terrain, detection of tracks caused by the vehicle, or estimation of crop yields over time.

Acknowledgments. This project was supported by a BBSRC SPARK Award, in collaboration with Househam Sprayers Ltd.

References

1. MRPT framework. Website (2013). http://www.mrpt.org
2. Anthonis, J., Audenaert, J., Ramon, H.: Design optimisation for the vertical suspension of a crop sprayer boom. Biosyst. Eng. **90**(2), 153–160 (2005)
3. Biber, P., Weiss, U., Dorna, M., Albert, A.: Navigation system of the autonomous agricultural robot "BoniRob". In: Proceedings IROS Workshop on Agricultural Robotics (2012)
4. Jeon, H., Womac, A., Wilkerson, J., Hart, W.: Sprayer boom instrumentation for field use. Trans. ASAE **47**(3), 659–666 (2004)
5. Magnusson, M., Lilienthal, A., Duckett, T.: Scan registration for autonomous mining vehicles using 3D-NDT. J. Field Robot. **24**(10), 803–827 (2007)
6. Endalew, A.M., Debaer, C., Rutten, N., Vercammen, J., Delele, M.A., Ramon, H., Nicolaï, B.M., Verboven, P.: Modelling the effect of tree foliage on sprayer airflow in orchards. Bound.-Layer Meteorol. **138**, 139–162 (2011)
7. Moorehead, S.J., Wellington, C.K., Gilmore, B.J., Vallespi, C.: Automating orchards: A system of autonomous tractors for orchard maintenance. In: Proceedings IROS Workshop on Agricultural Robotics (2012)
8. Moreno, F.-A., Gonzalez-Jimenez, J., Blanco, J.L., Esteban, A.: An instrumented vehicle for efficient and accurate 3D mapping of roads. Comput. Aided Civ. Infrastruct. Eng. **28**, 403–419 (2013)
9. Nüchter, A., Lingemann, K., Hertzberg, J., Surmann, H.: 6d SLAM - 3d mapping outdoor environments. J. Field Robot. **24**(8–9), 699–722 (2007)
10. Pedersen, S., Fountas, S., Have, H., Blackmore, B.: Agricultural robots - system analysis and economic feasibility. Precis. Agric. **7**(4), 295–308 (2006)
11. Rovira-Más, F., Zhang, Q., Reid, J.F.: Stereo vision three-dimensional terrain maps for precision agriculture. Comput. Electron. Agric. **60**(2), 133–143 (2008)
12. Sun, J., Miao, Y.: Simulation and controller design for an agricultural sprayer boom leveling system. In: Third International Conference on Measuring Technology and Mechatronics Automation (ICMTMA), vol. 3, pp. 245–248 (2011)
13. Thrun, S., et al.: Stanley: the robot that won the DARPA grand challenge. In: Buehler, M., Iagnemma, K., Singh, S. (eds.) The 2005 DARPA Grand Challenge. Springer Tracts in Advanced Robotics, vol. 36, pp. 1–43. Springer, Heidelberg (2007)
14. Walklate, P., Cross, J., Richards, G., Murray, R., Baker, D.: Comparison of different spray volume deposition models using lidar measurements of apple orchards. Biosyst. Eng. **82**(3), 253–267 (2002)
15. Wei, J., Salyani, M.: Development of a laser scanner for measuring tree canopy characteristics, phase 2: foliage density measurement. Trans. ASABE **48**(4), 1595–1601 (2005)
16. Wellington, C., Campoy, J. abd Khot, L., Ehsani, R.: Orchard tree modeling for advanced sprayer control and automatic tree inventory. In: Proceedings of IROS Workshop on Agricultural Robotics (2012)

Autonomous Coverage Expansion of Mobile Agents via Cooperative Control and Cooperative Communication

Said Al-Abri$^{(\boxtimes)}$ and Zhihua Qu$^{(\boxtimes)}$

Department of EECS, University of Central Florida, Orlando, FL 32816, USA
saidalabri@knights.ucf.edu, qu@eecs.ucf.edu

Abstract. In this paper, a distributed extremum seeking and cooperative control is designed for mobile agents to disperse themselves optimally in maintaining communication quality and maximizing their coverage. The agents locally form a virtual MIMO communication system, and they communicate among them by using the decode and forward cooperative communication. Outage probability is used as a measure of communication quality which can be estimated real-time. A general performance index balancing outage probability and spatial dispersion is chosen for the overall system. Extremum seeking control approach is used to estimate the optimal values specified by the performance index, and cooperative formation control is applied to move the agents to the optimal locations by using only the locally-available information. The network connectivity and coverage are much improved when compared to either non-cooperative communication approaches or other existing control results. Simulation analysis is carried out to demonstrate the performance and robustness of the proposal methodology, and simulation is done to illustrate its effectiveness.

Keywords: Cooperative control · Cooperative communication · Extremum seeking · Coverage · Connectivity

1 Introduction

Cooperative control is a new class of control systems which have been an active area of research since the last decade [6]. One of its early application is formation control of vehicles [3,7]. In [3], many linear and nonlinear cooperative control approaches are investigated and analyzed. In addition, cooperative control is used to maintain network connectivity of wireless sensor networks. Example works are shown in [8,9]. In [8], the connectivity of a network of mobile robots is controlled based on a proximity communication model like discs or a uniformly fading signal strength of communication links. However, in many situations,

This work is part of a Master Thesis done by Said Al-Abri under supervision of professor Zhihua Qu at University of Central Florida. Said's master study is funded by Sultan Qaboos University, Muscat, Oman.

A. Natraj et al. (Eds.): TAROS 2013, LNAI 8069, pp. 222–234, 2014.
DOI: 10.1007/978-3-662-43645-5_23, © Springer-Verlag Berlin Heidelberg 2014

these proximity communication models may not be practical due to dynamic environments of the robots which may add many communication noises that are not captured by the proximity model. In [10], a theoretic information based performance index is designed to ensure both connectivity and coverage. This performance index is a function of the outage probability which can be measured in real time. Hence, it captures any drop in the channel that may happened due to shadowing or any other fading source. Since this performance index is varying and cannot be known exactly in advance, the author of [10] used an extremum seeking control to search for the optimum locations of the nodes specified by the performance index. The algorithm and stability analysis of extremum seeking control can be found in [11].

Cooperative communication is a new technique that utilizes the benefits of Multiple Input Multiple Output (MIMO) communication system and relaying to enhance the quality of wireless networks [12,13]. Intuitively, cooperative communication can be viewed as a virtual MIMO communication system where the cooperating nodes behave like virtual antennas for the system [14]. The cooperating nodes add diversity to the network and some protocols to utilize this diversity are presented in [18]. In these protocols, the performance is evaluated by the outage probability which, in turn, is derived from the Shannon capacity. Some exact formulas of outage probabilities and the corresponding performance analysis are shown in [1]. The locations of the cooperating nodes and network structure play a major role in the performance of cooperative communication systems. It is shown in [15,16] that the communication coverage can be expanded further by carefully selecting the locations of the cooperating nodes. Some relay assignment protocols used to extend the coverage are shown in [17].

The objective of this paper is to design a distributed extremum seeking and cooperative control law for networked mobile agents such that a desired communication quality level is achieved while expanding the coverage to the possible maximum. The information between the networked mobile agents are shared via a decode and forward cooperative communication approach. To the best of our knowledge, this is the first time cooperative communication is used with cooperative control and extremum seeking in the problem of communication quality and coverage area of wireless sensor networks. As an extension to the performance index of direct transmission in [10], a new performance index is designed to capture the effects of a decode and forward cooperative communication. In addition, the design and stability analysis of a one variable extremum seeking control in [11] is extended to a two variable extremum seeking control. Furthermore, a cooperative control law similar to the one in [10] is designed with some modifications to accommodate with the new parameters added by the use of cooperative communication.

The rest of this paper is organized as follows. Section 2 presents the problem formulation and stability analysis of cooperative formation control. The outage analysis of decode and forward cooperative communication and performance index to maximize communication coverage for a predefined communication quality is presented in Sect. 3. In Sect. 4, the design and stability of two variable extremum seeking control integrated with cooperative control is emphasized. Simulation results are shown in Sect. 5. Finally, a conclusion of the work done and future work suggestion is given in Sect. 6.

2   Cooperative Formation Control

Consider a group of n networked mobile agents of the following dynamics [10]:

$$\dot{x}_i = u_i, \tag{1}$$

where $x_i \in \Re^m$ is the coordinate of the ith agent, $u_i \in \Re^m$ is the control to be designed. The connectivity between all the agents is described by a piecewise-constant binary matrix S(t). As shown in [3], there is a time sequence: $\{t_k : k \in \aleph\}$ such that $S(t) = S(t_k)$ for all $t \in [t_k, t_{k+1})$ where $\aleph = \{0, 1, ..., \infty\}$, and

$$S(t_k) = \begin{bmatrix} 1 & s_{12}(t_k) & \cdots & s_{1n}(t_k) \\ s_{21}(t_k) & 1 & \cdots & s_{2n}(t_k) \\ \vdots & \vdots & \ddots & \vdots \\ s_{n1}(t_k) & s_{n2}(t_k) & \cdots & 1 \end{bmatrix}, \tag{2}$$

where $s_{ij} = 1$, if the information of $x_j(t)$ is received by the ith agent, and $s_{ij} = 0$ if otherwise. In order to achieve a certain formation, the following nonlinear cooperative control is used to the ith agent:

$$u_i = \mu \sum_{j \in N_i} \frac{s_{ij}\, \varphi_{ij}}{\sum_{l=1}^{n} s_{il}\, \varphi_{il}} (1 - \frac{r_{ij}^*}{r_{ij}})(x_j - x_i) \triangleq \mu \sum_{j \in N_i} p_{ij}(1 - \frac{r_{ij}^*}{r_{ij}})(x_j - x_i), \tag{3}$$

where $\mu \geqslant 1$ is the control gain, N_i is the set of neighboring agents of ith agent, φ_{ij} are constant gains. Matrix $D = [p_{ij}]$ is nonnegative and row-stochastic, r_{ij}^* and $r_{ij} \triangleq ||x_j - x_i||$ are the desired and current distances between the ith and jth agents, respectively. Note that if $r_{ij} > r_{ij}^*$, the term $(1 - \frac{r_{ij}^*}{r_{ij}})$ in (3) behaves like an attractive force (i.e, the ith and jth agents move toward each other). The term becomes a repulsive force if $r_{ij} < r_{ij}^*$ (i.e, the ith and jth move away from each other). Thus, the desired separation distance r_{ij}^* for all $j \in N_i$ can be ensured distributively and asymptotically, provided that matrix $S(t)$ or (its corresponding graph) is cumulatively connected [3].

In order to prove the stability of system (1) under control (3), lets define the normalized error of $e_{ij} = \frac{x_j - x_i}{r_{ij}}$. Hence, the closed loop system becomes:

$$\dot{x}_i = \mu \sum_{j \in N_i} p_{ij}(x_j - x_i - r_{ij}^* e_{ij}) \triangleq \mu \sum_{j \in N_i} p_{ij}(\hat{x}_j - x_i), \tag{4}$$

where $\hat{x}_j = x_j - r_{ij}^* e_{ij}$. We need to show $\hat{x}_j \to x_i$, turn implies $x_j - x_i \to r_{ij}^* e_{ij}$. It is clear that

$$\mu x_i \sum_{j \in N_i} p_{ij}(\hat{x}_j - x_i) \leqslant 0 \tag{5}$$

for all values of $x(t)$ with $|x_i(t)| = \max_{1 \leqslant j \leqslant n} |\hat{x}_j(t)|$. Also, it follows

$$\mu(x_i - x_k) \left[\sum_{j \in N_i} p_{ij}(\hat{x}_j - x_i) - \sum_{j \in N_k} p_{kj}(\hat{x}_j - x_k) \right] \leqslant 0 \qquad (6)$$

for all values of $x(t)$ with $x_i(t) = \max_{1 \leqslant j \leqslant n} \hat{x}_j(t)$ and $x_k(t) = \min_{1 \leqslant j \leqslant n} \hat{x}_j(t)$. Should $D(t)$ be uniformly sequentially complete, we know from Theorem 6.8 of [3], that given (5) and (6), system (4) is both uniformly Lyapunov stable and asymptotically cooperatively stable. That is r_{ij} asymptotically converges to r_{ij}^*.

It is important to mention that for the problem under investigation, the desired separation distance r_{ij}^* is not known a priori and needs to be estimated online. Its value depends upon the level of communication quality as well as the desired coverage. It is time varying due to both the dynamic environment and the motion of the agents. Accordingly, a real time optimization algorithm will be developed in Sect. 4 to estimate the optimal separation distance r_{ij}^*. The optimal locations are specified by a performance index which will be designed in Sect. 3 based on a decode and forward cooperative communication. Once r_{ij}^* is being estimated, cooperative control law (3) moves the agents to the optimal r_{ij}^* in a way that is distributively and asymptotically.

3   Decode and Forward Cooperative Communication

In cooperative communication, a source cooperates with one or more relays to transmit data to a destination. Examples of one relay and two relays' models are shown in Figs. 1 and 2 respectively. The cooperating relays can be manipulated to create independently faded multipaths that can be utilized by the channel to achieve cooperative diversity [18]. Capacity of a communication channel, as initially motivated by Claude Shannon, is the maximum achievable data rate that can be transmitted over a channel with arbitrary tolerable small error probabilities that can be driven to zero using some coding techniques [2,5]. A communication system is said to be in outage if the source sends data in a spectral efficiency R (bits/s/Hz) that is larger than the channel capacity.

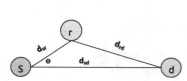

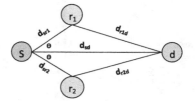

Fig. 1. Cooperative system of one source (s), one destination (d) and one relay (r) [1]

Fig. 2. Cooperative system of one source (s), one destination (d) and two relays (r_1,r_2) [1]

For direct transmission, the outage probability of Rayleigh faded and Additive White Gaussian Noise (AWGN) channel can be calculated as [2]:

$$P_{DT}^{out} = P(C < R) = 1 - \exp\left(-\frac{2^R - 1}{SNR}\left(\frac{d_{sd}}{d_0}\right)^\nu\right),   \tag{7}$$

where C is the channel capacity, SNR is the signal to noise ratio, ν is the path loss exponent, d_{sd} and d_0 are the source-destination and relative distances, respectively. Focusing on d_{sd}, it is clear that as d_{sd} increases, outage probability increases and vice versa. For m-relay decode and forward cooperative communication, the outage probability can be calculated by Eq. (10a) given in [1] which is omitted here to save space.

We are concentrating on the optimal locations and number of the relaying nodes to achieve the best communication coverage for a predefined level of communication quality measured by the outage probability. To do this, we need to understand the outage behaviour of a decode and forward cooperative communication described by Eqs. (7) and (10a) in [1]. Matlab simulations were carried out for different scenarios. Figures 3 and 4 show the outage probability versus source-relay distance, d_{sr} and angle, θ for one relay and two relays, respectively. Here, d_{sd} is fixed to be 600 m, $R = 1$ bit/s/Hertz, $\nu = 3$, $d_0 = 1$ m and $SNR = 100$ dB. It is clear that cooperating with two relays achieves lower P_{out} compared to one relay cooperation. In addition, the least P_{out} is achieved when $\theta = 0$, i.e. when the relay is on the line of sight between the source and destination. This is due to the shortest distances of d_{sr} and d_{rd} when $\theta = 0$. However, as θ increases, the decoding error at the relays increases making the channel to fall in an outage with higher probability. Furthermore, the relays tend to be closer to the source than to the destination which is due to the fact the relays need to decode the signal with minimum error before forwarding the signal to the destination.

In addition to the question of where to locate the relays, another question is regarding the effects of the number of the relays to the overall performance. Figures 5 and 6 show comparisons between direct transmission, one relay and two relays cooperation for $d_{sd} = 600$ and $d_{sd} = 1000$, respectively. For $d_{sd} = 600$, it

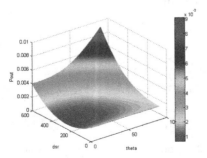

Fig. 3. Outage probability of one-relay cooperative system for $d_{sd} = 600$

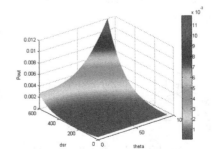

Fig. 4. Outage probability of two-relays cooperative system for $d_{sd} = 600$

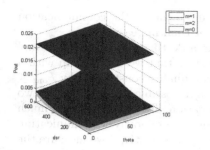

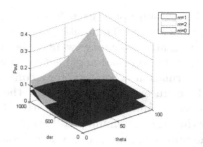

Fig. 5. Comparison of outage probability for none, one and two relays for $d_{sd} = 600$

Fig. 6. Comparison of outage probability for none, one and two relays for $d_{sd} = 1000$

is clear that there is a big improvement on the outage probability when using cooperation compared to the noncooperative communication. Furthermore, the two relay cooperation is better than the one relay cooperation for all d_{sr} and θ. However, as d_{sd} increases, the performance gaps between all the communication schemes start to decrease. This is because as d_{sd} becomes larger, the fading paths will become more similar to each other and increasing the number of the cooperative relays does not advantages to the system. Consequently, if the source is far away from the destination, decreasing the number of relays leads to a better performance.

Since the destination may also need to transmit data to the source, the best location of the relays to be at the middle between them. Therefore, in Figs. 7 and 8, we made the assumption that $d_{sr} = d_{rd}$ and then plotted the maximum d_{sd} that can be achieved for a predefined threshold on P_{out}. It is clear that the two relay cooperative communication achieves better d_{sd} than the one relay cooperative communication for low outage probabilities. In other words, the two relay cooperative communication achieves better communication quality and expands d_{sd}. As the outage probability threshold increases, the one relay cooperative communication and the noncooperative communication starts to become better than the two relay cooperative communication. This means

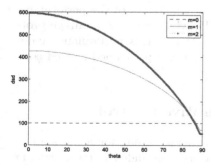

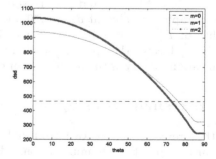

Fig. 7. Enclosed coverage area for $P_{out} \leq 0.0001$

Fig. 8. Enclosed coverage area for $P_{out} \leq 0.01$

that if a certain application has less strict requirement on the communication quality, decreasing the number of relays is better than increasing them.

Now, if any agent may want to be a source, a destination or a relay at any time, then the optimal angle is $\theta = 45$ which means that we will end up with a square formation. Therefore, for a communication quality constrained by $P_{out} \leq$ 0.0001, as an example, using Fig. 7 the maximum possible d_{sd} is 470 m of two relays decode and forward cooperative communication. Significantly, this is much larger than the maximum possible $d_{sd} = 100$ m of noncooperative communication for the same communication quality constraint. Consequently, if we measure the communication coverage as the enclosed area by the agents, then the coverage area of cooperative case is $A_{CC} = 334^2 = 111556 \, \mathrm{m}^2$ which is almost eleven times the area of the noncooperative case, $A_{DT} = 100^2 = 10000 \, \mathrm{m}^2$.

In a control objective, we need to have a performance index that take into account the previous analysis and is able to specify the optimal nodes' locations such that a predefined communication quality is achieved while expanding the coverage (i.e. the nodes' separation distances) to the maximum possible. To achieve this, the following performance index is proposed for a decode and forward cooperative communication:

$$
J\left(d_{sd}, \theta\right) = \left[1 - P_{out}^{DF}\right] \left[1 - exp\left(-\left(\frac{d_{sd}}{d_{sd}^{min}}\right)^{\upsilon}\right)\right] \cdot
$$

$$
\left[1 - exp\left(-\left(\frac{\theta}{\theta^{min}}\right)^{\upsilon}\right)\right] exp\left(-\left(\frac{d_{sd}}{d_{sd}^{max}}\right)^{\upsilon}\right) exp\left(-\left(\frac{\theta}{\theta^{max}}\right)^{\upsilon}\right), \quad (8)
$$

where the first term accounts for the outage probability given by equation (10a) in [1]. The second and third terms account for minimum source-destination distance, d_{sd} and angle, θ to ensure a minimum coverage area and for mobile agents safety. The last two terms are to ensure a predefined communication quality. This means that d_{sd}^{max} and θ^{max} are selected such that $P_{out}^{DF} \leqslant \zeta$, where ζ is a predefined required outage probability constraint. Note that υ is a tuning parameter to sharpen the curve of the performance index. Figure 9 shows a plot of the performance index (8) to achieve a communication quality of $P_{out}^{DF} \leqslant 0.0001$ of two cooperative relays. The parameters are selected as: $d_{sd}^{min} = 400$, $d_{sd}^{max} = 620$, $\theta^{min} = 40$, $\theta^{max} = 55$ and $\upsilon = 4$. Note that the maximum of the performance index graph is located at $d_{sd} = 470$ and $\theta = 45$, which gives an a outage probability of $P_{out}^{DF} < 0.0001$ as shown in Fig. 10. Hence, this performance index efficiently locates the optimal location which ensures a predefined desired communication quality with the maximum coverage possible.

4   Extremum Seeking and Cooperative Control

In the previous section we introduced a performance index for a decode and forward cooperative communication. This performance index is maximized at the optimal locations where all agents fairly utilize the resources in the system to achieve a predefined level of communication quality and coverage expansion.

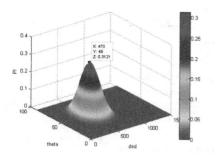

Fig. 9. Performance index of 2-relays system

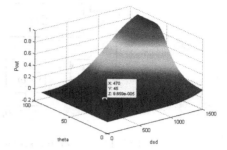

Fig. 10. Resulted outage probability

However, in reality we do not have the exact values of the outage probability which are changing due to the dynamic nature of the channel. Hence we cannot use classical optimal control approaches where a cost function needs to be known as a priori. Rather we need a real-time optimization process that can seek the optimal locations using local available information. Even though the outage probability is not known exactly in real time, however, it is measured or estimated based on the probability of a certain bit error rate which can be measured in the physical layer of the communication system. In this section, we extend the one variable extremum seeking control approach shown in [11] to two variables (i.e. d_{sd} and θ) extremum seeking control and derive the corresponding stability analysis. As shown in Fig. 9, the performance index, J in (8), is maximized at (d_{sd}^*, θ^*). Hence:

$$\frac{\partial J}{\partial \theta_1}(\theta_1^*, \theta_2^*) = 0, \ \frac{\partial J}{\partial \theta_2}(\theta_1^*, \theta_2^*) = 0,$$

$$\frac{\partial^2 J}{\partial \theta_1^2}(\theta_1^*, \theta_2^*)\frac{\partial^2 J}{\partial \theta_2^2}(\theta_1^*, \theta_2^*) - \left(\frac{\partial^2 J}{\partial \theta_1 \partial \theta_2}(\theta_1^*, \theta_2^*)\right)^2 > 0 \ \text{ and } \ \frac{\partial^2 J}{\partial \theta_1^2}(\theta_1^*, \theta_2^*) < 0. \quad (9)$$

In addition, cooperative control (3) is shown in Sect. 2 to be lyapunov stable and cooperatively asymptotically stable for any r_{ij}^* (i.e. $r_{ij}^* = d_{ij}^*$ if node j is a destination for source i, and $r_{ij}^* = d_{ij}^*/(2\cos\theta^*)$ if node j is a relay for source i). Consequently, cooperative control law (3) can be intuitively applied as an inner loop for the extremum seeking control algorithm of [11] as shown in Fig. 11. The dynamic equations of the overall system become:

$$\dot{x}_i = \mu \sum_{j \in N_i} p_{ij} \left[1 - \hat{r}_{ij}/r_{ij}\right](x_j - x_i) \quad (10)$$

$$\dot{\hat{d}}_{ij} = k_1 \xi_{ij}^1,$$

$$\dot{\hat{\theta}}_{ij} = k_2 \xi_{ij}^2,$$

$$\dot{\xi}_{ij}^1 = -\omega_{L_l} \xi_{ij}^1 + \omega_{L_l} \left[J(\hat{d}_{ij} + a\sin\omega_1 t, \hat{\theta}_{ij} + a\sin\omega_2 t) - \eta_{ij}\right] a\sin\omega_1 t,$$

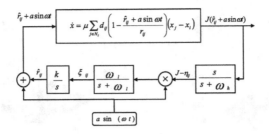

Fig. 11. Extremum seeking for cooperative control

$$\dot{\xi}_{ij}^2 = -\omega_{L_2}\,\xi_{ij}^2 + \omega_{L_2}\left[J(\hat{d}_{ij} + a\sin\omega_1 t, \hat{\theta}_{ij} + a\sin\omega_2 t) - \eta_{ij}\right]a\sin\omega_2 t,$$

$$\dot{\eta}_{ij} = -\omega_h\,\eta_{ij} + \omega_h\,J(\hat{d}_{ij} + a\sin\omega_1 t, \hat{\theta}_{ij} + a\sin\omega_2 t), \tag{11}$$

where we are using two perturbation frequencies, ω_1 for d_{ij} and ω_2 for θ_{ij}. In addition, ω_{L_l}, ω_{L_l} and ω_h are the frequencies of the low and high pass filters respectively. The constants a, k_1 and k_2 are the gains of the perturbation signal and integrators respectively. Let $\omega_1 = g\omega_2 = g\,\omega$ where g is an integer. Then, following the stability analysis in [11] and after lengthy calculations and several manipulations, the reduced averaged equilibrium of system (11) is:

$$
\begin{bmatrix}
\tilde{d}_{ij_r}^{a,e} \\
\tilde{\theta}_{ij_r}^{a,e} \\
\xi_{ij_{1r}}^{ae} \\
\zeta_{ij_{1r}}^{ae} \\
\tilde{\eta}_{ij_r}^{a,e}
\end{bmatrix}
=
\begin{bmatrix}
-\frac{\frac{1}{8}(m_3 m_4 - m_2 m_5) + \frac{1}{4}(m_3 m_6 - m_2 m_7)}{m_1 m_3 - m_2^2}a^2 + O(a^3) \\
-\frac{\frac{1}{8}(m_1 m_5 - m_2 m_4) + \frac{1}{4}(m_1 m_7 - m_2 m_6)}{m_1 m_3 - m_2^2}a^2 + O(a^3) \\
0 \\
0 \\
\frac{1}{4}(m_1 + m_3)a^2 + O(a^3)
\end{bmatrix},
\tag{12}
$$

where: $m_1 = \frac{\partial^2 \nu}{\partial x_1^2}, m_2 = \frac{\partial^2 \nu}{\partial x_1 \partial x_2}, m_3 = \frac{\partial^2 \nu}{\partial x_2^2}, m_4 = \frac{\partial^3 \nu}{\partial x_1^3}, m_5 = \frac{\partial^3 \nu}{\partial x_2^3},$
$m_6 = \frac{\partial^3 \nu}{\partial x_1 \partial x_2^2}, m_7 = \frac{\partial^3 \nu}{\partial x_1^2 \partial x_2}$, evaluated at $(0,0)$, and:

$$\nu(x_1, x_2) = \nu(\tilde{d}_{ij_r} + a\sin g\tau \ , \ \tilde{\theta}_{ij_r} + a\sin\tau)$$
$$= J(d_{ij}^* + \tilde{d}_{ij_1} + a\sin g\tau \ , \ \theta_{ij}^* + \tilde{\theta}_{ij} + a\sin\tau) - J(d_{ij}^*, \theta_{ij}^*). \tag{13}$$

Taking the Jacobian of (12), and using the Routh-Hurwitz criterion, it can be shown that the average equilibrium (12) is exponentially stable if:

$$\frac{1}{4}\,\omega_{L_1}'\,\omega_{L_2}'\left[\frac{\partial^2 \nu}{\partial x_1^2}(0,0)\frac{\partial^2 \nu}{\partial x_2^2}(0,0) - \left(\frac{\partial^2 \nu}{\partial x_1 \partial x_2}(0,0)\right)^2\right]a^4 + O(a^5) > 0, \tag{14}$$

and:

$$\frac{1}{2}\,\omega_{L_1}'^2\,\omega_{L_2}'^2(\omega_{L_1}' + \omega_{L_2}')\left(K_1'\frac{\partial^2 \nu}{\partial x_1^2}(0,0) + K_2'\frac{\partial^2 \nu}{\partial x_2^2}(0,0)\right) + O(a^2) < 0, \tag{15}$$

which are satisfied using assumption (9) in view of (13) and a is sufficiently small. Using Theorem 10.4 in [4] of the averaged systems, and following Theorem 4.1 in [11], the averaged system of (11) has a unique exponentially stable periodic solution $\left(\tilde{d}_{ij_r}(\tau), \tilde{\theta}_{ij_r}(\tau), \xi_{ij_{1r}}(\tau), \xi_{ij_{2r}}(\tau), \tilde{\eta}_{ij_r}(\tau)\right)$ of period 2π and this solution satisfies:

$$
\left\| \begin{bmatrix} \tilde{d}_{ij_r}^{2\pi}(\tau) + \frac{\frac{1}{8}(m_3 m_4 - m_2 m_5) + \frac{1}{4}(m_3 m_6 - m_2 m_7)}{m_1 m_3 - m_2^2} a^2 \\ \tilde{\theta}_{ij_r}^{2\pi}(\tau) + \frac{\frac{1}{8}(m_1 m_5 - m_2 m_4) + \frac{1}{4}(m_1 m_7 - m_2 m_6)}{m_1 m_3 - m_2^2} a^2 \\ \xi_{ij_{1r}}^{2\pi}(\tau) \\ \xi_{ij_{2r}}^{2\pi}(\tau) \\ \tilde{\eta}_{ij_r}^{2\pi}(\tau) - \frac{1}{4}(m_1 + m_3)a^2 \end{bmatrix} \right\| \le O(\delta) + O(a^3), \quad \forall \tau \ge 0.
$$

(16)

This means that all the solutions $\left(\tilde{d}_{ij_r}(\tau), \tilde{\theta}_{ij_r}(\tau), \xi_{ij_{1r}}(\tau), \xi_{ij_{2r}}(\tau), \tilde{\eta}_{ij_r}(\tau)\right)$ converge to an $O(\delta) + O(a^3)$ neighborhood of the origin. And if we choose $\delta = \frac{1}{\omega}$, then the extremum seeking error can be made arbitrarily small by selecting $\omega \gg 1$ and a is sufficiently small.

5 Simulation Results

Cooperative control system (10) with extremum seeking (11) was simulated first for four agents. Each agent is assumed to behave as a source, relay or destination at any time. The initial positions of the agents are: $x_1 = [0 \quad 0]^T$, $x_2 = [10 \quad -2]^T$, $x_3 = [6 \quad -1]^T$ and $x_4 = [-12 \quad 1]^T$. The performance index parameters are of the same values as of Fig. 9 and are assumed to ba same for all the channels. Moreover, the following constant parameters are selected: $\mu = 60$, $\omega_1 = 400$, $\omega_2 = 100$, $\omega_l = 4$, $\omega_h = 25$, $a = 0.05$, $k_1 = 80$ and $k_2 = 40$. The topology is assumed to be fixed for simplicity. The evolution of formation movement is shown in Fig. 12. Note that, compared to the noncooperative case[1], the connectivity is enhanced and the coverage is expanded largely. The estimated distances between the agents are plotted over time in Fig. 13. It is clear that the control reaches to the optimal separation distances effectively in less than 5 s.

Next the simulation was done for a group of seven agents. Each four agents are assumed to be as a cooperative communication subsystem as shown previously. Thus, one agent must serve for both of the subsystems to bridge them. We are doing this in order to make it easy for both design and implementation process when the system is further extended to a large number of agents. The system is simulated for the same parameters as before and the three new agents are initially located at: $x_5 = [10 \quad 20]^T$, $x_6 = [5 \quad -1]^T$ and $x_7 = [3 \quad -10]^T$. The simulation result is shown in Fig. 14. Unfortunately, the three new agents

---

[1] The result of the noncooperative case is not shown due to space limitation. However, the structure and evolution are similar to the cooperative case except the coverage is much more less and the connectivity is decreased.

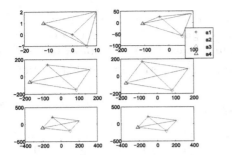

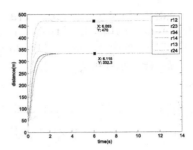

Fig. 12. Formation of the agents

Fig. 13. Estimated r_{ij}^*

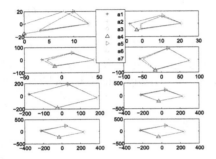

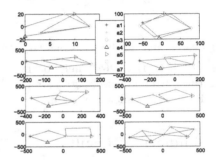

Fig. 14. Without virtual leader

Fig. 15. With virtual leader

coincide with the previous nodes. The reason behind this is that the added new agents are only linked with the bridging agent and are free from the other agents.

To solve this problem and to ensure that the navigation of the two subsystems are in a way that expands the coverage to the possible maximum, we introduced two virtual leaders located at the centers of mass of each subsystem. The dynamics of both of them are implemented in the bridging agent and designed to be:

$$\dot{x}_{v_1} = 0.5\mu \left(1 - \frac{r_{v_1 v_2}^*}{r_{v_1 v_2}}\right)(x_{v_2} - x_{v_1}), \quad \dot{x}_{v_2} = 0.5\mu \left(1 - \frac{r_{v_2 v_1}^*}{r_{v_2 v_1}}\right)(x_{v_1} - x_{v_2}), \quad (17)$$

where $r_{v_i v_j}^*$ is selected to be the same as r_{ij}^* for the bridging agent in each subsystem. Since the virtual leaders are located at the centers of each subsystem, the distance between any agent to the leader is selected to be half of the source-destination distance for any agent in any subsystem. Hence the new dynamics become:

$$\dot{x}_i = \mu \sum_{j \in N_i} d_{ij} \left[1 - \frac{\hat{r}_{ij} + a \sin \omega t}{r_{ij}}\right](x_j - x_i)$$

$$+ \mu d_{ik} \left[1 - \frac{0.5(\hat{r}_{ij} + a \sin \omega_1 t)}{r_{iv_k}}\right](x_{v_k} - x_i), \quad (18)$$

where $k = 1, 2$ representing the two virtual leaders. The remaining extremum seeking dynamics are same as before. The evolution of formation is shown in Fig. 15 and distance estimations is similar to Fig. 13 which is omitted here to save space. The results reveal the efficiency of the proposed control law.

6 Conclusion

In this paper, the cooperative formation control problem of mobile agents was solved such that the proposed control cooperatively and exponentially moves the agents to the optimal locations. The effects of using cooperative communication on the communication coverage and quality were studied and the results showed promised achievements to the communication coverage and quality of the sensors network.

Possible future work is to upgrade the methodology to tackle variable topologies instead of fixed topologies. Furthermore, other cooperative communication techniques may provides better performances in term of coverage and quality and hence they need to be studied to better solve the problem.

References

1. Beaulieu, N.C., Hu, A.: A closed-form expression for the outage probability of decode-and-forward relaying in dissimilar Rayleigh fading channels. IEEE Commun. Lett. **10**, 81–815 (2006)
2. Tse, D., Viswanath, P.: Fundamentals of Wireless Communication. Cambride University Press, Cambridge (2005)
3. Qu, Z.: Cooperative Control of Dynamical Systems. Springer, London (2009)
4. Khalil, H.K.: Nonlinear Systems. Prentice Hall, Upper Saddle River (2002)
5. Goldsmith, A.: Wireless Communications. Cambride University Press, Cambridge (2005)
6. Murray., R.M.: Recent research in cooperative control of multi-vehicle systems. ASME J. Dyn. Syst. Meas. Contrl **129**, 571–583 (2007)
7. Fax, J.A., Murray, R.M.: Information flow and cooperative control of vehicle formations. IEEE Trans. Autom. Control **49**, 571–583 (2004)
8. Zavlanos, M., Egerstedt, M., Pappas, G.: Graph theoretic connectivity control of mobile networks. Proc. IEEE **99**, 1525–1540 (2010)
9. Qu, Z., Li, C., Lewis, F.: Cooperative control based on distributed connectivity estimation of directed networks. In: American Control Conference, San Francisco, USA, pp. 3441–3446 (2011)
10. Li, C., Qu, Z., Ingram, M. A.: Distributed extremum seeking and cooperative control for mobile communication. In: 50th IEEE Conference on Decision and Control and European Control Conference (CDC-ECC), Orlando, USA, pp. 4510–4515 (2011)
11. Krstic, M., Wang, H.: Stability of extremum seeking feedback for general nonlinear dynamic systems. Automatica **36**, 595–601 (2000)
12. Nosratinia, A., Hunter, T.E., Hedayat, A.: Cooperative communication in wireless networks. IEEE Commun. Mag. **42**, 74–80 (2004)

13. Scaglione, A., Goeckel, D.L., Laneman, J.N.: Cooperative communications in mobile ad hoc networks. IEEE Signal Process. Mag. **23**, 18–29 (2006)
14. Coso, A., Savazzi, S., Spagnolini, U., Ibras, C.: Virtual MIMO channels in cooperative multi-hop wireless sensor networks. In: 40th Annual Conference on Information Science and Systems, pp. 75–80 (2006)
15. Gurrala, K.K., Das, S.: Impact of relay location on the performance of multi-relay cooperative communication. Int. J. Comput. Netw. Wirel. Commun. **2**, 2250–3501 (2012)
16. Cho, W., Oh, H.S., Kwak, D.Y.: Effect of relay locations in cooperative networks. In: 1st International Conference on Wireless Communication, Vehicular Technology, Information Theory, Aerospace and Electronic Systems Technology, pp. 737–741 (2009)
17. Sadek, A.K., Han, K., Liu, J.: Distributed Relay assignment protocols for coverage expansion in cooperative wireless networks. IEEE Wirel. Trans. Mob. Comput. **9**, 505–515 (2010)
18. Laneman, J.N., Tse, D.N.C., Wornell, W.: Cooperative diversity in wireless networks: efficient protocols and outage behavior. IEEE Trans. Inf. Theory **50**, 3062–3080 (2004)

Estimation of Contact Forces in a Backdrivable Linkage for Cognitive Robot Research

Richard Thomas and William Harwin[⊠]

Cybernetics Research Group, School of Systems Engineering,
University of Reading, Berkshire RG6 6AY, UK
rthomas101@hotmail.co.uk, W.S.Harwin@reading.ac.uk

Abstract. Most robot manipulators are poor at measuring and controlling forces, especially force of contact. At the innermost control loops these force estimates are needed to control compliance. At a more general level identifying when contact occurs, and the forces involved, are elemental requirements in building cognitive information that allow robots to adapt to an unstructured environment.

This paper shows that reliable internal force sensing can be achieved by using information that is already available in the control loop, and using methods based on system identification to estimate the force at the joint level, and hence by implication the end point force. The method relies on the concept of backdrivable linkage transmissions, a linkage design method that is well established in the field of haptic interface design. It is argued that these internal models should be more widely exploited in robotics and are no different to the internal models built by animals and humans that enable us to adapt rapidly in an unstructured world.

1 Introduction

It is well known that animals build internal models to compare sensory information to a belief of what should happen. Wider use of internal models should be the norm in robotics, extending beyond just the use of mathematical models to support the control law derivation. A particular problem in robotics is to ensure that forces of contact are built into the cognition and intelligence. Until this is done modern robots will be restricted to jerky movements or tasks that do not require contact such as waving. One example of the poor abilities robots have to adapt to their environment can be observed by considering the process of writing. The clockwork technology of Pierre Jaquet-Droz allowed one of his automata, the scribe, to write any sequence of letters, all be it at a slow and jerky pace. When observing most modern robots do the same tasks the same non-smooth movement can be readily observed.

Although a measure of force is not required for variable compliant joints (for example the dual nonlinear springs used by Laurin et al. [1]), most research interest is in producing compliant robots with a means of measuring force. The usual mechanism for compliant control is to control compliance at the joint

A. Natraj et al. (Eds.): TAROS 2013, LNAI 8069, pp. 235–246, 2014.
DOI: 10.1007/978-3-662-43645-5_24, © Springer-Verlag Berlin Heidelberg 2014

level and then relate this to end point compliance. It is relatively easy to show that given the joint admittance matrix A_{joint} the end point admittance will be $A_{end} = J A_{joint} J^T$. Equally the equation can be expressed as $A_{joint}^{-1} = J^T A_{joint}^{-1} J$ where J is the Jacobian of the linkage [2].

These equations assume that the linkage is backdrivable, which in this paper will be assumed to mean that firstly the force Jacobian equation $\tau = J^T F$ can be applied to the linkage where τ is the vector of joint forces/torques and F is the vector of end point forces/torques. Additionally it is assume that the transmission driving the joint has an efficiency close to 1 so that the joint torque/force can be related to the actuator torque/force via the transmission ratio. The above arguments also assume that the linkage dynamics do not have a significant effect on the forces and torques (quasi-static) however the methods described later can also be used to monitor, and react to, forces due to the acceleration and deceleration of the linkages.

Quality control of endpoint forces and torques has been well researched in respect of haptics interfaces and the control concepts presented there are equally applicable in robotics see [3,4]. Likewise estimating external forces from knowledge of the robot dynamics coupled with measurements of joint torques has also been demonstrated in simulation [5–7], however there is little attempt to link this work to an internal model outside those used for low level control, neither is there any experimental demonstration on backdrivable hardware.

Techniques of system identification are more apposite to this work since it allows the internal model to adapt. The method is general but is widely used in signal processing to adapt channel filters, in control engineering to identify complex plant dynamics, as well as a mechanism for fault detection [8–10]. A challenge in applying system identification techniques in robotics is the potentially dangerous results of applying random inputs into a robot so some care is needed.

2   Theory

This paper will consider initially each single degree of freedom joint and relate this to Cartesian space via the appropriate Jacobian. This assumption simplifies the mathematics and is appropriate for slow movements (where so called Coriolis forces are unlikely to be significant). Methods to compute the equations for the full dynamics of linkages are often expressed in the form

$$\tau = M(\dot{q}, q)\ddot{q} + B(\dot{q}, q) + G(q) \tag{1}$$

where τ are the joint torques and q and its derivatives are the joint states e.g. [11,12] however these methods result in complex equations where it is difficult to determine the relative importance of each sub term, and a calibration stage is needed to minimise modelling errors. A better insight can be achieved by considering joint level methods, and assuming quasi-static dynamics.

The block diagram representation for a single joint is given in Fig. 1. The blocks inside the dotted box are real physical components. The controller C

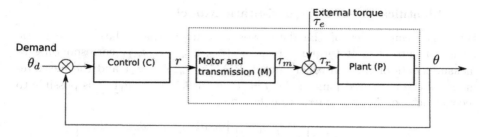

Fig. 1. Block diagram of a single joint. The components within the dotted box represent the physical components. The values within the controller (C) along with the command can be used to set the joint compliance

can be of any form and is assumed to be stable, but in most cases it will be a PID controller which then allows tuning of any standard methods for example Ziegler–Nichols (see most standard control texts e.g. /citeOgata2002modern). In the experiments that follow Zieger-Nichols was used to determine a conservative single set of PID values. Sets of PID values can then be found to allow differing levels of joint stiffness or compliance. For this study the tuning was necessarily conservative since for proximal joints are affected by the configuration of distal joints. M and P are the Motor along with the transmission, and the plant respectively.

From Fig. 1, it can be seen that by considering the combined torques at the joint that

$$\tau_r = \tau_e + \tau_m \tag{2}$$

where τ_e is external torque (that is the torque arising from the external forces operating through the torque Jacobian plus any forces due to the linkage dynamics), and τ_m is the torque from the joint actuator, usually assumed to be an electric motor and transmission.

$$\tau_m = MC(\theta_d - \theta) = Mr$$

where θ_d is demand angle (command) and the measured angle is θ.

If the transfer function of the plant P (i.e. the robot distal to the joint) is invertible then

$$\tau_r = P^{-1}\theta_c \tag{3}$$

By substituting Eqs. (2) and (3) into Eq. (1), an equation for τ_e is found as

$$\tau_e = (P^{-1} + MC)\theta - MC\theta_d = P^{-1}\theta - Mr \tag{4}$$

In considering Eq. (4) it can be observed that the output of the control block (r) and the joint angle (θ) are both available in the controller.

Equation (4) can be expressed as a vector multiplication

$$\tau_e = \begin{bmatrix} P^{-1} & -M \end{bmatrix} \begin{bmatrix} \theta_c \\ r \end{bmatrix} \tag{5}$$

2.1   Identification of the Quasi-Static Model

By considering a set of the data over n samples that relates some static measurements of joint torque data $T = \begin{bmatrix} \tau_{e,n} & \tau_{e,n-1} \cdots \tau_{e,1} \end{bmatrix}$ corresponding to measurements of the controller output $R = \begin{bmatrix} r_n & r_{n-1} \cdots r_1 \end{bmatrix}$ and the measurements of the measured joint angle $Q = \begin{bmatrix} \theta_n & \theta_{n-1} & \theta_{n-2} \cdots & \theta_1 \end{bmatrix}$, it is possible to reconstruct Eq. (5) as a matrix calculation.

$$\begin{bmatrix} \tau_{e,n} & \tau_{e,n-1} \cdots \tau_{e,1} \end{bmatrix} = \begin{bmatrix} P^{-1} & M \end{bmatrix} \begin{bmatrix} \theta_n & \theta_{n-1} & \theta_{n-2} & \theta_{n-3} \cdots \cdots \\ r_n & r_{n-1} & r_{n-2} & r_{n-3} \cdots \cdots \end{bmatrix} \tag{6}$$

or

$$T = \begin{bmatrix} P^{-1} & -M \end{bmatrix} \begin{bmatrix} Q \\ R \end{bmatrix} \tag{7}$$

This can be solved by a number of methods such as the pseudoinverse or singular value decomposition to fit a model of the Plant and Motor drives that can be used subsequently to reconstruct external torque from measured data from the controller and the joint angle displacement. Thus the fixed parameters of the inverse Plant and Motor can be computed as the pseudoinverse

$$\begin{bmatrix} P^{-1} & -M \end{bmatrix} = T \begin{bmatrix} Q^T & R^T \end{bmatrix} \left(\begin{bmatrix} Q \\ R \end{bmatrix} \begin{bmatrix} Q^T & R^T \end{bmatrix} \right)^{-1} \tag{8}$$

2.2   Identification of the Dynamic Model

System dynamics can be introduced into the models of the inverse plant and motor by using the delayed values of the input and output in a feedback loop. Models of this nature are often termed ARMAX (auto regressive moving average exogenous), and recursive methods such as those described by Soderstrom and Stoica [8] can be used. However for this work a direct approach is used where a single model is computed and used for subsequent analysis. The formulation of the problem is similar to that used in the quasi-static method above, although the matrix of input data is supplemented with the delayed version of the input (providing a reduced ARMAX model known as a FIR (finite impulse response) model. This approach allows the torque estimate to be determined without recourse to delayed estimates of the torque output.

The equation for a FIR model with two delays in each of the two inputs (r and θ) is formed as

$$\begin{bmatrix} \tau_{e,n} & \tau_{e,n-1} \cdots \tau_{e,1} \end{bmatrix} = \begin{bmatrix} p_1^{-1} & p_2^{-1} & p_3^{-1} & -m_1 & -m_2 & -m_3 \end{bmatrix} \begin{bmatrix} \theta_n & \theta_{n-1} & \theta_{n-2} & \theta_{n-3} \cdots \cdots \\ 0 & \theta_n & \theta_{n-1} & \theta_{n-2} \cdots \cdots \\ 0 & 0 & \theta_n & \theta_{n-1} \cdots \cdots \\ r_n & r_{n-1} & r_{n-2} & r_{n-3} \cdots \cdots \\ 0 & r_n & r_{n-1} & r_{n-2} \cdots \cdots \\ 0 & 0 & r_n & r_{n-1} \cdots \cdots \end{bmatrix} \tag{9}$$

For this configuration three plant model gain terms $(p_1^{-1} \ p_2^{-1} \ p_3^{-1})$ and motor/transmission terms $(-m_1 - m_2 - m_3)$ can be estimated. The pseudoinverse is again used to estimate these terms.

3   Setup and Experiments

The arm used for these experiments is a Phantom 1.5 Haptic arm, with three degrees of freedom although the two end joints are partially coupled together. The arm is commonly used as a haptic interface and the kinematics of the Phantom series of haptic device are described in Çavuşoğlu et al. [13]. The Phantom 1.5 is unmodified, however the encoders were interfaced directly to the quadrature inputs of a Quanser Q4 data acquisition board, and the Phantom 'lunchbox' was modified to take outputs directly from the analogue outputs of the same board. The board is mounted in an old PC computer running Matlab XPC-target, which allows rapid prototyping of controllers, flexible data acquisition for off line analysis. The Quanser board and XPC-Target operate at a loop time of 1 ms and manage the DoA converters, quadrature encoders, data logging. The Phantom 1.5, like many haptic devices, are designed to be backdrivable allowing the assumption to be made that there is a loss-less relationship between the joint torques and the motor torques.

This arrangement has the added advantage that the XPC-Target machine can run independently of the host PC, and also display real-time data on its designated screen. Logging of data occurs in real-time on the XPC-Target, but is only recovered once the control loop is stopped where upon it can be downloaded to the PC, saved and analysed at a later time.

Experiments are run by attaching a force gauge (RSFG-5000 from RS components) to a variable stack of elastic bands to either the distal or proximal link of the Phantom 1.5. Forces are applied approximately perpendicular to the link so that they equate directly to joint torques. This arrangement is shown in Fig. 2.

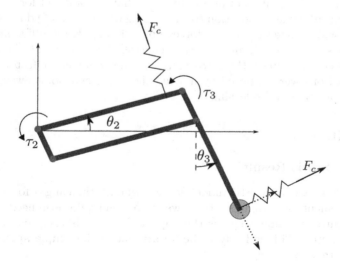

Fig. 2. Experimental setup of Phantom 1.5. Forces F_c are applied individually to a point near the end of the distal or proximal links as shown, via a spring-like connection, in this case a stack of elastic bands.

3.1   Quasi-Static Experiments

A static external force was applied to induce a torque on a joint by joint basis. The force was applied perpendicular to joint axis and the associated link so that the torque could be considered as directly proportional, thus force and torque are considered interchangeable via a per axis constant.

The force gauge was used to apply 0, 200, 400, 600 and 800 g of force to each joint. θ_c and C for the corresponding external force were saved into Matlab variables at the end of the simulation. From these variables, the model parameters are estimated using Eq. (8) and the resulting models are tested against the actual forces using the reconstruction Eq. (5). This process was applied to all three joints independently. This was monitored real-time with a scope displayed on the xPC-Target screen. The force applied via force gauge was compared visually to the recovered force.

After this proved successful, the PID controller values for the arm were altered so that the stiffness was altered and then the with values for P^{-1} and M for each joint, forces could be recovered when applied to the arm. Each joint was tested with a variety of forces, applied to the same areas as previous testing. The PID control values for the arm were changed, to make the arm stiffer. The arm was then re-tested to evaluate the force prediction capabilities of the $[P^{-1} \quad M]$ model in relation to the stiffness's of the arm i.e. was a correct model for the Plant and Motor's independent of the control system.

3.2   Dynamics Experiments

The force gauge was modified so that it could be attached to the ADC input of the xPC-Target via a differential amplifier so that the applied forces could be logged during run-time, in addition to the logging of the R and Q data sets. Each joint had a wide range of external forces applied, with three different stiffness configurations for each. Stiffnesses were controlled via a set of elastic bands that were used to connect the force gauge to the robot end-effector and step like perturbations were imposed by moving the robot end point away from the equilibrium position and releasing.

4   Results

4.1   Quasi-Static Results

Although the quasi-static FIR model worked when estimating a force from the joint displacement measurements there was a constant gain term needed to bring the results into the same range as the applied force as differing controller gain terms were tested. This is likely to be an artifact of the simple approach used during this stage of testing.

Figure 3 shows the recovered force from joint 3 in a stiff configuration using the stable PM model. Using the force gauge, forces were applied in 100 increments and held for a few seconds. As can be seen in Fig. 3, and Table 1 the

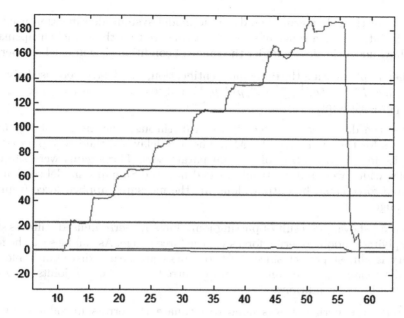

Fig. 3. Results for Joint 3 in stiff configuration force recovery, given 100 g intervals force input (red). Gridlines at multiples of 22.85 g force. Joints 1 and 2 estimates are also shown (green and blue) (color figure online)

Table 1. Results of quasi-static tests showing constant scaling error

Applied force (g)	Estimated force (g)
100	21
200	41
300	64
400	90
500	114
600	134
700	164
800	186

recovered force is showing 21 g, 41 g, 64 g, etc rather than the applied 100 g, 200 g, 300 g. This represents a constant scaling value of about 4.5 between measured and estimated forces, and plots of this recovered force shows that the force prediction is linear over the range of the joint, up to the point where the motor saturates.

4.2 Dynamic Model

The system identification methods used can only identify behaviours that are logged by the system, thus the collected data needs to be representative of the dynamics of the robot arm. The dynamic model system identification used the dynamic data across a range of stiffness's to ensure that the model encapsulated the behaviour of the device under the range of conditions it will need to operate.

When implementing the dynamic identification, five delays were used giving a total of five $P^{-1} = (p_1^{-1} \; p_2^{-1} \; p_3^{-1} \; p_4^{-1} \; p_5^{-1})$ and five $M = -(m_1 \; m_2 \; m_3 \; m_4 \; m_5)$ values per each joint.

After the data was collected, it was run through the inverse of Eq. (6) to get values for the model ($[P^{-1} \quad M]$). The five delays on the two inputs (r and θ) results in the estimation of 12 gain parameters. These gains were saved as variables which were automatically loaded into the Simulink model so that the estimated force could be output alongside the measured applied force from the force gauge.

Figure 4 shows the result of putting joint 3 in a hitherto unused stiffness state and applying random external forces via the force gauge. As can be seen the force recovery is almost perfect since the two signals are near indistinguishable, bar small discrepancies. The same results occurred for the other joints, in many different stiffness conditions.

A further experiment was done to evaluate the errors introduced by the linkage and the Jacobian when applying forces in a Cartesian frame. The force gauge was used to apply forces to the end effector of the arm via office elastic bands with the joints set to specific stiffness values. This allows for a compliant interaction and hence using the model above, together with the force Jacobian,

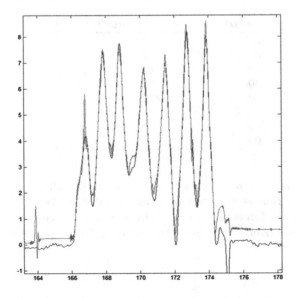

Fig. 4. Joint 3 predicted force (green) and measured external applied force (blue) (color figure online)

allows for an estimation of the tip forces. These Cartesian forces were summed using Pythagorean Theorem and plotted as a single force.

Figure 5 shows the outcome of the tip resultant force. This result relies the precomputed Jacobian and uses the combination of torque estimates from all three joints. The blue line represents the force gauge, the green the recovered Cartesian force. To reiterate, although the results here appear to show a single degree of freedom, it is in Cartesian space and relies on torque estimations from all three joints. Although the lines do not always match, the general shape and response is accurate. The recovered force with no external force, does show a small amount of force being applied. This makes sense as to be the effect of keeping the arm in a held state against gravity's pull.

5   Discussion

Although it is not immediately apparent from the block diagram, because we are considering the plant P in its inverse form, the system identification aspects of this work treats r and θ as inputs to the model, although from a control aspect r is the input and θ is the output. The models discussed are limited to considering only these input values and hence classed as finite impulse response (FIR) models. Future work could look at using the previous values of the estimated force/torque output, (for example ARMAX models), however the predictive capabilities of the FIR class is good.

The quasi-static model is evidently insufficient for accurate estimate of force estimation although it is clear that it is able to make a prediction of the level of force. The prediction is scaled by a constant factor of 4.5. It is not obvious why

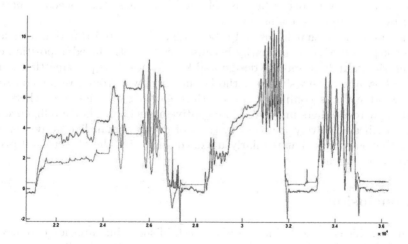

Fig. 5. Recovered Cartesian force (green) and external applied force (blue) (color figure online)

this scaling factor is not closer to 1. However given that the scaling is constant it can be seen that the method does provide a good estimate of externally applied forces via the estimate of joint torques. In practice this constant scaling should be simply incorporated into the inverse equation (Eq. 7/8).

The technique for joint level dynamic system identification proved more robust and effective at modelling the plant and motor. Using the standard joint Jacobian for the arm to convert the joint forces into Cartesian forces prove less effective, although the resultant magnitude was similar. This suggests that the Jacobian used may have been a poor model of the true relationship between endpoint force and joint torque. Calibration of the Jacobian is one way forward but are possibly over complex and their iterative nature does not guarantee an improvement [14]. Since the purpose of this work is to explore force control in the context of hand-eye coordination in robot cognition, estimating the endpoint forces in a Cartesian coordinate frame may not be appropriate since it is evident that humans do not rely on orthogonal, or well defined axes. It is possible that extension algorithms such as recursion, and least squares with forgetting factors, might be used to simulate reliable force control that can adapt to situations such as changing dynamics, for example as objects are grasped and released thereby adding or reducing the effective inertia of the final link. Although this technique assumes that joint positions are measured via quadrature encoders, it is equally applicable to other methods of estimating the joint state variable such as accelerometer arrays [15].

The method outlined should scale up to larger robots and haptic devices where the control structure relies on the concept of a strong link between joint torque and endpoint force (via the Jacobian). An immediate concern would be that of safety, since this is a learning method so it is likely that some method will be needed to constrain movements during the learning and evaluation stages. Problems are likely to occur because of the larger mass that needs to be supported by the actuators against gravity.

It is unfortunate that industrial robotics has dominated thinking for such a significant period of time, primarily because this has introduced a 'position control' centric attitude to linkage design and kinematics. It is pleasing that haptic devices allow a relatively simple method of moving away from position control to enable good stiffness/compliant control thus allowing robotics researchers, and in particular researchers interested in cognitive aspects of robot intelligence and control, with a relatively low cost method of doing human like activities with robots, that is robots that regularly make contact with the objects and people in their environment.

6   Conclusion

System identification techniques are well established, but are not common on mechanical devices. This work demonstrates a method that relies on so called 'backdrivable linkages' where the end point contact forces reflecting through the linkage to the joints and hence via the transmission to the closed look control of the actuator. Small haptic devices make ideal testbed systems for this approach since most haptic interfaces rely on the backdrivable concept. Problems associated with applying these ideas to larger linkages are discussed. The method described makes better use of the linkage since only the output of the controller and the joint sensor(s) are needed to estimate these forces. Contact force is often the 'elephant in the room' in robotics research and is particularly important to cognitive robotics where force must be an integrated part of the task and predictive models are an important area for research. Extending these methods to recursive and forgetting methods of system identification are the next obvious step, however parameter constraints will be an important part to ensure device (and person) safety. The approach of using haptic devices as commercial off the shelf hardware (COTS) is appropriate to cognitive robotics where most robots have poor force control and are expensive and rare.

Acknowledgments. This work was conducted as a placement in the University of Reading Summer Research opportunities scheme (UROP). The authors are grateful for the chance to explore 'blue skys' research topics that are provided and encouraged by these seed corn grants.

References

1. Laurin-Kovitz, K., Colgate, J., Carnes, S.: Design of components for programmable passive impedance. In: Proceedings of the 1991 IEEE International Conference on Robotics and Automation, 1991, pp. 1476–1481. IEEE (2002)
2. Salisbury, J.: Active stiffness control of a manipulator in cartesian coordinates. In: 19th IEEE Conference on Decision and Control including the Symposium on Adaptive Processes, 1980, vol. 19, pp. 95–100. IEEE (1980)
3. Carignan, C., Akin, D.: Using robots for astronaut training. Control Systems Magazine, vol. 23. IEEE (2003)
4. Colgate, J., Schenkel, G.: Passivity of a class of sampled-data systems: application to haptic interfaces. J. robot. syst. **14**(1), 37–47 (1997)
5. Van Damme, M., Beyl, P., Vanderborght, B., Grosu, V., Van Ham, R., Vanderniepen, I., Matthys, A., Lefeber, D.: Estimating robot end-effector force from noisy actuator torque measurements. In: 2011 IEEE International Conference on Robotics and Automation (ICRA), pp. 1108–1113. IEEE (2011)
6. De Luca, A., Mattone, R.: Sensorless robot collision detection and hybrid force/motion control. In: Proceedings of the 2005 IEEE International Conference on Robotics and Automation, 2005, ICRA 2005, pp. 999–1004. IEEE (2005)
7. Hacksel, P., Salcudean, S.: Estimation of environment forces and rigid-body velocities using observers. In: Proceedings of the 1994 IEEE International Conference on Robotics and Automation, 1994, pp. 931–936. IEEE (1994)
8. Söderström, T., Stoica, P.: System Identification. Prentice-Hall Inc, Englewood Cliffs (1988)
9. Ljung, L., et al.: Theory and Practice of Recursive Identification. The MIT Press, Cambridge (1983)
10. Ljung, L.: System Identification. Wiley Online Library, New York (1999)
11. Craig, J.J.: Introduction to Robotics: Mechanics and Control. Addison-Wesley, Reading (1989). ISBN 0-201-09528-9. UR call 629.892-CRA

12. Spong, M.W., Hutchinson, S., Vidyasagar, M.: Robot Modeling and Control. Wiley, New York (2006)
13. Çavuşoğlu, M.C., Feygin, D., Tendick, F.: A critical study of the mechanical and electrical properties of the phantom haptic interface and improvements for high performance control. Presence: Teleoperators Virtual Environ. **11**(6), 555–568 (2002)
14. Ouerfelli, M., Kumar, V., Harwin, W.: Kinematic modeling of head-neck movements. IEEE Trans. Syst. Man Cybern. Part A: Syst. Hum. **29**(6), 604–615 (1999)
15. Madgwick, S.O.H., Harrison, A.J.L., Sharkey, P.M., Vaidyanathan, R., Harwin, W.S.: Measuring motion with kinematically redundant accelerometer arrays: theory, simulation and implementation. Mechatronics **23**, 518–529 (2013)

A Simple Drive Load-Balancing Technique for Multi-wheeled Planetary Rovers

James C. Finnis[✉] and Mark Neal

Computer Science,
Aberystwyth University, Penglais, Aberystwyth SY23 3DB, UK
{jcf1,mjn}@aber.ac.uk

Abstract. A simple method for balancing the motor driver load across a six-wheeled rover is presented. This method uses the concept of inflammation to model the load on each motor driver by its temperature, decreasing the load as the local temperature increases. The method is compared with both the base case, where all motors run at a fixed load; and a relatively unintelligent method involving shutting off all motors while the temperature is above a given level. We show that load balancing in this manner has the beneficial effect of avoiding overheating in individual motors, while neither overly decreasing the traversed distance nor increasing the energy used.

Keywords: Robot · Rover · Temperature · PID · Load balancing · Inflammation

1 Introduction

Autonomous robots need to be able to function for extremely long periods of time without any operator intervention. This applies particularly to planetary rovers, which cannot be repaired or maintained after their initial deployment.

Faults can develop over time which can cause unforeseen variations across duplicate components. For example, the drive motors of multiwheeled rovers should ideally respond in an identical manner, but variations will occur — from minor changes in response to outright failure. This is compounded by the inevitable slight variations in manufacture between the components.

An example of such a failure occurred on the Mars Exploration Rover Spirit in 2004, when the right front drive motor began to draw approximately twice as much current as the other wheels due to a problem with the distribution of lubricant [1]. This problem persisted intermittently and slowly worsened, despite attempts to exacerbate the load by dragging the wheel, until it finally failed in March 2006 [2].

This case demonstrates one key area in which variations can occur: motor load. A typical planetary rover has six wheels, each of which may develop faults which could cause the load at a given speed to change — lubrication problems, gear wear and so on. In addition, each wheel is positioned at different heights on

A. Natraj et al. (Eds.): TAROS 2013, LNAI 8069, pp. 247–258, 2014.
DOI: 10.1007/978-3-662-43645-5_25, © Springer-Verlag Berlin Heidelberg 2014

a different patch of terrain with different properties. This can have a very large effect on the friction and slip experienced by each wheel, and therefore the load on that wheel's motor and its driver.

Such variations can cause one motor to become overloaded, which can eventually cause damage due to thermal and mechanical stress on the components. This can be exacerbated by the harsh environments in which planetary rovers must function: a lunar environment has no atmosphere to provide convective cooling, and no "cleaning events" in the form of fortuitous winds to clear away dust particles, which block radiative cooling. This led to the loss of the early Lunokhod-2 lunar rover [3]. The low-pressure Martian environment is little better for convective cooling.

The method described in this paper is designed to limit overloading individual motor drivers by using an analogue to a biological response to damage: inflammation. As part of the inflammatory response, substances called "cytokines" are released by damaged cells. These are small, short-lived proteins released by cells as signalling agents which are used for many purposes throughout the body, typically acting on nearby cells (i.e. in a *paracrine* manner). They are complex to classify and study: one task may be performed by several cytokines (redundancy), and one cytokine can have many unrelated functions (pleiotropy).

One fairly well-established effect, however, is the role of certain cytokines in the body's response to damage [5]. Cytokines are released by damaged cells, causing a complex series of changes leading to inflammation — the most important, perhaps, being the migration of white blood cells out of the bloodstream and into the injured area, where they can deal with potential infection and remove the damaged cells.

Another cytokinetic inflammatory effect directly changes the behaviour of the animal. Cytokines bind to pain receptors, sensitizing them so that normally painful stimuli become more painful (hyperalgesia) and those not normally experienced as painful evoke pain (allodynia.) This offers the organism a clear advantage: it will avoid exercising the damaged part of the body [7].

Although cytokine receptors such as the primary thermal nociceptor TRPV1 have sigmoid activation functions [4, 9, 11], we are currently using a ramp activation function. This is justified because we are simply using the temperature as a convenient analogue to a cytokine-like quantity, not as temperature *per se*. If we did not have temperature sensors, we could instead use the current to increment an exponentially decaying value in software, to the same effect. Thus we do not use temperature as the input to the receptor, but use it to change the threshold of the receptor so that more non-noxious stimuli become painful.

Our rover has low-resolution temperature sensors attached to the motor driver chips, which respond in an analogous way to cytokine release in the inflammatory response: as the motor is overloaded and "damage" increases, the temperature slowly increases, in the same way that cytokine concentrations would increase in a biological organism. Similarly, if the load is removed or decreased, the temperature falls; just as in biology the cytokine concentrations fall as the damage is repaired.

This similarity occurs because both processes can be modelled by a recurrence relation of the form

$$x_{n+1} = ax_n + b \tag{1}$$

In both cases, a quantity is being added to a value which is subject to an exponential decay [10]. We can therefore say that we are using the temperature of the driver as a physical analogue of the concept of a cytokine-like quantity: an exponentially decaying value which increases with load.

It might be thought more useful and direct to use the temperature directly, perhaps passing it through a sigmoid activation function such as would exist in a biological nociceptor (neuron sensitive to painful stimuli) [4]. However, the temperatures we are monitoring are non-noxious, well below any putative "pain threshold" for the rover. Again, we are simply using the temperature as a convenient analogue of a cytokine.

In our system, this temperature is used to sensitize a notional receptor, to the extent that allodynic pain might occur. If this sensitization is linear, we can argue that the *allodynic* pain response of nociceptors to the temperature is also linear, because the proportion of normally non-noxious stimuli which now cause pain will rise linearly. To model this, the temperature of each driver is passed through a simple linear activation function to produce the amount of allodynic pain.

With the pain response increased at the driver, the system will attempt to decrease the pain by decreasing the load. In our system, a short series of experiments showed that simply decreasing the required speed of a given motor slightly relative to others reduced the load on the driver enough to allow the temperature to start to fall.

2 Hardware

The rover used in the experiments is shown in Fig. 1, positioned on the simulated Martian regolith (see Sect. 3).

A half-sized ExoMars Concept-E chassis forms the basis of the rover. Each of the six wheels has three degrees of freedom: drive, steer and lift. The steer and lift motors are not used in these experiments and are commanded to the centre position. The numbering of the wheels and the organisation of the control system is shown in Fig. 2. The Concept-E chassis uses a "three-module concept," in which there are three independent suspension modules with two wheels each, each module having a central freely rotating pivot point [8] as shown in Fig. 3.

Our control system consists of nine off-the-shelf motor controller units based around an ATMega328p microcontroller and an L298 dual H-bridge driver. Each controller drives two motors. These controllers are themselves controlled by another ATMega328p, in the form of the very popular Arduino Uno board, via an I^2C bus. This "master controller" receives commands from a small on-board Linux PC (a commodity netbook) via USB.

The motor control PID loops run on the individual motor controller units. Commands from the PC, sent via the master controller, change parameters,

Fig. 1. The rover used in the experiments

required speed, etc. and can also read registers holding actual speed, odometry and similar data. Three of the motor controllers also read chassis orientation potentiometers.

Temperature monitoring is done via a network of DS-1820 1-Wire® sensors, all read by the master controller. Each sensor has an accuracy of ±0.5 °C, and is read every 10 s (to allow for parasitic power usage). One sensor was placed against each of the L298 driver chips, while a tenth sensor was mounted outside the enclosure to monitor ambient temperature. Since each controller is responsible either for one drive and one steer motor, or two lift motors; and the lift and steer motors are not loaded, any temperature increase must be from the drive loads.

2.1 Control

The drive motors are controlled using proportional control only — tuning a complex system of six wheels on a highly variable terrain proved intractable in the time available, particularly considering the complex interactions between the different wheels through the body of the rover. Using standard techniques such as Ziegler-Nichols did little to reduce the oscillations while maintaining a good response time to setpoint changes.

3   Environment and Experimental Setup

All experiments took place in Aberystwyth's Planetary Analogue Terrain laboratory, on a surface of Mars Soil Simulant-D from DLR Germany, which is geophysically analogous to Martian regolith. This surface is similar to talcum powder in consistency, and has a particular tendency to cause wheel slip.

A track was established on the simulant terrain, consisting of a short (200 distance units, 200 × 256 odometer ticks) — approximately 4 m) run, with a slight incline of about 10° for just over half its length. In all experiments, the rover was

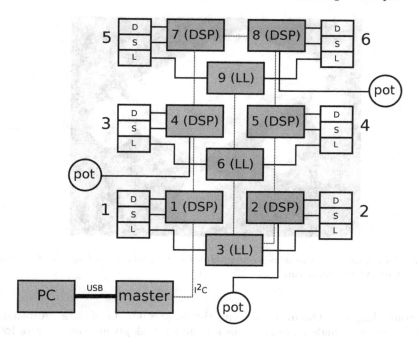

Fig. 2. Rover block diagram. Each numbered unit is a wheel, with 5 and 6 at the front. Each wheel has three motors for three degrees of freedom: Drive, Steer and Lift. Each pair of wheels has three controllers: one for each wheel's drive and steer motors (with optional bogey angle potentiometer), and one for both wheels' lift motors.

programmed to change direction every 200 units by negating the desired speed at the end of each length, with the speed being set according to the algorithm being run. Each experiment ran for one hour, and was repeated five times.

Three different control methods were compared:

- a baseline, where the motor speed was set to 700 ticks per second with no variation;
- a so-called "dumb" method, where all the rover motors were shut down for a minimum of 5 s if any motor driver temperature rose above a threshold level of 16 °C above ambient (as measured by the external sensor);
- the method which is the subject of this paper, which was parameterised to keep all driver temperatures below 17 °C above ambient.

4   Establishing a Baseline

Our first task was to establish a baseline heating curve. As stated above, this was done by simply driving the rover at the arbitrary maximum of 700 ticks per second. The results of all five runs are plotted in Fig. 5.

Fig. 3. Configuration of the suspension modules and wheels in the ExoMars Concept-E chassis, with wheel numbering.

Summarizing the Data. Processing the large amounts of data produced by several runs of a single technique was a complex task given that we were interested in the maximum driver temperature each technique achieved, and that each run produced its results at different timepoints, due to the variability of the loop length in the top-level PC control software.

To deal with this, we first took each run and found the maximum temperature across the drivers at each time point; and then we took the maxima for each run, extracted all the time points, and for each time point interpolated the values of all the runs' maxima. We then calculated the mean and standard deviation of the maxima thus found. This procedure is shown diagrammatically in Fig. 4.

Therefore, Fig. 5 shows the mean of the maximum temperature at each time for each run (since it is the maximum temperature we are trying to limit), with error bars showing 1 standard deviation. A curve of the form

$$y = \frac{a^x((a-1)c+b) - b}{a-1} \qquad (2)$$

(the solution to the recurrence given in (1)) was fitted to all the temperature maxima, not just the mean, using Levenberg-Marquardt non-linear least squares and plotted using matplotlib [6].

5   Testing the Load Reduction Method

Once this was done, tests were made to ensure that the cooling idea would work — that running a motor at a slightly lower speed relative to the others would reduce the load. The effects we see require that the terrain should have a high degree of slip (as the simulant terrain does).

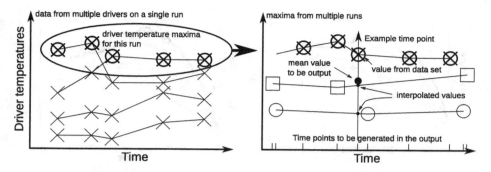

Fig. 4. Summarizing the data. In each run, the maxima of all driver temperatures is found at every given timepoint. Then, a set of all timepoints in all runs is found, and the mean and standard deviation of the maxima for all runs at all these timepoints is found, using linear interpolation where necessary since the intervals between the timepoints vary across the runs.

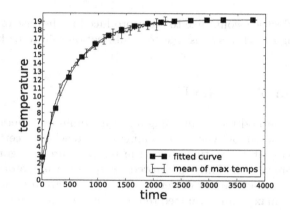

Fig. 5. An "baseline" run with no cooling algorithm — the motors always run at the set speed. The data is processed according to the data processing technique described in Sect. 4.

The test itself was straightforward: the rover was run as in the baseline test, and after a reasonably high temperature is reached, the set point of a hot motor was set to 0.7 of the set point of the other motors ($700 \times 0.7 = 490$). Two typical runs on different motors are shown below in Fig. 6, indicating a prompt reduction in temperature when a motor is commanded to run at a lower required speed. Motor 3 is the centre left motor, while motor 2 is the back right motor. There was no observed deviation in heading when any motor was set to a lower speed, over a distance of approximately 4 m repeated >50 times, as indicated by the wheel tracks in the soil simulant.

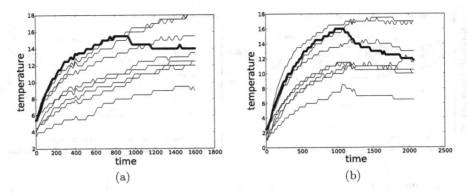

Fig. 6. Two runs testing the load reduction method. The temperature of the test motor is shown in bold. This motor is commanded from 700 to 490 after (a) ∼850 s and (b) ∼1000 s, leading to fairly prompt reductions in temperature through load reduction. In (a) it is motor 3, in (b) it is motor 2.

This is the effect we would like to see reproduced in the case of a motor with a persistently high load, perhaps due to a lubrication failure or terrain-related problem.

6   The "Dumb" Method

The most direct method for maintaining a low maximum temperature is simply to stop moving the robot when the temperature reaches a certain threshold level, only allowing it to continue once the temperature has gone below that level. Our experiment constantly monitored the temperature, stopping the rover if the temperature of any motor driver exceeded 16 °C, given that we arbitrarily set 16.5 °C as our maximum permitted temperature and our temperature sensors had a long latency (up to 10 s). The rover was only permitted to continue once no driver had been above 16 °C for at least 5 s. The purpose of this method was to ensure that the "obvious" solution to the problem of overheating was no better that the final method.

7   The "Inflammatory Model" Method

The Method. The desired outcome is a maximum temperature not exceeding about 16.5 °C. To obtain this, we used a linear ramp from 12 to 18 °C and a maximum reduction by a factor of 0.4 in the desired speed of each motor:

$$r(t, a, b) = \begin{cases} t \le a & 0 \\ t \ge b & 1 \\ \text{otherwise} & \frac{t-a}{b-a} \end{cases} \tag{3}$$

$$f(t) = 1 - 0.4r(t, 12, 18) \tag{4}$$

where $r(t)$ is the ramp function between edge values a and b, giving the pain response to temperature t; and f is the multiple of the required speed to be fed to motor controller i for a given temperature. Given that the experiment has a constant required speed of 700, we determine the motor speed with

$$s(t) = 700f(t) \tag{5}$$

This function is shown in Fig. 7.

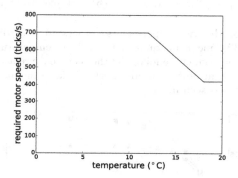

Fig. 7. The linear function mapping temperature to required motor speed, given a required vehicle speed of 700 ticks/s.

8   Results

The results of the two methods — the "dumb" method and the inflammatory method — were summarized in a similar manner to how the baseline was processed in Sect. 4, and exponential curves were fitted to the temperature results for all three experiments. The results are shown in Figs. 8 and 9.

The "dumb" method — stopping the rover whenever the temperature becomes too high — keeps the temperature lowest, as one would expect. However, the load balancing method which is the subject of this paper (labelled here as "simple") does succeed in maintaining a fairly low temperature. While it does not quite remain below 16.5 °C, it is certainly lower than the baseline. Better tuning of the parameters would give a better result.

These results should be viewed in conjunction with Fig. 10, which shows graphs of distance and energy against time for the three experiments. Figure 10a shows that the strategy of just stopping for a brief time, while it does keep the temperature down, results in an unacceptable drop in distance traversed; whereas the biologically inspired technique travels almost as far as the baseline in the same time. Figure 10b shows the mean energy usage of each method over time, showing that the energy usage of our method is nearly identical to

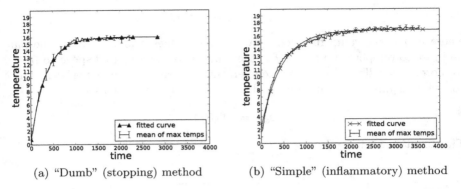

(a) "Dumb" (stopping) method   (b) "Simple" (inflammatory) method

Fig. 8. The results of the two experiments, processed according to the method presented in Sect. 4. The mean maxima of the driver temperatures are shown with error bars, and an exponential curve following the recurrence relation in Eq. 1 has been fitted. The graphs show that both methods effectively keep the maximum temperature down, well below the levels seen in Fig. 5.

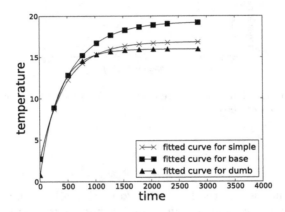

Fig. 9. The results of the baseline and the two methods, fitted to exponential curves of the form given in Eq. 2. These show that the inflammatory "simple" technique keeps temperature down almost as well as the "dumb" technique of stopping when a motor gets hot.

baseline, although it is considerably higher than the "dumb" method. This is to be expected, since it is merely a method for load balancing, not load reduction.

It's worth noting that in the actual experiments, motor 3 was always the first motor to show a rise in temperature and was therefore the first to be set to a lower speed. This likely to be because motor 3 had recently had some maintenance which required disassembly and reassembly. It should also be noted that after this initial phase, other motors begin to develop a high temperature: after about 2500 s, typically three motors are running hot and are slowed by the algorithm. Only motors 5 and 6 did not run hot consistently.

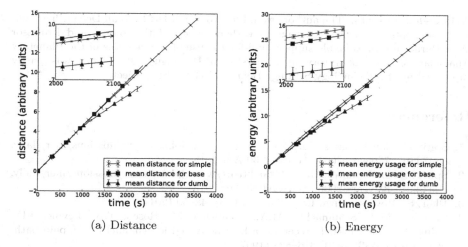

(a) Distance   (b) Energy

Fig. 10. Distance and energy against time for the baseline and both load reduction methods. These show that the distance traversed by the inflammatory technique ("simple") is much better than using the stopping ("dumb") technique and only slightly worse than the baseline (no temperature mitigation at all); while the energy usage for the inflammatory technique is only slightly higher than the "dumb" stopping technique.

9 Conclusions

By considering the biological phenomenon of inflammation, we have developed a very simple method of load-balancing for multiwheeled robots on high-slip surface. Although the method is simple, it maintains a low temperature while not sacrificing total distance travelled, and expending no more energy. This shows that biologically inspired techniques need not be complex or dependent upon emergent behaviour to provide real benefits. Further work will:

1. develop a more accurate model of the sensitisation phenomenon rather than the current ramp function, to see if this improves the behaviour;
2. look at the paracrine behaviour of cytokines, to see if a high cytokine concentration at one wheel can be diffused to another wheel to produce an effect there even if that wheel does not yet have a problem;
3. look at the endocrine behaviour of cytokines, to see if a slow buildup of cytokine concentration at a central location can produce useful behaviour;
4. develop a mechanical model of the rover and its interactions with the surface so that multiple simulated runs can be made simultaneously, speeding up the development of new techniques.

Acknowledgements. The authors would like to thank Dr. Hannah Dee for her valuable comments and suggestions on an earlier version of this paper, and Professor Dave Barnes for his valuable support and for the use of the facilities of the Planetary Analogue Terrain Laboratory at Aberystwyth University. This work was partly supported by the Technology Strategy Board's ENDOVER project.

References

1. Spirit Mission Updates, July 2004. http://www.jpl.nasa.gov/missions/mer/daily. cfm?date=7&year=2004. Accessed 30 Apr 2013
2. Spirit Mission Updates, March 2006. http://www.jpl.nasa.gov/missions/mer/daily. cfm?date=3&year=2006. Accessed 30 Apr 2013
3. Blair, S.: Rovers return. Eng. Technol. **6**(3), 48–50 (2011)
4. Caterina, M.J., Schumacher, M.A., Tominaga, M., Rosen, T.A., Levine, J.D., Julius, D.: The capsaicin receptor: a heat-activated ion channel in the pain pathway. Nature **389**(6653), 816–824 (1997)
5. Dinarello, C.A.: Proinflammatory cytokines. CHEST J. **118**(2), 503–508 (2000). http://dx.doi.org/10.1378/chest.118.2.503
6. Hunter, J.D.: Matplotlib: a 2D graphics environment. Comput. Sci. Eng. **9**(3), 90–95 (2007)
7. Kress, M.: Nociceptor sensitization by pro-inflammatory cytokines and chemokines. Open Pain J. **3**, 97–106 (2010)
8. Kucherenko, V., Bogatchev, A., Van Winnendael, M.: Chassis concepts for the ExoMars rover. In: Proceedings of the 8th ESA Workshop on Advanced Space Technologies for Robotics and Automation, Noordwijk, The Netherlands (2004)
9. Macpherson, L.J., Geierstanger, B.H., Viswanath, V., Bandell, M., Eid, S.R., Hwang, S., Patapoutian, A.: The pungency of garlic: activation of TRPA1 and TRPV1 in response to allicin. Curr. Biol. **15**(10), 929–934 (2005)
10. Medeiros, I.M., Castelo, A., Salomao, R.: Presence of circulating levels of interferon-g, interleukin-10 and tumor necrosis factor-a in patients with visceral leishmaniasis. Rev. Inst. Med. Trop. São Paulo **40**(1), 31–34 (1998)
11. Montell, C., Birnbaumer, L., Flockerzi, V., Bindels, R.J., Bruford, E.A., Caterina, M.J., Clapham, D.E., Harteneck, C., Heller, S., Julius, D.: Others: a unified nomenclature for the superfamily of TRP cation channels. Mol. Cell **9**(2), 229 (2002)

Control-Oriented Nonlinear Dynamic Modelling of Dielectric Electro-Active Polymers

Will Jacobs[1]([envelope]), Emma D. Wilson[1], Tareq Assaf[2], Jonathan M. Rossiter[2], Tony J. Dodd[1], John Porrill[1], and Sean R. Anderson[1]([envelope])

[1] Sheffield Centre for Robotics (SCENTRO), University of Sheffield, Sheffield, UK
{w.jacobs,s.anderson}@sheffield.ac.uk
[2] Bristol Robotics Laboratory (BRL), University of Bristol
and University of the West of England, Bristol, UK

Electro-Active Polymers (EAPs) are a rapidly developing actuation technology in the field of soft robotics. These soft actuators have the potential to replace existing hard actuator technologies for many applications, combining desirable features such as relatively large actuation strain, low mass, high response speed and compliance [1]. The characteristics of EAPs have drawn comparison to biological muscle [2], generating interest from the robotics community because of the potential for emulating the many versatile ways muscle is used in nature: as motors, brakes, springs and struts.

One of the key challenges in the development of EAP technology is control design [3]. However, like biological muscles that are typically nonlinear over a wide operating range [4], EAP artificial muscles also display nonlinear and time-varying characteristics. New application areas of EAPs in robotics require the development of control algorithms suited to these nonlinear, time-varying properties. An important goal to achieve, therefore, is the identification of dynamic models of EAP behaviour in forms that will facilitate control design and analysis, i.e. control-oriented dynamic models.

Control-oriented models should be simple, compact descriptions for efficient implementation and tractable analysis, as well as accurate over the operational range of interest. Many current dynamic models of EAP behaviour have focused on first principle descriptions [5], which can often be unwieldy and overly-complex for use in control design. Control-oriented models are typically obtained in compact forms using system identification techniques. The nonlinear, auto-regressive moving average with exogenous inputs (NARMAX) model is one such popular input-output description for use in nonlinear system identification [6,7].

In this investigation we applied the NARMAX modelling and analysis framework to the identification of control-oriented models of dielectric EAPs (DEAPs). In the experimental setup, input-output signals (voltage and displacement) were obtained by actuating a mass resting on a circular DEAP. A laser displacement sensor (Keyence LK-G152) was used to measure the vertical displacement of the mass (a 3 g sphere). The DEAPs were custom fabricated from acrylic elastomer acrylic elastomer 3M VHB 4905, with an initial thickness of 0.5 mm. Input voltages in the range 1.1–3.75 V were passed through a potentiostat and amplified with ratio of 15 V:12 kV then applied to the DEAP (Fig. 1A).

A. Natraj et al. (Eds.): TAROS 2013, LNAI 8069, pp. 259–260, 2014.
DOI: 10.1007/978-3-662-43645-5_26, © Springer-Verlag Berlin Heidelberg 2014

260 W. Jacobs et al.

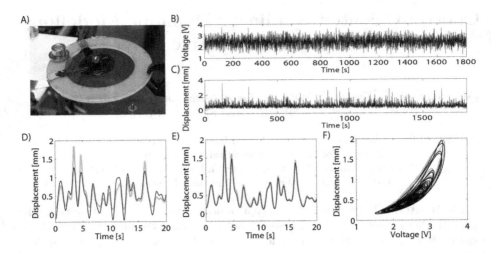

Fig. 1. (A) DEAP Actuator. (B) Voltage input prior to amplification. (C) Vertical displacement response of the DEAP. (D) Measured (_) and simulated (_) response of the DEAP actuator, modelled with a linear ARX model. (E, F) Measured (_) and simulated (_) response of the DEAP actuator, modelled with a NARX model.

We identified a NARX model from the voltage-displacement (input-output) signals and benchmarked it against a linear dynamic ARX model as well as a Hammerstein model (i.e. linear dynamics plus a static input nonlinearity). We found that the NARX model was comparatively compact and accurate, supporting its use in facilitating control design for DEAPs and providing a ready path for the development of soft robots and practical robotic devices.

References

1. Bar-Cohen, Y. (ed.): Electroactive Polymer (EAP) Actuators as Artificial Muscles: Reality, Potential, and Challenges. Society of Photo Optical, Bellingham (2004)
2. Meijer, K., Rosenthal, M., Full, R.J.: Muscle-like actuators? A comparison between three electroactive polymers. Proc. SPIE **4329**, 7–15 (2001)
3. Wilson, E., et al.: Bioinspired control of electro-active polymers for next generation soft robots. In: Herrmann, G., Studley, M., Pearson, M., Conn, A., Melhuish, C., Witkowski, M., Kim, J.-H., Vadakkepat, P. (eds.) TAROS-FIRA 2012. LNCS, vol. 7429, pp. 424–425. Springer, Heidelberg (2012)
4. Anderson, S.R., Lepora, N.F., Porrill, J., Dean, P.: Nonlinear dynamic modeling of isometric force production in primate eye muscle. IEEE Trans. Biomed. Eng. **57**(7), 1554–1567 (2010)
5. Wissler, M., Mazza, E.: Modeling and simulation of dielectric elastomer actuators. Smart Mater. Struct. **14**(6), 1396 (2005)
6. Chen, S., Billings, S.A.: Representations of non-linear systems: the NARMAX model. Int. J. Control **49**(3), 1013–1032 (1989)
7. Baldacchino, T., Anderson, S.R., Kadirkamanathan, V.: Structure detection and parameter estimation for NARX models in a unified EM framework. Automatica **48**(5), 857–865 (2012)

Emotionally Driven Robot Control Architecture for Human-Robot Interaction

Jekaterina Novikova[(✉)], Swen Gaudl, and Joanna Bryson

University of Bath, Bath, UK
{j.novikova,s.gaudl,j.j.bryson}@bath.ac.uk

With the increasing demand for robots to assist humans in shared workspaces and environments designed for humans, research on human-robot interaction (HRI) gains more and more importance. Robots in shared environments must be safe and act in a way understandable by humans, through the way they interact and move. As visual cues such as facial expressions are important in human-human communication, research on emotion recognition, expression, and emotionally enriched communication is of great importance to HRI and has also gained increasing attention during the last two decades [1,3,4,6,8]. Most of the existing work focuses on the recognition of human emotions or mimicking their expression [1,6] and emotional action selection [3,5]. Less work is done on the role of emotions in influencing human behaviour in HRI [4].

The goal of our work-in-progress is to further explore the idea of emotion as a strong influence of action selection and to analyse the role of emotion in HRI. We present here a model for incorporating emotions and emotional expressions into POSH dynamic plans [2] based on Behaviour Oriented Design. Emotions are represented as a factor for dynamic action selection, and an emotional expression is used in conjunction as a visual cue for communicating the current emotional state to a human before and during the execution of actions. We hypothesize that robots expressing emotions in a human-understandable way will improve the quality of communication and therefore collaboration between a robot and a human. We thus hypothesize that such emotional expressions can empower robots to influence humans behaviour in an HRI task.

The first phase of the emotional action selection includes detecting specific internal and/or external conditions that comprise the robots emotive response and thus are able to influence the emotional state of a robot. We use a list of conditions proposed by Breazeal [1]: presence of an undesired stimulus, presence of a desired stimulus, a sudden stimulus, delay in achieving goal. For determining an appropriate emotional state we use a two-dimensional representation for expressing basic human emotional states, proposed by Russell [8]. Here, *arousal* represents the strength of a stimulus, and the *valence* shows a positive/negative value of a stimulus. All the detected conditions influence both valence and arousal of a robots emotive response. We also use *intensity* as an additional property of an emotion. Intensity depends on time, number of detected stimuli, and an impact factor of an executed behaviour. We use impact factor as a property of a behaviour that depresses the intensity of the emotion this behaviour was triggered by.

A. Natraj et al. (Eds.): TAROS 2013, LNAI 8069, pp. 261–263, 2014.
DOI: 10.1007/978-3-662-43645-5_27, © Springer-Verlag Berlin Heidelberg 2014

Each emotion calls a specific behaviour of a POSH plan. While the selected behaviour is being executed it inhibits the intensity of the emotion it was triggered by, i.e. intensity of an emotion is a function of a behavioural impact over time. Emotional intensity in our model is an internal state of an agent, which is changed dynamically while robot is experiencing an emotion. 'Feeling' an emotion is a latched process [7], during which an intensity of the emotion is increasing over time from zero value until the maximum threshold of 100, and is reducing back to zero after the executing behaviour inhibits it. The expression of emotion starts after an increasing intensity of the emotion reaches the specified level and stops when the same level is reached while the intensity is decreasing. Emotions are expressed by a robot using two basic movements of its body: moving the 'neck' forward/backward, and raising/lowering 'eyebrows'. The execution of the selected behaviour starts when the intensity of an emotion reaches a specific level which is above the level of the start of expressing the emotion and below the maximal intensity level. The execution of behaviour, if not interrupted, stops when intensity of the emotion is zero.

We have collected data showing that the set of these two simple movements in a robot are recognised by people as expression of several basic emotions. In our future experiment we plan to apply the described model of emotional action selection for a human-robot interaction task where the robot must solicit assistance to achieve its goal. We intend to measure all three conditions: emotional action selection with and without expression and emotion free action selection, to observe whether emotional communication will empower the robot to influence a humans behaviour.

References

1. Breazeal, C.: Emotion and sociable humanoid robots. Int. J. Hum. Comput. Interact. **59**(1–2), 119–155 (2003). (Applications of Affective Computing in Human-Computer Interaction)
2. Bryson, J.: The behaviour-oriented design of modular agent intelligence. In: Proceedings of the NODe 2002 Agent-Related Conference on Agent Technologies, Infrastructures, Tools, and Applications for E-services (NODe'02) (2002)
3. Bryson, J., Tanguy, E.: Simplifying the design of human-like behaviour: emotions as durative dynamic state for action selection. Int. J. Synth. Emot. **1**(1), 30–50 (2010)
4. Gonsior, B., Sosnowski, S., Buss, M., Kuhmlenz, K.: An emotional adaption approach to increase helpfulness towards a robot. In: Proceedings of International Conference on Intelligent Robots and Systems, Vilamoura, Algarve, Portugal (2012)
5. Johansson, A., Dell'Acqua, P.: Introducing time in emotional behaviour networks. In: 2010 IEEE Symposium on Computational Intelligence and Games (CIG), pp. 297-304, 18–21 Aug 2010
6. Kawamura, K., Wilkes, D., Pack, T., Bishay, M., Barile, J.: Humanoids: future robots for home and factory. In: Proceedings of the First International Symposium on Humanoid Robots (HURO96), Tokyo, Japan, pp. 53–62 (1996)

7. Rohlfshagen, P., Bryson, J.: Flexible latching: a biologically-inspired mechanism for improving the management of homeostatic goals. Cogn. Comput. **2**(3), 230–241 (2010)
8. Russell, J.: Reading emotions from and into faces: resurrecting a dimensional-contextual perspective. In: Russell, J., Fernandez-Dols, J. (eds.) The Psychology of Facial Expression, pp. 295–320. Cambridge University Press, Cambridge (1997)

Adaptive Control of Robot System
of up to a Half Passive Joints

Chenguang Yang[1,2]($\boxtimes$), Jing Li[3], Zhijun Li[2], Weisheng Chen[3],
and Rongxin Cui[4]

[1] School of Computing and Mathematics, Centre for Robotics and Neural Systems,
Plymouth University, Plymouth, UK
cyang@ieee.org
[2] MOE Key Lab of Autonomous Systems and Networked Control, College of
Automation Science and Engineering, South China University of Technology,
Guangzhou, China
[3] Department of Mathematics, Xidian University, Xi'an, China
[4] School of Marine Science and Technology, Northwestern Polytechnical University,
Xi'an, China

Abstract. In this paper, we study adaptive control of a robot system of
$2n$ joints with up to n joints being passive. By exploiting the dynamics
couplings between the active joints and the passive joints, we have developed a method to use desired trajectories of active joints to indirectly
"control" the motion of the passive joints. Optimal control techniques
have been employed to control the active joints with smooth motion of
minimized acceleration. Neural network (NN) has been used for block
function approximation, in order to generate ideal desired trajectory of
active joints. It has been theoretically established that under the developed adaptive controller and NN based trajectory generator, the passive
joints can be effectively controlled to follow the predefined trajectory.

Keywords: Passive joint · Optimization · LQR · Model reference
control

1 Introduction

The control problem of underactuated robots has received considerable research
attention in the last two decades. Conventional robot usually has one actuator for
each joint, such that its degree of freedom (DOF) equals the number of actuators.
An underactuated robot has passive joints equipped with no actuators. Passive
joints without actuators could help to reduce weight, energy consumption, and
cost for robot manipulators. They also play useful roles in many applications,
e.g., hitting or hammering an object. The compliance of joint could actually
improve efficiency in these applications [1]. It is also noted that in many cases
full actuated system is not necessary, e.g., in the gymnastic swing up motion, the
sportman's joints on the upper limb are in fact passive [2]. The popular personal

A. Natraj et al. (Eds.): TAROS 2013, LNAI 8069, pp. 264–275, 2014.
DOI: 10.1007/978-3-662-43645-5_28, © Springer-Verlag Berlin Heidelberg 2014

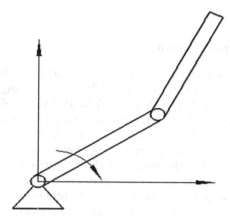

Fig. 1. Example of a two joints robot with last joint passive

transportation wheeled robot Segway HT system [3,4], is also underactuated as there are only two motors but three DOF. Similarly, most underwater vehicles also have less number of actuators than the number of DOF [5]. This configuration does not reduce the performance of Segway but make its riding more convenient.

While for more generalized active/passive hybrid joints robot, we would not lose control of the purposely made passive joint. Due to the dynamics coupling between the active and passive joint, we could control these passive joints indirectly by active joints. To illustrate this point, let us consider a simple two joints robot as shown in Fig. 1. The second joint is passive but we could still be able to control its motion by using the first joint. A typical example of this kind of two joints robot is the two link inverted pendulum.

As we know, the zero torque at the passive joints leads to a second-order nonholonomic constraint, whereas nonholonomic constraints is defined as nonintegrable differential equations with time-derivatives of the generalized coordinates (e.g., joint velocity). The nonholonomic mechanism could allow one to control a multiple joints robot with a reduced number of inputs [6].

Motivated by the above discussion, in this paper we consider adaptive control of a robot system with $2n$ joints/DOFs with up to n joints passive. It is worth to mention that the robot system dynamics is considered uncertain in this work. In practice, it is hard to obtain an accurate model, and there are also unavoidable uncertain loads. From this point of view, the development of adaptive control that can compensate for system uncertainties become an important issue. In this work, the overall $2n$ order system is decomposed into an up to an nth order passive subsystem and an nth order active subsystem. The uncertainty of the active subsystem is handled by adaptive control techniques while the uncertainties of the passive subsystems is dealt with neural network (NN) approximation, which has been widely used to compensate for nonlinear uncertainties [7–9].

2   Preliminaries

2.1   RBFNN Approximation

In this paper, an unknown smooth nonlinear function $\phi(W, z) : R^{l+m} \to R$ will be approximated on a compact set D by the following radial basis function neural network (RBFNN) [10] $\phi(W, z) = W^T S(z) + \delta(z)$, where $S(z) = [s_1(z), \cdots, s_l(z)]^T : D \to R^l$, is a known smooth vector function with the NN node number $l > 1$. Basis function $s_i(z)$, $1 \le i \le l$, is chosen as the commonly used Gaussian function with the form $s_i(z) = \exp[-(z - \mu_i)^T (z - \mu_i)/\eta_i^2]$, where $\mu_i = [\mu_{i1}, \cdots, \mu_{im}]^T \in D$ is the center of the receptive field and $\eta_i > 0$ is the width of the Gaussian function. The optimal weight vector $W = [w_1, \cdots, w_l]^T$ is a constant vector and $\delta(z)$ denotes the NN inherent approximation error, which satisfies $|\delta(z)| \le \epsilon$, for a small ϵ. In many previous published works, the approximation error is assumed to be bounded by a fixed constant.

For convenience of approximation of a nonlinear matrix function with each element a unknown scalar function, we use the following block matrix operators as introduced in [11], which defines block matrices W and S as follows:

$$
W \triangleq \begin{bmatrix} W_{11} & W_{12} & \cdots & W_{1m} \\ W_{21} & W_{22} & \cdots & W_{2m} \\ \vdots & \vdots & \ddots & \vdots \\ W_{n1} & W_{n2} & \cdots & W_{nm} \end{bmatrix} \in R^{nl \times m} \quad S \triangleq \begin{bmatrix} S_{11} & S_{12} & \cdots & S_{1m} \\ S_{21} & S_{22} & \cdots & S_{2m} \\ \vdots & \vdots & \ddots & \vdots \\ S_{n1} & S_{n2} & \cdots & S_{nm} \end{bmatrix} \in R^{nl \times m}
$$

with each block $W_{ij} \in \mathbf{R}^l$ a column vector of NN weight and $S_{ij} \in \mathbf{R}^l$ a column vector of NN basis function. A "transpose" operation $\langle T \rangle$ of the block matrix W is defined in such a way

$$
W^{\langle T \rangle} \triangleq \begin{bmatrix} W_{11}^T & W_{12}^T & \cdots & W_{1m}^T \\ W_{21}^T & W_{22}^T & \cdots & W_{2m}^T \\ \vdots & \vdots & \ddots & \vdots \\ W_{n1}^T & W_{n2}^T & \cdots & W_{nm}^T \end{bmatrix} \in R^{n \times ml}
$$

that each block of the column vector is transposed to a row vector while the relative location of each block in the matrix W is not changed. Furthermore, we define block matrix multiplication between $W^{\langle T \rangle}$ and S as

$$
[W^{\langle T \rangle} \langle \cdot \rangle S] \triangleq \begin{bmatrix} W_{11}^T S_{11} & W_{12}^T S_{12} & \cdots & W_{1m}^T S_{1m} \\ W_{21}^T S_{21} & W_{22}^T S_{22} & \cdots & W_{2m}^T S_{2m} \\ \vdots & \vdots & \ddots & \vdots \\ W_{n1}^T S_{n1} & W_{n2}^T S_{n2} & \cdots & W_{nm}^T S_{nm} \end{bmatrix} \in R^{n \times m}
$$

In this manner, we could easily use $[W^{\langle T \rangle} \langle \cdot \rangle S]$ to emulate a unknown matrix of dimension $n \times m$.

2.2   Finite Time Linear Quadratic Regulator (LQR)

Consider the following linear system with input u

$$\dot{x} = Ax + Bu, \quad x(t_0) = x_0, \quad x \in R^n, \quad u \in R^n \tag{1}$$

where $[A, B]$ is completely stabilizable pair. The optimal control $u^*(t)$, $t > 0$, that minimizes the following performance index

$$J = \int_{t_0}^{t_f} ((x - x_d)^T Q(x - x_d) + u^T Ru)dt, \quad R = R^T \geq 0, \quad Q = Q^T > 0 \tag{2}$$

is given by $u^* = -R^{-1}B^T(Px + s)$ with P and s the solutions of the following Riccati Equations [12]

$$\begin{aligned}
-\dot{P} &= PA + A^T P - PBR^{-1}B^T P + Q, \quad P(t_f) = 0_{[n,n]} \\
-\dot{s} &= (A - BR^{-1}B^T P)^T s - Qx_d, \quad s(t_f) = 0_{[n]}
\end{aligned} \tag{3}$$

3   Dynamics of Underactuated Robot Manipulator

Without loss of generality, we perform partition of generalized coordinate vector q as $q = [q_a^T, \; q_b^T]^T \in R^{2n}$ with q_a being the active joints and q_b being the passive joints. Then, the dynamics model of robot manipulator with both active and passive joints can be described as follows:

$$\begin{bmatrix} m_a & m_{ab} \\ m_{ba} & m_b \end{bmatrix} \begin{bmatrix} \ddot{q}_a \\ \ddot{q}_b \end{bmatrix} + \begin{bmatrix} c_a & c_{ab} \\ c_{ba} & c_b \end{bmatrix} \begin{bmatrix} \dot{q}_a \\ \dot{q}_b \end{bmatrix} + \begin{bmatrix} g_a \\ g_b \end{bmatrix} + \begin{bmatrix} \tau_{d_a} \\ \tau_{d_b} \end{bmatrix} = \begin{bmatrix} \tau_a \\ 0 \end{bmatrix} \tag{4}$$

Define

$$M = \begin{bmatrix} m_a & m_{ab} \\ m_{ba} & m_b \end{bmatrix} \quad C = \begin{bmatrix} c_a & c_{ab} \\ c_{ba} & c_b \end{bmatrix} \quad G = \begin{bmatrix} g_a \\ g_b \end{bmatrix} \quad \tau_d = \begin{bmatrix} \tau_{d_a} \\ \tau_{d_b} \end{bmatrix} \tag{5}$$

as the inertia matrix, Centripetal/Coriolis matrix, gravity torque vector and disturbance torque vector, respectively. As mentioned above, at most a half of the joints can be passive, Without loss of generality, in the partition $q = [q_a, q_b]$, let us assume that $q_a \in R^n$ and $q_a \in R^n$ such that there exist a matrix

$$I_0 = [I_{[n,n]}, 0_{[n,n]}] \in R^{n \times 2n} \tag{6}$$

such that $q_a = I_0 q$ and $\tau_a = [\tau_1, \tau_2, \dots, \tau_n]$. Then (4) can be written in a compact form as

$$M\ddot{q} + C\dot{q} + G + \tau_d = I_0^T \tau_a \tag{7}$$

The following property are well known for the Lagrange-Euler formulation of robotic dynamics:

Property 1. [13] The matrix M is symmetric and positive definite. Accordingly, the sub-matrix M_a and M_b must be positive definite as well. In addition, we assume that M_{ab} and M_{ba} are also positive definite. The inverse of matrix M must exist and is

$$M^{-1} = \begin{bmatrix} s_b^{-1} & -m_a^{-1} m_{ab} s_a^{-1} \\ m_b^{-1} m_{ba} s_b^{-1} & s_a^{-1} \end{bmatrix} \tag{8}$$

where S_a and S_b are Schur complements of m_a and m_b, respectively, defined as

$$s_a = m_b - m_{ba} m_a^{-1} m_{ab} \quad s_b = m_a - m_{ab} m_b^{-1} m_{ba} \tag{9}$$

Multiplying $I_0 M^{-1}$ on both sides of (7) gives us

$$\ddot{q}_a + I_0 M^{-1} C \dot{q} + I_0 M^{-1} G + I_0 M^{-1} \tau_d I_0 M^{-1} = I_0^T \tau_a = s_b^{-1} \tau_a \tag{10}$$

Then, multiplying s_b on both sides of the above equation, we have

$$s_b \ddot{q}_a + s_b I_0 M^{-1} C \dot{q} + S_b I_0 M^{-1} g + S_b I_0 M^{-1} \tau_d = \tau_a \tag{11}$$

Define $M_a \triangleq s_b \in \mathbf{R}^{n \times n}$, $C_{ab} \triangleq s_b I_0 M^{-1} C = [C_a, C_b] \in \mathbf{R}^{n \times 2n}$ with $C_a \in \mathbf{R}^{n \times n}$, $C_b \in \mathbf{R}^{n \times n}$, and $G_a = (S_b I_0 M^{-1} g + S_b I_0 M^{-1} \tau_{da})$, then, we have Σ_a-subsystems of active joints q_a as follows:

$$\Sigma_a : M_a \ddot{q}_a + C_a \dot{q}_a + C_b \dot{q}_b + g_a = \tau_a \tag{12}$$

while the dynamics of passive joints q_b can be written as

$$\Sigma_b : \ddot{q}_b = -M_b^{-1} C_b \dot{q}_b - M_b^{-1} g_b - M_b^{-1} \tau_{db} - M_b^{-1} M_{ba} \ddot{q}_{ad} - M_b^{-1} C_{ba} \dot{q}_{ad} \tag{13}$$

It is well known that matrix $2C_a - \dot{M}_a$ is a skew-symmetric matrix [14], and there exist positive scalars θ_M, θ_C, θ_G such that $\|M_a\| \leq \theta_M$, $\|C_a\| \leq \theta_C \|\dot{q}\|$ and $\|G_a\| \leq \theta_G$ [14]. It should be mentioned that due to the unknown system parameters in the above dynamics formulation, the dynamics matrices are actually unknown for control design.

4   Optimal Model Reference Control of Subsystem Σ_a

4.1   Optimal Reference Model

Given the desired joint trajectories of the active joints q_a, denoted as q_{ad}, which is to be designed later, we construct a reference model for the dynamics (12) as follows:

$$M_d \ddot{q}_a + C_d \dot{q}_a + K_d q_a = -F_\eta(q_{ad}, \dot{q}_{ad}) \tag{14}$$

which can be regarded as a virtual impedance model with M_d, C_d, K_d the virtual inertia, damping and stiffness matrices, respectively, and F_η a virtual force.

Due to the difference between the initial value of actual joint position q_a and desired joint position q_d, we cannot expect to make q_a follow q_d immediately. In that way, too much acceleration will be caused and gear and actuator wear as well as safety issue would arise. Therefore, not only tracking error but also acceleration should be taken into consideration for minimization of the following performance index

$$I_P = \int_{t_0}^{t_f} \left(e^T (q_a - q_{ad})^T Q_p (q_a - q_{ad}) \right.$$
$$\left. + (\dot{q}_a - \dot{q}_{ad})^T Q_v (\dot{q}_a - \dot{q}_{ad}) + \ddot{q}_a^T M_d \ddot{q}_a \right) dt. \tag{15}$$

where $[t_0, t_f]$ is a predefined time span from start of operation, Q_p and Q_v are predefined non-negative weight matrix for position tracking error and velocity tracking error, separately. Given that weight matrix of joint acceleration, which is the same positive definite matrix M_d used in (14), now let us consider how to minimize the performance index I_P by suitably designing C_d, K_d and F_η using LQR technique detailed in Sect. 2.2. For this purpose, we rewrite the reference model (14) as

$$\dot{\bar{q}}_a = A\bar{q}_a + Bu \tag{16}$$

with

$$\bar{q}_a = [q_a, \dot{q}_a], \quad \bar{q}_{ad} = [q_{ad}, \dot{q}_{ad}] \tag{17}$$

$$A = \begin{bmatrix} 0_{n \times n} & I_{n \times n} \\ 0_{n \times n} & 0_{n \times n} \end{bmatrix}, \quad B = \begin{bmatrix} 0_{n \times n}, I_{n \times n} \end{bmatrix}^T, \quad Q_p = \begin{bmatrix} q_p & 0_{[n \times n]} \\ 0_{[n \times n]} & q_v \end{bmatrix} \tag{18}$$

$$u = -M_d^{-1} [K_d, \ C_d]\bar{q} - M_d^{-1} F_\eta (q_d, \dot{q}_d) \in R^n \tag{19}$$

Introducing an Q defined as

$$Q = \begin{bmatrix} Q_p & 0_{n \times n} \\ 0_{n \times n} & Q_v \end{bmatrix} \tag{20}$$

and noting that $u \equiv \ddot{q}$ according to the definition (38), we can then rewrite the performance index (15) as

$$\bar{I}_P = \int_{t_0}^{t_f} \left((\bar{q}_a - \bar{q}_{ad})^T \bar{Q} (\bar{q}_a - \bar{q}_{ad}) + u^T M_d u \right) dt \tag{21}$$

If we regard u as the control input to system (16), then the minimization of (21) subject to dynamics constraint (16) becomes a typical LQR control design problem. According to the LQR optimal control technique reviewed in Sect. 2.2, the solution of u that minimizes (21) is

$$u = -M_d^{-1} B^T P\bar{q} - M_d^{-1} B^T s \tag{22}$$

where P is the solution of the following differential equation

$$-\dot{P} = PA + A^T P - PBM_d^{-1}B^T P + \bar{Q}, P(t_f) = 0_{2n \times 2n} \tag{23}$$

and s is the solution of the following differential equation

$$-\dot{s} = (A - BM_d^{-1}B^T P)^T s - \bar{Q}\bar{q}_d, \quad s(t_f) = 0_{2n\times 1} \tag{24}$$

Comparing Eqs. (38) and (22), we can see that the matrices K_d and C_d can be calculated in the following manner:

$$[K_d, C_d] = B^T P, \quad F_\eta = B^T s, \quad t_0 \le t \le t_f \tag{25}$$

Remark 1. As $P(t_f) = 0_{2n\times 2n}$ and $s(t_f) = 0_{n\times 1}$, we see that $K_d = C_d = 0_{n\times n}$ when $t = t_f$. During $[t_0, t_f]$, active joints q_a have been controlled to be very close to q_{ad}. For $t > t_f$, we simply keep $K_d = C_d = 0_{n\times n}$, and let $F_\eta = -M_d\ddot{q}_{ad}$ such that $\ddot{q}_a \equiv \ddot{q}_{ad}$ for $t > t_f$.

4.2 Model Matching Errors

By using $e = q_a - q_{ad}$, the reference model (14) can be rewritten as

$$M_d\ddot{e} + C_d\dot{e} + K_d e = -\eta \tag{26}$$

where $\eta = F_\eta + M_d\ddot{q}_{ad} + C_d\dot{q}_{ad} + K_d q_{ad}$. We can see that the objective of the control design inclines to seek a proper control input torque τ_a in (12), such that the dynamics (12) match the desired reference model dynamics (26). In order to measure the difference between the closed-loop dynamics of (12) and the reference model dynamics (26), we employ a matching error as in [4] $w = M_d\ddot{e} + C_d\dot{e} + K_d e + \eta$ such that if $w(t) \equiv 0_{n\times 1}$ can be guaranteed, the dynamics (12) will exactly match the desired reference model dynamics (26).

For the convenience of the following analysis, we define an augmented matching error as

$$\bar{w} = K_\eta w = \ddot{e} + C_m\dot{e} + K_m e + K_\eta\eta \tag{27}$$

where $C_m = M_d^{-1}C_d$, $K_m = M_d^{-1}K_d$ and $K_\eta = M_d^{-1}$.

Remark 2. The virtual mass matrix M_d is always chosen to be positive definite such that it is invertible and $\bar{w}$ in (27) is well defined.

Denote L and L^{-1} as Laplace transformation operator and inverse Laplace transformation operator, respectively. We design a filtered matching error z as:

$$z = L^{-1}\left\{\left(1 - \frac{\Gamma}{s+\Gamma}\right)L\{\dot{e}\} + \frac{1}{s+\Gamma}L\{\varepsilon\}\right\} \tag{28}$$

where ε is the practical implementation of $\bar{w}$, due to acceleration measurement being normally unavailable:

$$\varepsilon = C_m\dot{e} + K_m e + K_\eta\eta \tag{29}$$

For convenience of following computation we rearrange z:

$$z = \dot{e} - e_h + \varepsilon_l \tag{30}$$

where $e_h = L^{-1}\{\frac{\Gamma s}{s+\Gamma}L\{e\}\}$, $\varepsilon_l = L^{-1}\{\frac{1}{s+\Gamma}L\{\varepsilon\}\}$, i.e. high-pass and low-pass filtered respectively, we see that the augmented matching error $\bar{w}$ can be written as

$$\bar{w} = \dot{z} + \Gamma z \tag{31}$$

which implies that z could be obtained by passing $\bar{w}$ through a filter. From (31) and (27), we see that $z = 0_{n\times1}$ and subsequently $\dot{z} = 0_{n\times1}$ will lead to $w = 0_{n\times1}$, i.e., matching error minimized. Then, the closed-loop dynamics (12) would exactly match the reference model (14). After a finite time t_f, we will have $q = q_d$. In the next subsection, we will design an adaptive controller which guarantees that there exists a finite time $t_z \ll t_f$ such that $z(t) = 0_{n\times1}, \forall t > t_z$.

4.3   Adaptive Controller Design

Denote the desired trajectories for subsystem Σ_a as $\dot{q}_{ad}$ and $\ddot{q}_{ad}$ which are defined as below:

$$\dot{q}_{ad} = \dot{q}_d + e_h - \varepsilon_l, \quad \ddot{q}_{ad} = \ddot{q}_d + \dot{e}_h - \dot{\varepsilon}_l \tag{32}$$

Now let us design the adaptive controller for subsystem Σ_a (12). It consist of three terms as below:

$$\tau_a = -Y(\ddot{q}_{ad}, \dot{q}_{ad}, \dot{q}, q, z)\hat{\Theta} - K_s \operatorname{sgn}(z) - K_z \tag{33}$$

where the first term is the computed torque part that is used to compensate system dynamics, and the last two terms are feedback control part that regulate model matching error. The parameters K_1 and K_2 are diagonal positive definite matrix; $\operatorname{sgn}(\cdot)$ is element wise sign function; and $\hat{\Theta}$ is the estimate of

$$\Theta = [\theta_M, \theta_C, \theta_{f1}, \theta_G + \theta_{f2}]^T \tag{34}$$

and

$$Y(\ddot{q}_{ad}, \dot{q}_{ad}, z) = [\|\ddot{q}_{ad}\| \operatorname{sgn}(z), \|\dot{q}_a\|\|\dot{q}_{ad}\| \operatorname{sgn}(z), \|\dot{q}_a\|\|q_a\| \operatorname{sgn}(z), \operatorname{sgn}(z)]$$

For convenience, we use Y instead of $Y(\ddot{q}_{ad}, \dot{q}_{ad}, \dot{q}, q, z)$ where it does not result in any confusion. The following parameter estimation update law is used to generate $\hat{\Theta}$:

$$\dot{\hat{\Theta}} = \Gamma_\Theta^{-1} Y^T z \tag{35}$$

where Γ_Θ is a diagonal positive definite matrix.

Theorem 1. *Consider the closed-loop system combining subsystem dynamics (12) and the controller (33). We have the following results: (i) all the signals in the closed-loop are uniformly bounded, and (ii) there exists a finite time t_z such that $z(t) = 0_{n\times1}$ for $t > t_z$, where as the value of t_z can be tuned to be arbitrarily small by using large K_s and K_z.*

Proof. The proof is similar to the proof in [4] and is thus omitted.   ∎

5   Reference Trajectory Generator for Subsystem Σ_b

In the following, we are going to show that the n passive joints can be fully "controlled" by the active joints dynamics. As active joint acceleration $\ddot{q}_a \equiv \ddot{q}_{ad}$ for $t > t_f$, we could treat the desired joint acceleration $\pi \triangleq \ddot{q}_{ad}$ as "control variables". As above discussed, after finite time t_f, the dynamics (13) becomes as follows:

$$\ddot{q}_b = -M_b^{-1} C_b \dot{q}_b - M_b^{-1} g_b - M_b^{-1} \tau_{db} - M_b^{-1} M_{ba} \pi - M_b^{-1} C_{ba} \dot{q}_{ad} \qquad (36)$$

where π is defined as q_{ad}. Let $\varphi = [\varphi_1, \varphi_2]^T = [q_b, \dot{q}_b]^T$, $\phi = [\phi_1^T, \phi_2^T]^T = [q_{ad}^T, \dot{q}_{ad}^T]^T$. Consider the desired position/velocity trajectories of the passive joints as q_{bd} and $\dot{q}_{bd}$, respectively. Then, our design objective is to construct a π (subsequently ϕ_1 and ϕ_2) such that φ_1 and φ_2 of system (37) follow $\varphi_{1d} \triangleq q_{bd}$ and $\varphi_{2d} \triangleq \dot{q}_{bd}$. Let $\pi = v + u$, then, Eq. (36) can be rewritten as

$$\ddot{q}_b = f_v(\varphi, \phi, v) + f_u(\varphi, \phi, u)$$
$$f_v(\varphi, \phi, v) = -M_b^{-1} M_{ba} v - M_b^{-1} (C_b \varphi_2 + g_b + \tau_{db} + C_{ba} \phi_2)$$
$$f_u(\varphi, \phi, u) = -M_b^{-1} M_{ba} u \qquad (37)$$

Let us choose u as

$$u = K_1(\varphi_1 - \varphi_{1d}) + K_2(\varphi_2 - \varphi_{1d}) - \ddot{q}_{bd} \qquad (38)$$

with $K_1 \in R^{n \times n}$ and $K_2 \in R^{n \times n}$.

Because $\frac{\partial f_v}{\partial v} = -M_b^{-1} M_{ba} < 0$, according to implicit function theorem [15] based neural network design [16], there must exist an implicit function v^* such that

$$f_v(\varphi, \phi, v^*) + f_u(\varphi, \phi, u) + u = 0 \qquad (39)$$

Refer to Sect. 2.1, we can see that there exists ideal RBFNN weight vectors such that

$$v^* = [W_v^{*\langle T \rangle} \langle \cdot \rangle S_v(z)] + \epsilon_v, \quad z = [q_{bd}, \dot{q}_{bd}, \varphi^T, \phi^T, u]^T \qquad (40)$$

where $\epsilon_v \in \mathbf{R}^n$ is the neural network approximation error vector. Let us employ RBFNNs to approximate v^* as follows:

$$v = \hat{W}_v^{\langle T \rangle} \langle \cdot \rangle S_v(z) \in R^n \quad \hat{W}_v \in R^{nl \times 1}, \hat{W}_v^{\langle T \rangle} \in R^{n \times l} \quad S_v(z) \in R^{nl \times 1} \qquad (41)$$

where one should note that $\langle T \rangle$ is different from conventional notation of transpose, and each vector $\hat{W}_v i$ in $\hat{W}_v$ is of dimension l.

Substituting $\pi = u + v$ into (37), we have

$$\dot{\varphi}_1 = \varphi_2$$
$$\dot{\varphi}_2 = -K_1(\varphi_1 - \varphi_{1d}) - K_2(\varphi_2 - \varphi_{1d}) + \ddot{q}_{bd})$$
$$\quad - M_b^{-1} M_{ba}([\tilde{W}_v^{\langle T \rangle} \langle \cdot \rangle S_v(z)] - \epsilon_v \qquad (42)$$

where $\tilde{W}_v = \hat{W}_v - W_v^*$. Define $\tilde{\varphi}_1 = \varphi_1 - \varphi_{1d}$ and $\tilde{\varphi}_2 = \varphi_2 - \varphi_{2d}$ such that $\tilde{\varphi} = \hat{\varphi} - \varphi_d$. Then, from (42) we have

$$\dot{\tilde{\varphi}} = A_W\tilde{\varphi} - [0 \ \ I]^T M_b^{-1} M_{ba}([\tilde{W}_v^{\langle T\rangle}\langle\cdot\rangle S_v(z)] - \epsilon_v)$$

$$= A_W\tilde{\varphi} - [0 \ \ I]^T M_b^{-1} \sum_{i=1}^{n} M_{bai}(\tilde{W}_{vi}^T S_{vi}(z) - \epsilon_{vi}) \qquad (43)$$

where M_{bai} represents the i-th column of matrix M_{ba}, $S_{vi}(z)$ the i-th column of $S_v(z)$, $\tilde{W}_{vi}^T$ the i-th row of $\tilde{W}_v^T$, and $A_W = \begin{bmatrix} 0 & I \\ -K_1 & -K_2 \end{bmatrix}$ satisfies the Lyapunov equation

$$A_W^T P_W + P_W A_W = -Q_W$$

i.e., for any symmetric positive definite matrix Q_W, there exists a symmetric positive definite P_W satisfying the above equation.

Theorem 2. *Consider the following weight adaptation law for RBFNN employed in (41)*

$$\dot{\hat{W}}_{vi} = \Gamma_{vi} S_{vi}(z) M_{bai}^T M_b^{-1} [0 \ \ I] P_W^T \tilde{\varphi} - \sigma \Gamma_{vi} \hat{W}_{vi} \qquad (44)$$

where $\Gamma_{vi} \in \mathbf{R}^{l\times l}$ and σ are suitably chosen as a symmetric positive definite matrix and a positive scalar, respectively. Then, the tracking errors $\tilde{\varphi}_1$ and $\tilde{\varphi}_2$ will be eventually bounded into a small neighborhood around zero.

Proof. Considering the following Lyapunov function

$$V_2(t) = \tilde{\varphi}^T P_W \tilde{\varphi} + \sum_{i=1}^{n} \tilde{W}_{vi}^T \Gamma_{vi}^{-1} \tilde{W}_{vi}^T \qquad (45)$$

and the closed-loop dynamics (43) with the update law (44), we obtain

$$\dot{V}_2(t) \leq -\lambda_{Q_W}\|\tilde{\varphi}\|^2 - 2\sigma\sum_{i=1}^{n}\|\tilde{W}_{vi}\|^2 + \varepsilon^2\|\tilde{\varphi}\|^2 + \varepsilon^2\sum_{i=1}^{n}\|\tilde{W}_{vi}\|^2$$

$$+ \frac{1}{\varepsilon^2}\epsilon_0^2\|M_b^{-1}\|\sum_{i=1}^{n}\|M_{bai}\|^2\|\|P_W[0 \ \ 1]^T\|^2 + \frac{1}{\varepsilon^2}\sigma^2\sum_{i=1}^{n}\|W_{vi}^{*\langle T\rangle}\|^2$$

where $\max|\epsilon_{vi}| \leq \epsilon_0$, λ_{Q_W} is the minimum eigenvalue of Q_W, ε is any given positive constant and we can choose it sufficiently small. Furthermore, we can choose the suitable Q_W and σ making $\lambda_{Q_W} \geq \varepsilon^2, 2\sigma \geq \varepsilon^2$, and it follows that $\dot{V}_2(t) \leq 0$ in the complementary set of a set B defined as

$$B \triangleq \left\{ (\tilde{\varphi}, \tilde{W}) \ \middle| \ \frac{\sum_{i=1}^{n}\|\tilde{W}_{vi}\|^2}{\bar{a}^2} + \frac{\|\tilde{\varphi}\|^2}{\bar{b}^2} - 1 \leq 0 \right\}$$

with $\bar{a} = \frac{1}{\varepsilon} \sum\limits_{i=1}^{n} \sqrt{\epsilon_0^2 m_1 m_2 \|P_W[0, \quad 1]^T\|^2 + \sigma^2 \|W_{vi}^{*\langle T \rangle}\|^2} / \sqrt{\lambda_{Qw} - \varepsilon^2}$, and $\bar{b} =$

$\frac{1}{\varepsilon} \sum\limits_{i=1}^{n} \sqrt{\epsilon_0^2 m_1 m_2 \|P_W[0, \quad 1]^T\|^2 + \sigma^2 \|W_{vi}^{*\langle T \rangle}\|^2} / \sqrt{2\sigma - \varepsilon^2}$, with $m_1 = \sup\{\|M_b^{-1}\|\}$
and $m_2 = \sup\{\|M_{bai}\|\}$. Obviously, the set B defined above is compact. Hence, by LaSalle's theorem, it follows that all the solutions of (43) are bounded around a neighborhood of zeros. Thus, the proof is completed.

6   Conclusion

This work has developed a framework of controlling a robot system with up to a half joints passive. The active joints are controlled using adaptive control approach with optimized performance while the passive joints are indirectly "controlled" by the desired trajectories of active joints, which are generated using NN approximation techniques. The proposed control scheme has achieved smooth motion of active joints and guaranteed tracking of passive joints.

Acknowledgement. This work was supported in part by the EU Project (FP7-PEOPLE-2010-IIF-275078); the Foundation of Key Laboratory of Autonomous Systems and Networked Control (2012A04, 2013A04), Ministry of Education, P.R. China; the Natural Science Foundation of China under Grants (51209174, 61174045, 61111130208, 61203074).

References

1. Garabini, M., Passaglia, A., Belo, F., Salaris, P., Bicchi, A.: Optimality principles in variable stiffness control: The VSA hammer. In: 2011 IEEE/RSJ International Conference on Intelligent Robots and Systems (IROS), pp. 3770–3775. IEEE (2011)
2. Xin, X., Kaneda, M.: Swing-up control for a 3-DOF gymnastic robot with passive first joint: design and analysis. IEEE Trans. Robot. **23**(6), 1277–1285 (2007)
3. Li, Z., Yang, C.: Neural-adaptive output feedback control of a class of transportation vehicles based on wheeled inverted pendulum models. IEEE Trans. Control Syst. Technol. **20**(6), 1583–1591 (2012)
4. Yang, C., Li, Z., Li, J.: Trajectory planning and optimized adaptive control for a class of wheeled inverted pendulum vehicle models. IEEE Trans. Cybern. **43**(1), 24–35 (2013)
5. Cui, R., Ge, S.S., How, V.E.B., Choo, Y.S.: Leader-follower formation control of underactuated autonomous underwater vehicles. Ocean Eng. **37**(17–18), 1491–1502 (2010)
6. Ostrowski, J.P.: Controlling more with less using nonholonomic gears. In: Proceedings of IEEE/RSJ International Conference on Intelligent Robots and Systems, 2000 (IROS 2000), vol. 1, pp. 294–299. IEEE (2000)
7. Liu, Y.J., Chen, C.L.P., Wen, G.X., Tong, S.C.: Adaptive neural output feedback tracking control for a class of uncertain discrete-time nonlinear systems. IEEE Trans. Neural Netw. **22**(7), 1162–1167 (2011)
8. Chen, W., Jiao, L.: Adaptive tracking for periodically time-varying and nonlinearly parameterized systems using multilayer neural networks. IEEE Trans. Neural Netw. **21**(2), 345–351 (2010)

9. Chen, W., Jiao, L., Li, J., Li, R.: Adaptive NN backstepping output-feedback control for stochastic nonlinear strict-feedback systems with time-varying delays. IEEE Trans. Syst. Man Cybern. B Cybern. **40**(3), 939–950 (2010)

10. Ge, S.S., Hang, C.C., Lee, T.H., Zhang, T.: Stable Adaptive Neural Network Control. Kluwer Academic, Norwell (2001)

11. Ge, S.S., Lee, T.H., Harris, C.J.: Adaptive Neural Network Control of Robotic Manipulators. World Scientific, London (1998)

12. Anderson, B.D.O., Moore, J.B.: Optimal Control: Linear Quadratic Methods. Prentice Hall, Englewood Cliffs (1990)

13. Bernstein, D.S.: Matrix Mathematics. Princeton University Press, Princeton (2005). ISBN:0-691-11802-7

14. Tayebi, A.: Adaptive iterative learning control for robot manipulators. Automatica **40**(7), 1195–1203 (2004)

15. Munkres, J.R.: Analysis on Manifolds. Addison-Wesley, Reading (1991)

16. Yang, C., Ge, S.S., Xiang, C., Chai, T., Lee, T.H.: Output feedback NN control for two classes of discrete-time systems with unknown control directions in a unified approach. IEEE Trans. Neural Netw. **19**(11), 873–1886 (2008)

Realtime Simulation-in-the-Loop Control for Agile Ground Vehicles

Nima Keivan[(✉)] and Gabe Sibley

George Washington University, Washington, DC 20052, USA
{nimski,gsibley}@gwu.edu

Abstract. In this paper we present a system for real-time control of agile ground vehicles operating in rough 3D terrain replete with bumps, berms, loop-the-loops, skidding, banked-turns and large jumps. The proposed approach fuses local-planning and feedback trajectory-tracking in a unified, simulation-based framework that operates in real-time. Experimentally we find that fast physical simulation-in-the-loop enables impressive control over difficult 3D terrain. The success of the proposed method can be attributed to the fact that it takes advantage of the full expressiveness of the inherently non-linear, terrain-dependent, highly dynamic systems involved. Performance is experimentally validated in a motion capture lab on a high-speed non-holonomic vehicle navigating a 3D map provided by an offline perception system.

1 Introduction

We present a unified approach to both planning and control that leverages accurate *physical simulation* to perform both tasks jointly and in real-time. Physical simulation is beneficial because it can model not only complex vehicle dynamics, but also vehicle-terrain interaction. For instance, simulation can include complex

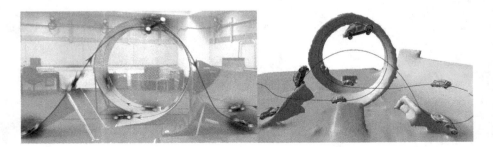

Fig. 1. Long jump and loop-the-loop experiment. A feasible trajectory is first solved through waypoints using a boundary value solver on a 3D scanned terrain model (right). The trajectory is then tracked and executed by the real-time feedback controller (left).

A. Natraj et al. (Eds.): TAROS 2013, LNAI 8069, pp. 276–287, 2014.
DOI: 10.1007/978-3-662-43645-5_29, © Springer-Verlag Berlin Heidelberg 2014

3D surface models, varying friction models, realistic contact models – indeed anything deemed necessary to more faithfully predict reality. Our approach is based on the insight that if simulation is sufficiently fast and accurate, it can be used to predict system state farther into the future. This allows us to servo on errors between desired future performance and simulated future performance. One may think of this as closing the loop using an accurate, long-range process model (Fig. 1).

For both feasible trajectory-generation and feedback-control we employ an optimization-based boundary-value solver. The solver takes as input a starting configuration, a terrain model, and a goal. The goal can be either a waypoint or a desired trajectory. The aim is to find a *command sequence, $c(\mathbf{p})$*, that will drive, in simulation *and in reality*, the vehicle from the start to the goal, over the intervening terrain. The command sequence itself is a low-degree-of-freedom generator parameterized by $\mathbf{p}$, such as a polynomial, spline, etc. The solver is used to find the parameters that yield a desired command sequence. We find that this approach allows accurate real-time planning and control in full 3D environments, with complex vehicle dynamics and terrain interaction, including slipping, jumping and driving upside-down.

The use of motion primitives [17] and stochastic search methods such as RRT and RRT* [12,13] and probabilistic methods such as PRMs [2,4] and POMDPs have resulted in algorithms that successfully navigate complex obstacle fields even in higher order configuration space. A major advantage of these methods is that they can employ nonlinear dynamics models thereby enabling physically accurate planning in complex environments without approximation or linearization. However, this advantage comes at a performance price as stochastic methods invariably sample infeasible trajectories. Conversely, optimization based methods [10] employ effective initial guesses and numerical or analytical optimization techniques to rapidly converge on optimal paths. However due to the reliance on the accuracy of the initial guess, these methods are susceptible to failure or suboptimal performance depending on the quality of this guess. The quality, optimality and methodology of the plans notwithstanding, their open loop performance in real robots is inevitably impaired by the existence of imperfections or extraneous inputs that may not have been included in the dynamics model. Therefore for real-life applications, some form of closed loop control is desired.

In contrast to simulation-in-the-loop or model-predictive control, traditional feedback systems use static and/or dynamic feedback of the state to determine the controls for the next time steps. To aid stability analysis, traditional approaches also typically model the system with simplified, analytically treatable, vehicle models. Recent developments have resulted in methods allowing the calculation of Lyapunov functions for nonlinear systems [11] and defining graphs of Lyapunov-stable regions around states as in the case of LQR-Trees [20]. These methods rely on the linearization of the state transfer function in order to obtain analytically tractable control policies, and the automatic recovery of stability regions. Linearized time-varying systems often entail substantial gain-tuning, which is known to be an arduous and time consuming process, especially due to their non-intuitive nature [6]. In contrast, with simulation-in-the-loop based approaches, we find that it is possible to automatically learn physically mean-ingful parameters that lead directly to improved system performance – that is, learning to make simulation match reality.

Recent developments in model predictive control (MPC) [9] and learning-based model predictive control (LBMPC) [3] both implement model-based con-trol schemes and infer the underlying model parameters. Likewise, our system relies on accurate models of the *physical parameters* that define the system. In our case, these models are learned beforehand using online non-linear regression [15]. We find that expressive models learned via online regression are advanta-geous as they allow accurate simulation, farther into the future, over difficult and challenging terrain.

2   Methodology

The proposed approach relies on simulating the effect of applying a sequence of linear and angular velocity commands which are defined by a low-degree of freedom control-function, $c(\mathbf{p})$. Feasible trajectories are generated by solving a boundary value problem for the parameterized control function that links two waypoints – i.e., the command sequence that will move the car between these two waypoints. To control the vehicle we repeatedly minimize the difference between a forward-simulated vehicle and the desired trajectory. This approach leverages accurate simulation of the vehicle and terrain dynamics at the expense of a model which is not analytically differentiable. Therefore, in order to generate a control law to optimize either the boundary or the trajectory cost, we will rely on finite-difference estimations of the gradient of the vehicle model, in a least-squares optimization setting.

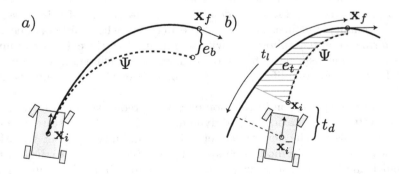

Fig. 2. (a) Planning to a waypoint: e_b is the boundary residual between a trajectory and the waypoint $\mathbf{x}_f$. (b) Trajectory tracking: e_t is the trajectory residual calculated between the simulation Ψ and the reference trajectory over a finite horizon t_l. Control delay compensation is shown where the delay is t_d (For more refer to Sect. 2.3).

2.1 Optimization

The boundary error as shown in Fig. 2a, is used when planning feasible trajectories between waypoints and is formulated as:

$$e_b = \left\| \mathbf{x}_{lf} \boxminus \Psi_l \left(\mathbf{c} \left(\mathbf{p} \right), \mathbf{x}_i, t_f \right) \right\|_2$$

We define the function $\Psi_l \left(\mathbf{c}, \mathbf{x}_i, t_f \right)$ as representing the numeric simulation, where $c \left(\mathbf{p} \right)$ is the control function defined by parameters $\mathbf{p}$, $\mathbf{x}_i$ is the starting state, and t_f is the time instant for which we desire the state of the vehicle. Ψ_l returns the state $\phi_{lt_f} = \begin{bmatrix} \mathbf{T}_{lv} & \mathbf{v} \end{bmatrix}$ where $\mathbf{v} \in \mathbb{R}^3$ is the velocity vector in local coordinates and $\mathbf{T}_{lv} \in \mathbb{R}^{4 \times 4}$ is the transformation matrix from vehicle to the local coordinates. The local coordinates are defined by the average of the initial and final waypoint terrain normals. $\mathbf{x}_f = \begin{bmatrix} \mathbf{T}_{lv} & \mathbf{v} \end{bmatrix}$ is the desired state at the end of the trajectory in local coordinates, set by a waypoint. The optimization minimizes the weighted square norm of the error vector by varying the parameter vector $\mathbf{p}$, and consequently the control function that drives the vehicle forward in time. The operator $\boxminus$ calculates the velocity and $se\left(3\right)$ pose error between two vehicle states.

The trajectory cost is shown in Fig. 2b and is used to track a reference trajectory. The error vector is taken between samples on the simulated trajectory (defined by Ψ_l) and a finite horizon over the reference trajectory, $\mathcal{X} = \{\mathbf{x}_0 ... \mathbf{x}_n\}$:

$$e_t = \left\| \sum_{j=0}^{n} w_j \mathbf{x}_{lj} \boxminus \Psi_l \left(\mathbf{c}, \mathbf{x}_i, t_j \right) + t_f \right\|_2$$

Where t_l is the duration of the finite horizon and is included in the error to favor time optimal solutions, that is the optimization will search for the shortest path that converges to the trajectory being tracked. This error is a weighted average between the reference trajectory set $\mathcal{X}$ and the simulated trajectory given by Ψ. The weighting imposed by the set $\mathcal{W} = \{w_0 ... w_n\}$, $\sum \mathcal{W} = 1$ is required to ensure more importance is placed at the end of the trajectory rather than the beginning, since due to the non-holonomic constraints, the initial cost cannot be minimized. In our implementation, a simple linear weighting scheme was used from the beginning to the end of the trajectory.

The optimization attempts to minimize the error vector by the weighted least squares optimization $\left(J^T W J \right) \mathbf{p} = J^T W \mathbf{r}$ where W is a diagonal matrix of weights for each individual row of the error vector and $\mathbf{r}$ is the error vector. The weight matrix is required as we are optimizing cartesian poses, angles and velocities in the same vector, and a rational weighting scheme based on the importance of the relative units is required for efficient optimization. These weights can also reflect the fact that certain degrees of freedom are more important to minimize than others (i.e. in-plane yaw, vs. pitch and roll).

The Jacobian is defined as $J = \partial \mathbf{e} / \partial \mathbf{p}$. Due to the inclusion of the simulation function Ψ in the error vector, the Jacobian cannot be evaluated analytically and must be obtained through finite differences. In order to obtain real-time performance, we calculate the Jacobian via a multi-threaded simulation scheme, whereby each column of the Jacobian (which corresponds to an evaluation of the simulation function Ψ) is undertaken in a separate thread. Due to the extreme nonlinearity caused by the terrain, a parallelized trust region damping scheme is used.

In the trajectory cost function, we have used the set $\mathcal{X}$ to represent the reference trajectory. The duration of this set, denoted as t_l, represents the finite horizon for which we optimize when tracking a feasible trajectory. As time itself is a cost in the optimization and t_l is allowed to change (due to the change in $\mathbf{x}_f$), the solutions will favor time-optimal trajectories, if they can be planned without increasing the trajectory cost. The choice of initial t_l however, has an impact on convergence. If t_l is too small, there will be insufficient time and space for the vehicle to be able to close the gap on the trajectory. In the pathological case, this can lead to a local minimum whereby t_l cannot be incrementally increased by the optimization, and convergence will fail. Conversely, if t_l is set too large, there may be insufficient expressivity in the command sequence to solve solutions that adhere to the reference trajectory. This problem has been studied in the literature and heuristics-based methods exist to find the optimal finite horizon [8]. In our solution, we have experimentally determined $t_l = 0.4s$ as a suitable horizon duration to initialize the solution. The optimization will then attempt to reduce (and may potentially increase) the finite horizon time.

2.2 Control Parametrization

Our vehicle model is controlled via steering and acceleration commands. As stated previously, we are interested in a low degree of freedom control function, $\mathbf{c} = C(\mathbf{p})$ which generates an expressive command sequence.

Here $\mathbf{p}$ is a vector which parametrizes the command sequence, which is then discretized for simulation. In our previous work [15] we had employed the use of cubic curvature polynomials [13] to parametrize steering. However, cubic curvature polynomials require expensive optimizations in order to generate command sequences given the boundary conditions. This necessitates the use of pre-calculated lookup tables and an additional optimization in order to produce an initial guess for real-time planning. Therefore, in the proposed methodology, we have employed the use of 2D bezier curves for steering, and a constant acceleration parameter to control velocity. Bezier curves have been used extensively in path planning for autonomous vehicles [5,14] due to the ease with which one can extract the path curvature given the control points and final pose. The desired quality for the chosen parameterization (cubic-curvature, bezier curves, etc.) is that (a) the resulting command sequence is "expressive" – i.e., it can actually cause the vehicle to move in a desired fashion, and (b) that it is parameterized by few degrees of freedom – to facilitate fast finite differences optimization. In Appendix A we detail how we have used bezier curves to this effect.

2.3 Feedfoward

An advantage of the simulation-in-the-loop approach is that it provides easy access to physically meaningful parameters like terrain surface and vehicle configuration.

These can be used for feedforward compensation to help the optimization avoid local minima that litter the cost landscape. We employed the use of three feedforward terms in our implementation: gravity, wheel friction and control-delay.

Gravity feedforward is useful to mitigate the effects of steep and undulating terrain. A simple gravity feedforward model was implemented as

$$a_g = \frac{1}{m} \sum_{i=1}^{4} \mathbf{n_i} F_i . \mathbf{v}_x$$

where $\mathbf{n_i}$ is the surface normal at the point of contact of wheel i, F_i is the normal force at contact for wheel i, $\mathbf{v}_x$ is the axis orientation vector for the vehicle chassis and m is the mass of the vehicle.

Friction feedforward is used to compensate for wheel resistance. The acceleration compensation for friction is calculated as

$$a_f = \frac{1}{m} \sum_{i=1}^{4} \mu F_i$$

where μ is the friction coefficient and F_i is the normal force of wheel i. Once a_g and a_f are calculated for the current time step, they are fed forward into the acceleration parameter a of the next time step. The final acceleration used in the simulation is given by $a_{totoal} = a + a_g + a_f$ where a is the parameter used in the optimization. This type of terrain-aware feed-forward compensation (which uses terms such as F_i and $\mathbf{n}_i$) is made possible by a full physical simulation of the vehicle *and* terrain dynamics.

Control delay feedforward is used to compensate for both delayed state estimation and delays in the control pipeline. Compensating for these phenomenon increases stability. In order to compensate for control delay, we make use of a *control queue* which serves to store all commands that are sent to the vehicle. The current state of the vehicle $\mathbf{x}_i^-$ can then be transformed to the compensated state $\mathbf{x}_i$ as follows

$$\mathbf{x}_i = \Psi \left(\mathbf{c}_q \left(t_i - t_d : t_i \right), \mathbf{x}_i^-, t_d \right)$$

Where Ψ is the simulation function as defined in Sect. 2.1. To compensate for control delay, we simulate the vehicle forward by the control delay duration t_d, using the control queue $\mathbf{c}_q \left(t_i - t_d : t_i \right)$, which contains the commands sent to the car from time $t_i - t_d$ up to the current time t_i. The augmented state $\mathbf{x}_i$ is then used to generate the next set of commands. In this fashion, we are predicting the pose of the vehicle at the time after which the current commands will begin executing $(t_i + t_d)$, using commands already in the execution pipeline $(\mathbf{c}_q)$.

2.4   Vehicle Modeling

Our experiments use a modified racing remote-control vehicle with rear-wheel electric propulsion, rack and pinion steering and spring/damper suspension. The parameters described in the models below are learnt automatically using an online nonlinear regression [15]. In order to accurately simulate the dynamics of our vehicle, and fully consider the uneven terrain, we use a double track vehicle model with suspension dynamics, as detailed in [7]. The chassis is modeled as a rigid body with forces and moments imparted due to wheel contacts and motor torques. The wheel forces are modeled using spring/damper dynamics as $F_w = kx + c\dot{x}$ where c is the damping coefficient, k is the spring stiffness and x is the deviation from the spring rest state. As F_w acts along the suspension axis, moments are induced about the center of gravity of the vehicle. To obtain x and $\dot{x}$, we leverage an off the shelf physics simulation engine [1] along with a scanned 3D model of the terrain, represented as a dense triangle mesh, to calculate the wheel position and velocities.

When accelerating or braking, the electric motor applies a torque to the rear wheels which is proportional to the current passing through the motor winding. This current is itself dependent on the voltage applied to the motor as well as the current rotor speed. The motor controller hardware employed applies a voltage across the rotor terminals for a given input signal. Therefore, the torque applied to the rear wheels is modeled as $T = V.C_{T_{stall}} - \omega.C_\omega$ where V is the voltage applied to the motor, $C_{T_{stall}}$ is a constant relating the stall torque of the motor to the applied voltage, ω is the motor speed and C_ω defines the slope with which the torque diminishes as the motor gains speed. The torque resulting from this equation is then applied to the rear wheel axles. As expected, this rear-wheel torque also induces an opposing torque about the CG, resulting in "wheelies" if the applied torque is too high.

To enforce non-holonomic constraints, sideways motion from the wheels must be eliminated. We therefore enforce a constraint on the relative velocities of the wheels and the terrain, which applies a force to enforce the non-holonomic constraints of the vehicle, and prevent lateral slip. However in practice, slip conditions can cause violations of non-holonomic constraints. To model these phenomena, we use the Magic Tire Model [16] which defines the maximum lateral force applicable as a function of the normal force F_w, the slip angle and 4 coefficients.

3   Results

The local planner and trajectory tracker were experimentally tested on a wide array of challenging environments including jumps, loop-the-loops and quarter pipes. The experiments were carried out on a small high-speed autonomous robot in a motion capture environment. In order to build an accurate model of the vehicle, a machine learning based approach to model identification was used in conjunction with the accurate simulation model in order to infer the difficult to measure parameters such as tire model coefficients, friction coefficients and electric motor torque parameters. For state estimation, global pose estimates from a Vicon motion capture system operating at 120 Hz were fused with IMU measurements at 400 Hz in a sliding window filter in order to obtain accurate pose and velocity estimates. The local planning and trajectory tracking were

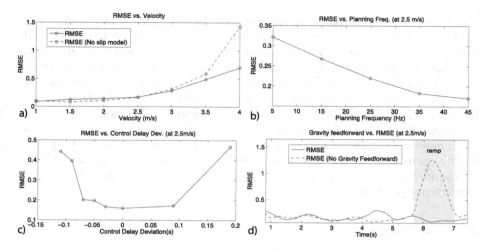

Fig. 3. All experiments are performed on the figure-8 trajectory in Fig. 4. The RMSE error is taken as the average trajectory residual taken over two consecutive laps of the figure-8 consisting of a weighted combination of translation, heading and velocity error. (a) Trajectory error with varying velocity setpoints, showing the effect of tire slip model at higher velocities. (b) Effects of planning frequency on the error. (c) Effects of incorrect control delay feedforward on RMSE. (d) The effects of gravity feedforward. Large errors are apparent for both solutions at the start of the trajectory due to initial convergence, and on the ramp when gravity feedforward has been turned off.

performed in real-time on an Intel i7 laptop with 4 hyper-threaded cores. Replanning rates with the specified hardware was between 40 and 60 Hz depending on the terrain.

For results on solving and tracking the aforementioned challenging environments, please refer to the video submission. In order to qualitatively explore the effects of parameters on the performance of the real-time trajectory tracker, a number of experiments were performed over the figure-8 trajectory shown in Fig. 4. A feasible trajectory was solved between a series of manually placed waypoints, and the controller was subsequently used to track the trajectory with the experimental vehicle.

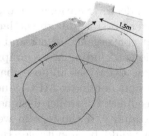

In Fig. 3a, the effects of the tire model described in Sect. 2.4 on the tracking error are shown. At higher speeds, there is significant slip between the plastic wheels and the carpet floor. Furthermore, the acceleration applied reduces the normal forces

Fig. 4. Experimental figure 8 over flat terrain and quarter pipe, showing boundary value solution trajectory

on the front wheels, increasing slip. The inclusion of the tire model allows the optimization to fold in these effects and correct the steering. Figure 3b shows the effect of planning frequency on the error. It is apparent that stable performance is observed down to re-planning rates of 5 Hz, however this is at the expensive of higher tracking error. Figure 3c demonstrates the effect of control delay feed-forward on tracking accuracy. The control delay of the system was determined via model identification to be 0.11s. The negative effect on tracking accuracy as the control delay parameter deviates from the true value is due to the inability to properly compensate the control inputs for delays. Figure 3d demonstrates the effect of gravity feedforward on tracking accuracy. The plot demonstrates two separate runs with sections taking place on the ramp. It can be seen that when gravity feedforward is not used, there is a large increase in tracking error on the ramp. Due to the increased acceleration required to maintain velocity on the incline, the lack of gravity feedforward leads to bad initialization and convergence problems.

4   Discussion

Our approach to local planning and control is different in that it does not make use of an analytically differentiable model to provide a feedback command-sequence. Rather, we solve an optimization problem involving the full simulation. This approach works not only for obtaining feasible paths given the boundary conditions (i.e. local planning), but also in the case of trajectory tracking. In the latter case, continuous re-planning based on an accurate vehicle and terrain model provides stability and accurate tracking even without an explicit feedback law. This tracking stability is observed down to very low re-planning frequencies.

However, that is not to say that continuous re-planning does not constitute feedback control. On the contrary, perturbations are folded into subsequent plans as they are sensed by the state estimation. However, the level of detail folded into the simulation ensures that unexpected perturbations are rare, as all extraneous effects of the terrain and vehicle model are already considered.

Indeed, traditional guarantees about stability, given a perturbation are no longer possible due to the lack of an analytical model, however our results show that convergence is fast and reliable for a wide range of terrain and vehicle conditions including large jumps, loop-the-loops, quarter pipes, bumps and berms. These situations all pose challenges to traditional control schemes including linear and nonlinear MPC, due to the non-analytical terrain term. Even though a feedback controller designed with an analytical model would yield guarantees of stability, if the terrain were to be sufficiently perturbed, divergence of control would be inevitable. However, with simulation-in-the-loop the knowledge obtained from simulating the future would provide powerful predictions of the failure modes, and would also facilitate triggering of abort maneuvers.

Furthermore, the use of an accurate simulation model replaces gains in traditional control schemes with physically meaningful model parameters. Whereas tuning LQR, PID and other feedback control schemes involves the calculation and setting of abstract gains relating to error and control amplitudes, the only parameters that need to be tuned in the proposed methodology are the model parameters, which have physical meanings.

5   Conclusions

We have presented a unified approach to planning and control based on fast physical simulation-in-the-loop. To find feasible trajectories between waypoints, we solve a boundary-value-problem that depends on a low degree-of-freedom but highly expressive control function that drives a simulated vehicle forward in time. To track the reference trajectory, a trajectory-cost is repeatedly minimized using the same optimization. Similar to receding horizon or model predictive control, constant re-planning provides feedback to mitigate deviation from the desired trajectory, however we avoid simplifying the underlying model. Simulation-in-the-loop can include any phenomenon worth modeling, such as 3D terrain, complex tire models, friction, slipping and jumping. Simulation also provides a means to fold in perturbations before they happen, and to preemptively adjust the control sequence accordingly. The results show that this methodology is able to control a high-speed autonomous vehicle through challenging terrain such as large jumps, skids, steep banks and loop-the-loops.

Appendix A

We define the bezier curve $\mathbf{P}(s)$, $s \in [0,1]$ as $\mathbf{P}(s) = \sum_{i=0}^{n} B_i^n(s)\mathbf{P}_i$ where $\mathbf{P}_i \in \mathbb{R}^2$ is the ith control point. We have initially parametrized the bezier curve with the variable s rather than the traditional t, as we desire the final curve to indeed be parametrized by time, which we represent as t. The relationship between s and t depends on the evolution of the vehicle velocity. As we have proposed a constant acceleration parameter a, we can parametrize the bezier curve w.r.t time as $\mathbf{P}(s(t))$ where $s(t) = t/t_{total} = t/(2d/(v_i + v_f)$, d is the distance along the bezier curve, and v_i and v_f are the initial and final velocities respectively. The curvature of the bezier curve at any value of t defined as $\kappa(t)$ can then be calculated using the analytical first and second derivatives of the curve in x and y [5]. The steering angle θ_s can then be obtained from κ, based on a simplified symmetric bicycle steering model with zero slip [18] $\theta_s = tan^{-1}(l\kappa)$ where l is the wheel base.

The order of the bezier curve is chosen such that sufficient expressivity exists to satisfy boundary conditions. For trajectory tracking, a major requirement is C^2 continuity. This is needed to ensure maximum feasibility of steering commands, as the actuators cannot be discontinuously controlled, when switching from one command sequence to the next. To independently constrain the curvature on the boundary of a bezier curve, three control points are required [19], therefore we use fifth degree bezier curves ($n = 6$) in order to independently satisfy the constraint at both ends of the curve.

Given the bezier curve control points, the parameter vector $\mathbf{p}'$ can be expressed as $\mathbf{p}' = \begin{bmatrix} \mathbf{P}_0, \dots, \mathbf{P}_6 \, a \end{bmatrix}^T$ where $\mathbf{P}_0, \dots, \mathbf{P}_6 \in \mathbb{R}^2$ are the control points. $\mathbf{p}'$ is defined by 13 parameters: 2 for each control point and 1 for acceleration. There are three gauge freedoms associated with transforms in $\mathbb{SE}2$, reducing the actual parameter space to 10. However, properties of the bezier curve can be exploited to further reduce the number of parameters. Given the initial projected 2D pose ($\mathbf{x}_{i_{proj}} = \begin{bmatrix} x_i \ y_i \ \theta_i \ \kappa_i \end{bmatrix}$, as described below), the position, heading and curvature of the vehicle can be constrained (the curvature is constrained by the current curvature κ_i). Consequently, the first three control points can be parametrized using only h_i and a_i (refer to [19]). Fixing x_i, y_i and θ_i also removes the gauge freedoms of the control points. At the other end of the bezier curve, $\mathbf{x}_{f_{proj}} = \begin{bmatrix} x_f \ y_f \ \theta_f \ \kappa_f \end{bmatrix}$ is used to reduce the parameter space of the final 3 control points from 6 to 3, as the end curvature is also constrained by k_f to enforce C^2 continuity. This leaves the parameters h_i, a_i, h_f and a_f. To further reduce the parameter space, these values were set as follows: $h_i = a_i = h_f = a_f = \frac{1}{5} \left\| \begin{bmatrix} x_f \ y_f \end{bmatrix} - \begin{bmatrix} x_i \ y_i \end{bmatrix} \right\|$ where the factor of 1/5 times the distance between the start and end points was empirically determined to result in adequate control laws for a wide range of start and goal positions, headings and curvatures. Therefore, the initial parameter space $\mathbf{p}$ can be obtained from the reduced parameter space: $\mathbf{p}' = \mathcal{B}(\mathbf{p}, \mathbf{x}_{i_{proj}}, \kappa_f)$ where $\mathbf{p} = \begin{bmatrix} x_f \ y_f \ \theta_f \ a \end{bmatrix}^T$, and $\mathcal{B}$ is a function which maps the reduced parameter space to the full 5th degree bezier control points. Note that $\mathcal{B}()$ requires the initial state $\mathbf{x}_i$ to constrain the first three control points, and the final curvature κ_f to constrain the last 3 points according to $\mathbf{p}$. Therefore the command-sequence used in the optimization is defined as $\mathbf{c} = C\left(\mathbf{x}_i, \mathcal{B}(\mathbf{p}, \mathbf{x}_{i_{proj}}, \kappa_f) \right)$.

The control law parametrization scheme detailed previously uses bezier curves in $\mathbb{R}^2$. While control commands generated from these curves would perfectly steer an ideal vehicle in $\mathbb{R}^2$, the same statement can not be made for a vehicle in $\mathbb{R}^3$ driving over rough terrain. For this reason, we are interested in initializing the steering command-sequence in a way which places us in the basin of attraction of the optimization that follows.

To initialize the bezier curve, we project the pose of the vehicle onto a plane in $\mathbb{R}^2$, the normal of which is calculated as the average of the terrain normals at the start and end of the trajectory. This serves to maximize the likelihood that following the curvature of the bezier curve will lead the vehicle to follow the shape of the curve. We define the projection as $\mathbf{x}_{proj} \in \mathbb{R}^4 = \mathcal{P}(\mathbf{T}_{pw}\mathbf{x}_w, \kappa)$. Where x_{proj} is the state vector in projected $\mathbb{R}^2$ coordinates defined as $\begin{bmatrix} x \ y \ \theta \ \kappa \end{bmatrix}$, $\mathbf{T}_{pw}$ is the $\mathbb{R}^{4 \times 4}$ transformation matrix from world coordinates to the projection plane coordinates (obtained from the terrain normals), and $\mathcal{P}$ is the projection function, which takes the x and y components, and projects the vehicle theta onto the plane. Once both $\mathbf{x}_{i_{proj}}$ and $\mathbf{x}_{f_{proj}}$ have been obtained, a bezier curve can be initialized using the aforementioned re-parametrization function $\mathcal{B}(\mathbf{p}', \mathbf{x}_{i_{proj}}, \kappa_\mathbf{f})$, which returns the initial bezier control points.

References

1. Bullet physics engine. http://bulletphysics.org. Accessed July 2012
2. Amato, N.M., Wu, Y.: A randomized roadmap method for path and manipulation planning. In: IEEE International Conference on Robotics and Automation, pp. 113–120 (1996)

3. Aswani, A., Bouffard, P., Tomlin, C.: Extensions of learning-based model predictive control for real-time application to a quadrotor helicopter. In: ACC 2012, Montreal, Canada, June 2012 (to appear)
4. Barraquand, J., Kavraki, L., Latombe, J.-C., Li, T.-Y., Motwani, R., Raghavan, P.: A random sampling scheme for path planning. Int. J. Rob. Res. **16**, 759–774 (1996)
5. Choi, J.-W., Curry, R., Elkaim, G.H.: Continuous curvature path generation based on bézier curves for autonomous vehicles. IAENG Int. J. Appl. Math. **40** (2010)
6. Divelbiss, A.W., Wen, J.T.: Trajectory tracking control of a car-trailer system. IEEE Trans. Control Syst. Technol. **5**(3), 269–278 (1997)
7. Genta, G.: Motor Vehicle Dynamics: Modeling and Simulation. Advances in Fuzzy Systems. World Scientific Publishing Company, Incorporated (1997)
8. Grieder, P., Borrelli, F., Torrisi, F., Morari, M.: Computation of the constrained infinite time linear quadratic regulator. In: Proceedings of the 2003 American Control Conference, vol. 6, pp. 4711–4716 (2003)
9. Howard, T.M., Green, C.J., Kelly, A.: Receding horizon model-predictive control for mobile robot navigation of intricate paths. In: Howard, A., Iagnemma, K., Kelly, A. (eds.) Field and Service Robotics. STAR, vol. 62, pp. 69–78. Springer, Heidelberg (2010)
10. Howard, T.M., Kelly, A.: Optimal rough terrain trajectory generation for wheeled mobile robots. Int. J. Rob. Res. **26**(2), 141–166 (2007)
11. Johansen, T.A., Norway, N.-T.: Computation of lyapunov functions for smooth nonlinear systems using convex optimization. Automatica **36**, 1617–1626 (1999)
12. Karaman, S., Frazzoli, E.: Incremental sampling-based algorithms for optimal motion planning. In: Robotics: Science and Systems (RSS), Zaragoza, Spain, June 2010
13. Lavalle, S.M.: Rapidly-exploring random trees: a new tool for path planning. Technical report (1998)
14. Miller, I., Lupashin, S., Zych, N., Moran, P., Schimpf, B., Nathan, A., Garcia, E.: Cornell university's 2005 darpa grand challenge entry. J. Field Robot. **23**(8), 625–652 (2006)
15. Lovegrove, S., Keivan, N., Sibley, G.: A holistic framework for planning, real-time control and model learning for high-speed ground vehicle navigation over rough 3d terrain. In: IROS (2012)
16. Pacejka, H.B., Society of Automotive Engineers: Tire and Vehicle Dynamics. SAE-R. Society of Automotive Engineers, Incorporated (2006)
17. Pivtoraiko, M., Kelly, A.: Kinodynamic motion planning with state lattice motion primitives. In: Proceedings of the IEEE/RSJ International Conference on Intelligent Robots and Systems (2011)
18. Rajamani, R.: Vehicle Dynamics and Control. Mechanical Engineering Series. Springer, New York (2011)
19. Sederberg, T.W.: Computer Aided Geometric Design. Brigham Young University, April 2007
20. Tedrake, R.: LQR-trees: feedback motion planning on sparse randomized trees. In: Proceedings of Robotics: Science and Systems, Seattle, USA, June 2009

Humanoid and Robotic Arm

Development of a Highly Dexterous Robotic Hand with Independent Finger Movements for Amputee Training

Ali H. Al-Timemy, Alexandre Brochard, Guido Bugmann[✉],
and Javier Escudero

School of Computing and Mathematics, Plymouth University, Plymouth, UK
alialtimemy2006@yahoo.com, brochard.al@gmail.com,
g.bugmann@plymouth.ac.uk, javier.escudero@ed.ac.uk

Around the world, there are thousands of people who lost a hand during war or as a consequence of an accident. Artificial hand prosthesis controlled by surface electromyography (EMG) signals offers a promising solution to improve the quality of life of amputees. As part of the process of prosthesis fitting, an occupational therapist will try to train the amputee with the help of a physical prosthesis [1] that is not actually fitted, but only displayed to provide visual feedback, but these are expensive (>£16,000). This training should be performed for long periods of time at the rehabilitation centre prior to the prosthesis fitting. It aims at improving the generation of nerve signals for capture by EMG probes, and at tuning of the EMG pattern recognition algorithm to the actions most suited for each amputee [2]. This part of the rehabilitation process can be made more efficient and more widely available through the use of a low-cost actuated hand with the same degrees of freedom as the prosthetic device to be fitted.

To address this need, a custom-built fully-articulated robotic hand has been developed at Plymouth University. The hand needed to have five fully articulated fingers and the ability to perform hand rotations in 2 directions. In addition, the cost of the hand should be within the margin of £250 in order for it to be affordable for academic and clinical purposes. The robotic hand consists of three main components, the mechanical hand, servo motors and the control unit.

(1) The mechanical component of the hand is implemented by rapid prototyping of a 3D open-source hand model (http://www.thingiverse.com/thing:14986). A total of 28 parts of the hand were printed by the 3D printer with a highly durable acrylonitrile butadiene styrene (ABS) material at Plymouth University. An extrinsic design was adopted, where the hand is actuated by tendons driven by servo motors located outside the hand.

(2) In order to actuate the hand, five tendons were constructed with high-strength (20.5 kg) and abrasion-resistant fishing line and were connected to the finger tip of each finger through a grove inside each finger. Highly durable dental rubber bands were used to passively return each finger to its rest position. After finishing the hand design and fabrication, the finger tips were covered with a black rubber material with to facilitate object gripping. Six servo motors (TowerPro MG995) were used to actuate the hand through six tendons. The motors were assembled into a custom-build metal framework and they were connected through the tendon

A. Natraj et al. (Eds.): TAROS 2013, LNAI 8069, pp. 291–293, 2014.
DOI: 10.1007/978-3-662-43645-5_30, © Springer-Verlag Berlin Heidelberg 2014

to each of the 5 fingers (little, ring, middle, index and thumb). A sixth motor was assembled at the base of the mount to rotate the hand in 2 directions. Figure 1(A) shows the complete robotic hand with the location of the six servo motors and the components of the control unit. After assembling the motors on the framework, the motors were calibrated to open and close each finger and the initial and final motor positions were recorded for future actuation. The hand can firmly hold objects of moderate weight, such as a cordless drill.

(3)   The control unit of the hand consisted of two boards, the microcontroller board and the power supply board for the motors. The Arduino Uno microcontroller receives hand posture commands from a PC via the serial port and converts them into commands for individual servo motors. A second board was designed to supply the 5 volts, with current limitation for force control, to the six servo motors. In total, 18 different hand and finger postures were pre-programmed and built into the Arduino. The robotic hand is controlled either with EMG signals via the Matlab interface on the PC [2] or via an electrical switch for a manual selection of the desired movement.

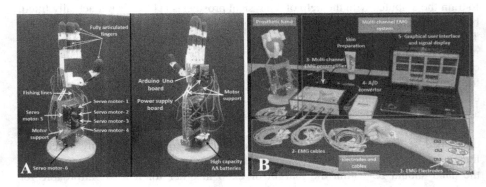

Fig. 1. (A) The components of the custom-build highly dextrous fully articulated robotic hand. (B) The complete custom-build robotic hand system controlled with multi-channel EMG.

The total cost of the 3 components of the hand was £232 which is below the estimated budget (£250). This cost is very low compared to the commercially available fully articulated hands and makes possible its widespread use possible for rehabilitation.

This hand can also be used by researchers who are working in the field of myoelectric control. Figure 1B illustrates its integration with a custom-build, multi-channel EMG system designed to identify and control a large number of finger and hand movements with EMG signals [2].

Acknowledgments. This work is supported by the Ministry of Higher Education scholarship/ Iraq.

References

1. Muzumdar, A.: Powered Upper Limb Prostheses: Control, Implementation and Clinical Application. Springer, Heidelberg (2004)
2. Al-Timemy, A.H., Bugmann, G., Escudero, J., Outram, N.: Classification of finger movements for the control of dexterous hand prosthesis with surface electromyogram. IEEE J. Biomed. Health Inf. 17(3), 608–618 (2013)

A Robotic Suit Controlled by the Human Brain for People Suffering from Quadriplegia

Alicia Casals[1], Pasquale Fedele[2], Tadeusz Marek[3], Rezia Molfino[4],
Giovanni Gerardo Muscolo[4,5(✉)], and Carmine Tommaso Recchiuto[6]

[1] Fundació Privada Institut de Bioenginyeria de Catalunya, Barcelona, Spain
[2] Liquidweb s.r.l., Siena, Italy
[3] Univwersytet Jagiellonski, Krakow, Poland
[4] PMAR Lab, Department of Applied Mechanics and Machine Design,
Scuola Politecnica, University of Genova, Genoa, Italy
info@humanot.it
[5] Creative and Visionary Design Lab, Humanot s.r.l., Prato, Italy
[6] Electro-Informatic lab, Humanot s.r.l., Prato, Italy

Abstract. The authors present an introductory work for the implementation of an international cooperative project aimed at designing, developing and validating a new generation of ergonomic robotic suits, wearable by the users and controlled by the human brain. The aim of the proposers is to allow the motion of people affected by paralysis or with reduced motor abilities. Therefore, the project will focus on the fusion between neuroergonomics and robotics, also by means of brain-machine interfaces. Breakthrough solutions will compose the advanced robotic suit, endowed with soft structures to increment safety and human comfort, and with an advanced real-time control that takes into account the interaction with the human body.

Keywords: Neuroergonomics · Brain computer interfaces · Robotics · Robotic suits · Compliant actuators · Exoskeleton · EEG · Dynamic balance control

1 Concept and Research Objectives

Quadriplegia and paraplegia are often caused by spinal cord injuries but can also be caused by nerve diseases such as multiple sclerosis and amyotrophic lateral sclerosis. Inactivity due to paraplegia and quadriplegia can cause additional problems like bedsores, spastic limbs, weakened bones and chronic pain. People with paraplegia and quadriplegia may also become depressed because of social isolation, lack of emotional support, and increased dependence on others [1]. Based on the recent advances in rehabilitation robotics and Brain-Computer Interfaces [2], the aim of the international consortium is to fuse together neuroergonomics and robotics by developing a novel robotic suit able to support the patient's body and controlled by the user's brain by means of an advanced Brain Machine Interface. The project will start with a study of the physiological characteristics of the body of persons suffering from quadriplegia. Early studies will be conducted with real patients and healthy subjects and these

A. Natraj et al. (Eds.): TAROS 2013, LNAI 8069, pp. 294–295, 2014.
DOI: 10.1007/978-3-662-43645-5_31, © Springer-Verlag Berlin Heidelberg 2014

studies will be primarily aimed at identifying the ergonomically correct zones to connect the robotic structure to the human body and a set of reference trajectories for upper and lower limbs motion. In parallel, we aim at measuring the brain neural signals related to the decision-making skills applying advanced EEG analysis techniques to identify neural markers of movement based on the availability of perceptual information arising from the movements. A dense array system designed to study neuronal activity by recording brain waves via 256 electrodes will allow performing the necessary neuropsychological tests. Moreover, an eye-tracker will provide data such as head position, eye position, direction of eye movement in all dimensions, eye closure and pupil dilation. EEG recording will be synchronized with eye movement tracking, and the precise timing of salient events in neuroergonomic tests will be marked on recorded files. All the analysis will be translated into neuroscientific models and design guidelines for the robotic suit. Moreover, the resulting models will be used as inputs of the robot through advanced Brain-Computer Interfaces. The new BCI will identify EEG patterns to detect the intent of the user to move, the direction of movement, stand up or sit down, perform simple and specific tasks, by managing 12 kinds of conscious thoughts of movement (left, right, up, down, push, and pull, rotations, etc.). A big effort will then be employed in the design and development of the robotic suit, able to walk, also in complex environment, such as obstacle, stairs, while carrying a human body; thus, the robot will be endowed with force and tactile sensors, accelerometers and gyroscopes, encoder joint sensors and it will make use of sensor fusion techniques to achieve a stable balancing, in order to have a high level of autonomy while moving in a unknown environment. High power density actuator systems will be necessary: elastic and compliant pneumatic actuators and series elastic actuators, based on the muscle's morphology, elasticity and structure will be the primary choice. Sensorized soft-intelligent materials will ensure safety and comfort in the interaction with the patient, while giving proprioceptive postural information. The suit control will combine proprioceptive postural information from the real-time measurement of the effective center of mass of the robot, calculated through an online processing of inertial and force sensors data, with the input of the BCI system that will guide the robotic suit in the environment. At the end of the project, the EEG recording system will be used also to analyze the effective impact of the robotic system as rehabilitation and assistive tool. The study of novel rehabilitation models of persons suffering from quadriplegia and the development of new BCI systems and new ergonomic robotic structure could open up new horizons in rehabilitation, neuroergonomics and robotics.

References

1. Manns, P.J., Chad, K.E.: Components of quality of life for persons with a quadriplegic and paraplegic spinal cord injury. Qual. Health Res. **11**(6), 795–811 (2001)
2. Daly, J.J., Wolpaw, J.R.: Brain–computer interfaces in neurological rehabilitation. Lancet Neurol. **7**(11), 1032–1043 (2008)

A Novel Acoustic Interface for Bionic Hand Control

Richard Woodward[1], Marcus Gardner[1(✉)], Paolo Angeles[1],
Sandra Shefelbine[2], and Ravi Vaidyanathan[1]

[1] Department of Mechanical Engineering, Imperial College London,
London, UK
m.gardner12@imperial.ac.uk
[2] Department of Mechanical and Industrial Engineering,
Northeastern University, Boston, USA

Abstract. Recent advances in prosthesis technology have led to the commercialization of several bionic hands for forearm amputees. Current methods of control involve the use of electromyography (EMG) to open and close the prosthetic hand with pre-defined grip patterns. Research into the use of mechanomyographic (MMG) signals have been successful in hand prosthesis control, and could provide a superior alternative to EMG sensors. This paper investigates the use of a novel MMG sensor for control of a commercial bionic hand.

Keywords: Mechanomyography · Prosthesis control · Bionic hand

1 Introduction

Advancements in prosthesis technology have led to improved functionality and control of bionic limbs for amputees. Current bionic hands use EMG signals from residual muscle in the forearm to produce open/close functions. MMG is not currently used to control prosthesis commercially; however, research has shown that they have the potential to replace EMG. MMG is the measurement of vibrations on the skins surface, or sound, as a result from muscle contraction [1], and are usually recorded with accelerometer or microphone based transducers. MMG provides many benefits over EMG: non-precise sensor placement; unaffected by changes in skin impedance e.g. sweat; significantly less expensive [1, 2]; doesn't require conductive gel or shaving of placement area. The aim of this paper is to test the suitability of a novel microphone-based MMG sensor for control of a Bebionic V2 hand by RSLSteeper.

2 Methodology

MMG is collected via a novel microphone based transducer. Computer communication can either be wired, sampling at 5 KHz, and down-sampled to 1 KHz, or wireless via Bluetooth, sampling at 512 Hz. Both have pros and cons; the wired method samples at a faster rate, but isn't portable like the wireless counterpart. Since harmonic data is

A. Natraj et al. (Eds.): TAROS 2013, LNAI 8069, pp. 296–297, 2014.
DOI: 10.1007/978-3-662-43645-5_32, © Springer-Verlag Berlin Heidelberg 2014

irrelevant in this application, the signal was filtered between 2 Hz and 50 Hz with a software 2nd order band-pass Butterworth filter. The sensor has a standing voltage of ~ 0.7 V, and a 1 V threshold is used to perform open/close actions when reached. This threshold value was chosen as it performed suitably across a number of test subjects without undesirable activation during movement or involuntary muscle activity.

Two MMG sensors are placed over the flexor and extensor digitorum muscles in the forearm, detecting finger flexion/extension (Fig. 1). Software was written to open the prosthetic hand on finger flexion, close on extension, and change grips when flexing twice, switching from a full-finger hand clench to a thumb-index finger pinch.

Fig. 1. The Bebionic V2 hand connected to MMG sensors on a subject's arm

3 Results

Preliminary tests have shown good results in function classification; however some alterations are required to remove external interference to improve accuracy. Further testing is required with multiple subjects to improve data reliability and assess differences in signal output.

4 Conclusion

Through initial testing, it was observed that the Bebionic hand could be controlled effectively using two MMG sensors. The sensor shows good initial results for the classification of signals for multi-functional control of a bionic hand. The added advantages of MMG over EMG and their compatibility with myoelectric hands suggest that the sensor has great potential to be used for hand prosthesis control.

Acknowledgements. The authors would like to thank RSLSteeper and Adam McEvoy for their support for the Bebionic V2 hand.

References

1. Orizio, C.: Muscle sound: bases for the introduction of a mechanomyographic signal in muscle studies. Crit. Rev. Biomed. Eng. **21**(3), 201 (1993)
2. Silva, J., Heim, W., Chau, T.: A self-contained, mechanomyography-driven externally powered prosthesis. Arch. Phys. Med. Rehabil. **86**(10), 2066–2070 (2005)

An Approach to Navigation for the Humanoid Robot Nao in Domestic Environments

Changyun Wei$^{(\boxtimes)}$, Junchao Xu, Chang Wang,
Pascal Wiggers, and Koen Hindriks

EEMCS, Delft University of Technology, Delft, The Netherlands
{c.wei,j.xu-1,c.wang-2,k.v.hindriks}@tudelft.nl, p.wiggers@hva.nl

Abstract. Humanoid robot navigation in domestic environments remains a challenging task. In this paper, we present an approach for navigating such environments for the humanoid robot Nao. We assume that a map of the environment is given and focus on the localization task. The approach is based on the use only of odometry and a single camera. The camera is used to correct for the drift of odometry estimates. Additionally, scene-classification is used to obtain information about the robot's position when it gets close to the destination. The approach is tested in an office environment to demonstrate that it can be reliably used for navigation in a domestic environment.

Keywords: Humanoid · Navigation · Localization

1 Introduction

Service robots need to coexist with and assist humans in domestic environments such as houses, offices and hospitals. Such robots might perform various tasks including, for example, fetch-and-carry tasks, cleaning tasks, or educational and entertaining tasks. To this end, robots have to be able to adapt to environments that are designed for humans. Humanoid robots appear to be very suitable in this setting because, among others, they navigate by means of biped walking, are able to make gestures, and can express emotions [1]. A key task that they would need to master first is navigating through human environments.

In our research, we use the humanoid robot Nao from Aldebaran Robotics, which is full-body functional, convenient to program, and affordable. The Nao platform is widely used in research on *human robot interaction*. In most research of this type, e.g., [2], however, the Nao is typically not supposed to navigate autonomously but subjects in experiments are supposed to interact with the robot that remains in a fixed position. In other settings where the Nao is supposed to navigate, additional sensors such as a laser head [3] or ceiling cameras [4] are used. An exception is the RoboCup where considerable progress has been made on navigating the Nao platform autonomously without the use of additional sensors, but the navigation techniques need the specific soccer field lines that are unusual in office environments.

A. Natraj et al. (Eds.): TAROS 2013, LNAI 8069, pp. 298–310, 2014.
DOI: 10.1007/978-3-662-43645-5_33, © Springer-Verlag Berlin Heidelberg 2014

It remains a challenging problem to have the Nao robot navigate in more general environments such as office environments without the use of additional sensors. In this paper, we propose an approach that supports navigation in such more general environments using only the built-in sensors of the Nao platform. We present a robot control strategy that enables the Nao to navigate in indoor environments. We use the virtual odometer based on Nao's built-in IMU (Inertial Measurement Unit) using the inertial and gyro sensors for tracking the robot's pose. An active localization method using SIFT-based landmark recognition is used to correct for any errors in the robot's pose estimated by the IMU. Finally, scene-classification is used to obtain additional information when the Nao arrives at its target position.

The paper is organized as follows. Section 2 discusses related work on navigation methods and in particular on related work that involves the Nao. In Sect. 3, we briefly describe the task scenario that we have used. Section 4 presents our navigation approach for the Nao platform. Section 5 presents initial results. Finally, Sect. 6 discusses future work and concludes the paper.

2   Related Work

Methods for localization and navigation provide techniques for estimating a robot's pose by interpreting sensor data and matching the information obtained through sensors with a map of the robot's environment. Significant progress has been made on the more general problem of *Simultaneous Localization and Mapping* (SLAM). The effectiveness of various SLAM methods, however, greatly depends on the available sensors. The most effective methods are applicable to robots equipped with laser range scanners. There are also effective methods for robots with stereo vision. Davison [5] even provides an approach to SLAM that only uses a monocular camera. The Nao also can use basically only one camera for navigation in indoor environments; however, monocular SLAM and visual SLAM more generally requires high-quality images at a high frame-rate neither of which is supported by the Nao platform.

An important issue is that by using only a single camera it is hard to perceive depth or distance between the robot and obstacles in its environment. Havlena et al. [6] presents a technique for measuring the depth using the camera of the Nao by processing two-frame image sequences using *Structure from Motion* (SfM) methods. However, the method proposed cannot be used for efficient, real-time navigation because it requires too much computational resources.

Another approach to navigation uses landmark-based localization. Oßwald et al. [7] presents a reinforcement learning method for navigating the Nao in a hallway. This work aims at an optimal balance between accurate localization and fast walking that uses the patterns present in a wooden floor. This approach can be considered as a land-mark based method that relies on a continuous presence of landmarks. Our method uses a quite sparse set of landmarks, namely, a limited set of pictures on walls. Even though landmarks may be used to simplify the localization task, a landmark recognition method still needs to deal with

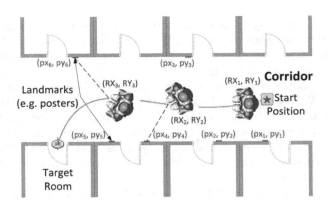

Fig. 1. An overview of the navigation task and environment.

issues such as occlusion or background clutter, lighting conditions, and variance in a robot's point of view. Elmogy and Zhang [8] present a landmark recognition and pose estimation method for humanoid robot navigation. Their method is based on image segmentation, two-step classification and stereo vision. For robust recognition, contour segmentation based on the Canny Filter is used to prevent the stored images from matching to the background. To increase the efficiency, a color histogram is used to roughly classifies the landmarks before SIFT is used for accurate classification. In contrast, our method avoids the segmentation stage by using RANSAC to effectively filter out the outliers on the background, and landmark positions are estimated using mono vision.

3   Task Scenario

In order to test our navigation approach, we have used a scenario where the Nao robot needs to deliver a message to a staff member in an office environment. An example of a message to be delivered is "It's time for coffee/lunch" which reminds a staff member to once in a while take some minutes off from work. In the scenario, the Nao is supposed to walk autonomously from its current location to a target room and has to navigate a corridor (see Fig. 1). The robot needs to recognize that it has arrived at the target room, needs to adjust its pose to be able to address the staff member, and deliver the message.

Indoor environments are structured places designed with specific constraints on the shape, size, spatial connections and orientation of navigational components such as doors, walls and corridors. It is an important competence for a mobile robot to discover and distinguish these components for navigation and performing context interpretation. We chose the office corridor because (i) it is similar to other general domestic environments in several ways, e.g., it has a similar spatial structure including doors and walls, and, e.g., pictures on walls, (ii) but otherwise it does not require continuous obstacle avoidance as, e.g.,

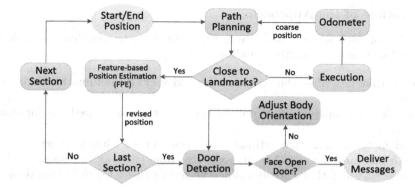

Fig. 2. The overall flow chart of the task.

would be required in a bedroom, and, finally, (iii) a corridor poses still some issues that need to be avoided such as disorientation and bumping into a wall.

4 Localization and Navigation

Our navigation approach for the Nao robot in domestic environments is based on using landmarks and uses odometry and a single camera to solve the localization problem. The recognition of landmarks provides sufficiently accurate pose estimates for navigation. In principle, this method can also solve the kidnapping problem. Landmarks also provide a tool to construct semantic maps. We use SIFT to recognize landmarks while a map that includes the landmarks, walls and doors assist in understanding the overall environment structure.

4.1 The Humanoid Robot Nao

The Nao is a 58 cm tall humanoid robot, and the academic version used in this study has 25 degrees of freedom. It is equipped with an AMD Geode 500 MHz CPU and 256 MB SDRAM that runs an embedded Linux system. Remote machines can connect to the system via WiFi. Two cameras with maximum resolution of 640 × 480 are located vertically on the forehead and on the mouth position respectively. The cameras do not have an overlapping field of view and, moreover, cannot be used at the same time. The cameras therefore do not support stereo vision. We have used the forehead camera for the vision-based localization component in our approach. The maximal nominal frame rate is 30 fps. However, this frame rate is only available for a lower resolution of 160 × 120. The maximum frame rate for the higher resolution is 4 fps. For our purposes, we used the higher resolution needed for feature extraction. The Nao's built-in walking behaviours (forward-backward, sideways and turning) are based on the *Zero-Moment Point* (ZMP) technique [9]. We used the default APIs to control these behaviors.

4.2   Outline of the Approach

An outline of the approach is shown in Fig. 2. In order to complete the task, we combined several components including:

- A map that describes the environment and includes landmarks, doors and walls;
- A localization module that revises the robot's pose based on landmark detection;
- An odometer module that estimates the robot's pose while walking;
- A path planning module that supports real-time path optimization;
- A scene classifier that classifies different scenes (used for detecting doors in our case).

A map is needed for localization and navigation. A map is a list of objects in the environment and their locations, and can be categorized as feature-based or location-based [10]. Feature-based maps only specify the shape of the environment at specific locations, namely the locations of the objects contained in the map, while location-based maps are volumetric and offer a label for any location in the world. In this paper, a feature-based grid map is used to describe the landmark positions, the doors and the walls, as shown in Fig. 7.

Localization and path planning are required for robot navigation. Localization determines the robot's current state in terms of spatial information (e.g., the coordinates and the orientation); Path planning determines the optimal path to the next destination according to the structure of the environment, where the path computed should be short and safe at the same time. An odometer is used to continuously provide a rough estimate of the robot's pose. Walking is interrupted occasionally and at those times a vision-based method is used to correct the robot's current pose estimate.

Odometry is not sufficient by itself for accurate pose estimation. Humanoid robots use a complex mechanism for biped walking that produces small but unpredictable errors due to leg joint backlash and foot slippage. These errors accumulate and become larger with every walking step. A vision-based method that uses landmarks cannot be applied for pose estimation while walking, however. The robot cannot always see landmarks while walking and, more importantly, it is not possible to continuously run a vision-based method because this is computationally too expensive.

An overview of the overall control flow is provided in Fig. 3. The odometry keeps track of the robot's position S and provides position estimation Z. Based on these position estimates, the robot then decides whether to trigger a feature-based position estimation module to obtain alternative data Z' for position estimation. A path planner continuously recalculates the optimal path and determines the next step the robot should take based on either Z or, if available, the improved position estimate Z'. Finally, the robot performs the action U that is computed by the path planner. This procedure is repeated until the estimated position is within a certain range of the target destination.

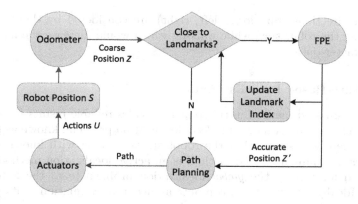

Fig. 3. The loop of localization strategy combining odometry and vision

When the robot has arrived at its target position, in our task, the robot needs to face a door and deliver a message (see Sect. 3). This part of the task requires a method for door detection. We use a trained classifier for differentiating between different states of the door. The results could then be used to determine the robot's position relative to the door.

4.3   Odometry and Real-Time Path Planning

Odometry is a pose estimation method frequently used for wheeled vehicles navigation. The odometer in the vehicles estimates the change of the vehicle position over time by tracking the revolutions of the wheels. By contrast, for biped humanoid robots that do not have wheels, the odometer is usually constructed by using inertial and gyro sensors. The current robot's position estimate, the velocity and acceleration data from the inertial sensor, and the orientation from the gyro sensor are used to estimate the robot's new pose after movement.

The raw sensor data can be interpreted as the coordinates of three different spaces that have been defined by the manufacturer. We used the so-called *World Space*. The origin of the World Space is fixed when the robot is booted and cannot be altered by any other means than rebooting the robot. Furthermore, when the robot's feet do not touch the floor, the sensors will not record any position change. Therefore, the World Space cannot be directly used in our task because its origin is not fixed physically with the environment. Instead, we have created a *Map Space* that is built and provides a blueprint of our office corridor. The World Space coordinates that correspond to sensor data readings then are transformed into Map Space coordinates and re-calibrated every time. For visualization of the robot trajectory on the GUI, these coordinates need to be further transformed into QtGUI coordinates (see Fig. 7).

The Path Planning module calculates the optimal path using A-star algorithm in the Map Space once the robot position is updated. For simplicity, only quadrantal directions (up, down, left, right) are considered by the planner to avoid having the robot turn while walking which would increase the inaccuracy of the odometer data.

4.4 Feature-Based Pose Estimation

The landmark-based method in this paper provides reliable results for localization. Basically, odometry solves the *local* localization problem, known as position tracking, where we assume that the robot starts walking with known starting coordinates. A landmark-based method is an active localization method, which can be used for solving the *global* localization problem (e.g., the kidnapping problem). Ideally, the robot can compute an accurate estimation of its position whenever the robot sees any known landmarks.

The robot interrupts walking whenever it believes it is close to a landmark. Closeness to a landmark is computed by using the current robot's position estimate provided by the odometer and the landmark coordinates on the 2D grid map. The robot stops walking in order to avoid image blur that is caused by the shaking of the robot's body when it is walking. When the robot stops walking it captures an image frame while standing still and then continues walking again. Relative position estimation is triggered if a pre-stored landmark is recognized.

The idea of the landmark methods is to use posters on the wall of a corridor (see Fig. 1) as landmarks because these can be considered as "natural" landmarks. It is common to show research work by posters hanging on the wall of the laboratory corridor, and they have been in our corridor for months now and have become part of the environment. These posters provide sufficient information for the vision-based robot localization. Moreover, using natural landmarks is of great importance for human-robot collaboration because the environments where the robots work with humans is made for humans by definition such as office and home and few changes of the environment should be made specifically for the robot. However, due to physical restrictions (e.g., the height of Nao and the width of the corridor), the Nao failed to capture good images of the real posters because they are positioned relatively high on the walls. Instead, we chose to use printed pictures that we pasted on the walls of the corridor.

A sufficient number of keypoints are required to recognize an object, and it should be possible to extract these keypoints from images observed from different angles and distances. Scale-invariant feature detection is crucial to landmark recognition because the actual position where the robot starts searching landmarks is unpredictable due to drift errors of the odometer. Several approaches such as SIFT [11], SURF [12], and their derivatives, Harris-SIFT [13] and Any-SURF [14], can be used. As a mature and feasible method, we have chosen SIFT even though it is relatively slow. Experiments showed that sufficient keypoints can be detected within the observed image via SIFT (see Fig. 4). In the implementation, the SIFT feature descriptors of each landmark image are calculated beforehand and stored in the database for real-time matching. The landmarks are recognized by matching local feature correspondences.

Fig. 4. The local features extracted by SIFT and matched by RANSAC.

The SIFT features extracted from the observed landmark image are used for relative pose estimation. Once a captured frame contains a recognizable landmark, the position estimation process starts. We used a method proposed by Fischler and Bolles called *RANSAC* that optimizes the correspondences matching process [15]. RANSAC can be applied and used for solving the Location Determination Problem (LDP) (i.e., calculating the viewpoint position from the observed images). This method uses at least four coplanar points on the object plane to find the viewpoint in 3D space. In our task, the robot camera is the viewpoint, and the SIFT features are the coplanar points.

4.5   Door Detection

In our task, the Nao needs to navigate towards the target door and stop right in front of the open door, facing the door (see Fig. 5(a)). However, due to the inaccurate odometer and the drift errors, in most cases the robot stops facing towards part of the open door or the wall (see Fig. 5(f)). In addition, there are other situations like facing the corridor (see Fig. 5(e)) or the wall (see Fig. 5(d)).

There are many ways to detect a closed door (see Fig. 5(c)) such as using adaptive boosting to train a strong classifier with several weak classifiers trained by utilizing a variety of features, including vertical lines, concavity, gap below the door, kick plate, color, texture, and vanishing point [16]. However, there is no general vision-only method to detect an open door which largely depends on the scenery inside the room. Low-level line features and further high-level processing can be combined for corridor and door structure detection and their hypothesis generation and verification [17]. But for the humanoid Nao, as its camera is swinging when walking, it is better to utilize global features, e.g., histogram, than local edge features to get robust results. Another advantage of the histogram is that it can be used as part of an appearance-based method for topologically localizing the robot [18].

In the cluttered environment, we need to find distinctive features of the open door to distinguish it from other scenes. Scene classification is used in our case because it is able to discriminate open doors from other indoor places, e.g., closed

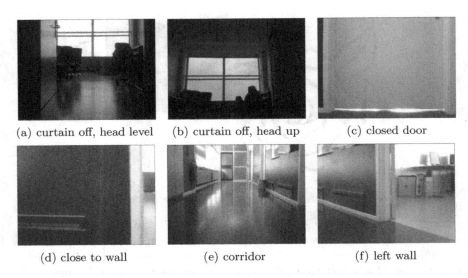

(a) curtain off, head level (b) curtain off, head up (c) closed door

(d) close to wall (e) corridor (f) left wall

Fig. 5. The classified scenes captured by NAO's camera for open door detection.

Table 1. A comparison of prediction rate between 2 and 5 classes

Result	2 classes	5 classes
True positive rate	0.37	0.99
False positive rate	0.25	0.12
Prediction rate	0.59	0.66

doors, corridors and walls. What makes an open door different here is the scene of the room such as strong light from the window (see Fig. 5(b)). A Support Vector Machine (SVM) has been chosen as the discriminative classifier for several visual recognition domains. SVMs have the advantage that they generalize well on difficult image classification problems where the only features are high dimensional histograms. The Nao, controlled by a GUI, walked around the corridor while 7 videos were recorded. Videos were converted into JPEG images which were labeled into different categories to train and test the classifier.

In the model training process, BGR channels and the intensity channel are used to compose the feature vector. At first, all the images from the first 6 videos are classified into 2 classes. Images are labelled as an open door if the Nao is facing towards a bright window. Then, a linear classifier is trained to predict unseen images as an open door or not. Then, the images taken near the open door are classified into 5 classes: open door, corridor, left wall, right wall and others. Images from the other video are used as a test set. A histogram is used for training the classifier. The results of open door detection are presented in Table 1. It shows that the prediction of true door rate is increased and false door rate decreased when the images are classified into 5 classes rather than

(a) landmark set 1   (b) landmark set 2   (c) landmark set 3

Fig. 6. Three example landmarks used in our test.

2 classes, which means the door is better classified with more careful manual labelling of the images. However, the overall prediction rate is not good enough for classifying non-separable places.

5   Results

The proposed method has been evaluated in the corridor of our office (see Fig. 1). There are eight rooms alongside the corridor. A 2D grid map (see Fig. 7) representing the environment Fig. 1 has been used for path planning. The grid cells used are squares of 250 mm by 250 mm. Grey grid cells denote the walls whereas landmarks (pictures pasted on the wall) are denoted by pink circles.

In the field test, the robot was supposed to walk from the starting point through the corridor and finally arrive at the door of the target room. Because the pose estimates based on odometry are not sufficiently reliable for localization during the entire journey of about 12 m, the corridor was partitioned into shorter sections with a length ranging from 175 cm to 250 cm (different section lengths were used because of physical constraints). These sections either contained landmarks (so-called *landmark sections* with lengths within the indicated range) or consisted of doors which are about 100 cm long (*door sections*). Different sets of landmarks (see Fig. 6) were placed near the endpoints of each of the landmark sections. Each set contained multiple, different landmarks to ensure that the robot would see at least one of them. We used black-and-white pictures for reasons of computational efficiency. Odometry was used to estimate robot position while the robot was walking from one endpoint of a section to the next one. Feature-based pose estimation was used to correct for errors when the robot believes it has reached an endpoint of a section. To search for landmarks, the robot first turned its head 45° and then further to a full 90°. If after turning its head the robot still can not detect any landmarks, the robot would turn its body 90° and repeat the head turning procedure. This strategy turned out to be sufficient in our tests.

Finally, to detect a door section, the robot first navigates to the target door following a planned path, and then estimates its position relative to the door by classifying the scene and adjusting its body direction until it faces the door.

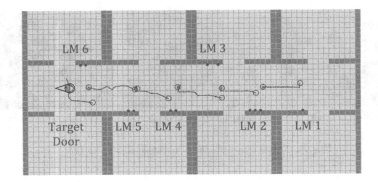

Fig. 7. The 2D Grid Map and the robot trajectory in the field test: Magenta dots denote the landmarks on the walls; Green dots denote the landmark section endpoints; Red dot lines denote the robot trajectories estimated by the odometer; The robot positions just before and after each vision-based correction are marked by red circles; LM 1 to 6 denote the landmark groups 1 to 6 (Color figure online).

The Nao succeeded in arriving at the endpoint of the last landmark section, as shown in Fig. 7. Overall, the odometry worked sufficiently well but occasionally the robot deviated quite far from the planned path and walked virtually towards the wall. As expected, our tests confirm that vision-based localization is needed for avoiding collisions with walls. The door detection method performed quite well and open doors were recognized using histogram and color features given similar illumination conditions as were used for training the classifier and the robot successfully adjusted its body direction to face the door.

6   Discussion and Conclusion

Our test setup and results demonstrate the feasibility of our proposed navigation method. However, there are various areas for improvement that are left for future work. For example, deviations from planned paths due to foot slippage are occasionally quite big and our approach is not sufficiently able to correct yet for these position estimate errors (see also Fig. 7).

The computational resources needed for the feature-based pose estimation, moreover, are quite expensive, and the method cannot be run during the entire journey of the robot. As a result, it may still be possible that error in the robot pose estimate becomes too high and cannot always be corrected by landmarks. It remains a challenge to construct more computationally efficient algorithms that can be used continuously for detecting landmarks. Of course, it is always possible to use more landmarks. Other natural landmarks than pictures such as marks on doors might also be used. This would, however, also require a more detailed map. For our tests, we have manually constructed the map but if more detailed maps are needed it would be useful to apply learning techniques that enable the robots to explore and automatically remember (new) landmarks.

Other problems that need more work include failure of landmark recognition due to too large viewing angles (which significantly reduces image quality) and variation in lighting conditions. For example, the dimmest region of the corridor is near LM3 (see Fig. 6(c) and Fig. 4). Here the auto-gain feature of the camera of the Nao boosted the CMOS sensor sensitivity, which also significantly reduced image quality. In those cases, SIFT may not extract enough useful features.

References

1. Brooks, R.A.: Humanoid robots. Commun. ACM **45**, 33–38 (2002)
2. Ros, R., Nalin, M., Wood, R., Baxter, P., Looije, R., Demiris, Y., Belpaeme, T., Giusti, A., Pozzi, C.: Child-robot interaction in the wild: advice to the aspiring experimenter. In: Proceedings of the 13th International Conference on Multimodal Interfaces, pp. 335–342. ACM (2011)
3. Hornung, A., Wurm, K.M., Bennewitz, M.: Humanoid robot localization in complex indoor environments. In: 2010 IEEE/RSJ International Conference on Intelligent Robots and Systems (IROS), pp. 1690–1695. IEEE (2010)
4. Yan, W., Weber, C., Wermter, S.: A hybrid probabilistic neural model for person tracking based on a ceiling-mounted camera. J. Ambient Intell. Smart Environ. **3**(3), 237–252 (2011)
5. Davison, A.: Real-time simultaneous localisation and mapping with a single camera. In: Proceedings of the Ninth IEEE International Conference on Computer Vision, 2003, vol. 2, pp. 1403–1410, Oct 2003
6. Havlena, M., Šimon Fojtů, Průša, D., Pajdla, T.: Towards robot localization and obstacle avoidance from Nao camera. Technical report, Czech Technical University in Prague, Nov 2010
7. Oßwald, S., Hornung, A., Bennewitz, M.: Learning reliable and efficient navigation with a humanoid. In: 2010 IEEE International Conference on Robotics and Automation (ICRA), pp. 2375–2380. IEEE (2010)
8. Elmogy, M., Zhang, J.: Robust real-time landmark recognition for humanoid robot navigation. In: IEEE International Conference on Robotics and Biomimetics, 2008. ROBIO 2008, pp. 572–577, Feb 2009
9. Vukobratovic, M., Borovac, B.: Zero-moment point - thirty five years of its life. I. J. Humanoid Robot. **1**(1), 157–173 (2004)
10. Thrun, S., Burgard, W., Fox, D.: Probabilistic Robotics. Intelligent Robotics and Autonomous Agents. The MIT Press, Cambridge (2005)
11. Lowe, D.G.: Distinctive image features from scale-invariant keypoints. Int. J. Comput. Vis. **60**(2), 91–110 (2004)
12. Bay, H., Tuytelaars, T., Van Gool, L.: SURF: speeded up robust features. In: Leonardis, A., Bischof, H., Pinz, A. (eds.) ECCV 2006, Part I. LNCS, vol. 3951, pp. 404–417. Springer, Heidelberg (2006)
13. Azad, P., Asfour, T., Dillmann, R.: Combining Harris interest points and the SIFT descriptor for fast scale-invariant object recognition. In: IEEE/RSJ International Conference on Intelligent Robots and Systems, pp. 4275–4280. IEEE, Oct 2009
14. Sadeh-Or, E., Kaminka, G.A.: AnySURF: flexible local features computation. In: Röfer, T., Mayer, N.M., Savage, J., Saranlı, U. (eds.) RoboCup 2011. LNCS, vol. 7416, pp. 174–185. Springer, Heidelberg (2012)
15. Fischler, M.A., Bolles, R.C.: Random sample consensus: a paradigm for model fitting with applications to image analysis and automated cartography. Commun. ACM **24**(6), 381–395 (1981)

16. Chen, Z., Birchfield, S.: Visual detection of lintel-occluded doors from a single image. In: IEEE Computer Society Conference on Computer Vision and Pattern Recognition Workshops, 2008. CVPRW '08, pp. 1–8. IEEE, June 2008

17. Shi, W., Samarabandu, J.: Investigating the performance of corridor and door detection algorithms in different environments. In: International Conference on Information and Automation, 2006. ICIA 2006, pp. 206–211. IEEE (2006)

18. Pronobis, A., Caputo, B., Jensfelt, P., Christensen, H.I.: A discriminative approach to robust visual place recognition. In: 2006 IEEE/RSJ International Conference on Intelligent Robots and Systems, pp. 3829–3836. IEEE (2006)

An Embodied-Simplexity Approach to Design Humanoid Robots Bioinspired by Taekwondo Athletes

Rezia Molfino[1], Giovanni Gerardo Muscolo[1,2(✉)], Domenec Puig[3],
Carmine Tommaso Recchiuto[4], Agusti Solanas[3],
and A. Mark Williams[5]

[1] PMAR laboratory, Department of Applied Mechanics and Machine Design,
Scuola Politecnica, University of Genova, Genoa, Italy
info@humanot.it
[2] Creative and Visionary Design laboratory, Humanot s.r.l., Prato, Italy
[3] Universitat Rovira i Virgili, Tarragona, Catalonia, Spain
[4] Electro-Informatic laboratory, Humanot s.r.l., Prato, Italy
[5] Brunel University, London, UK

Abstract. We are investigating new approaches to design and develop embodied humanoid robots with predictive behaviour in the real world. Our approaches are strongly based on the symbiotic interaction between the concepts of the embodied intelligence and simplexity that enables us to reproduce the artificial body in symbiosis with its intelligence. The research is based on the study of martial arts athletes and their planned movements. In particular, the taekwondo martial art competition where anticipation and adaptive behaviours are key points is used as a vehicle to address conceptual and design issues. The expected outcome will be the development of a robot that will have the capability to anticipate an opponent's actions by coordinating eyes, head and legs, with a stable body constituted by an embodied platform bioinspired by the taekwondo athletes.

Keywords: Bio-inspired robotics · Cognitive robotics · Humanoid robotics · Human-robot interaction · Robot vision · Anticipation · Simplexity · Embodied intelligence

1 Concept and Research Objectives: Beyond the State of the Art

Pfeifer and Bongard [1] demonstrated that through coordinated interaction with the environment, an agent can structure its own sensory input. Because physical interaction of an agent with its environment leads to structured patterns of sensory stimulation, neural processing is significantly facilitated, demonstrating the tight coupling of embodiment and cognition. Berthoz [2] introduced the concept of the simplexity in neuroscience as a set of solutions living organisms find, that enable them to deal with information and situations, while taking into account past experiences and anticipating future ones. Considering these two concepts (embodiment and simplexity), our research goal is to synthesise them in new embodied humanoid robots that can move, kick, punch

A. Natraj et al. (Eds.): TAROS 2013, LNAI 8069, pp. 311–312, 2014.
DOI: 10.1007/978-3-662-43645-5_34, © Springer-Verlag Berlin Heidelberg 2014

and jump in an autonomous way and interact with other robots, with a human opponent and with the environment, anticipating movements in a fight. In a broader sense, the research also aims to establish scientific foundations for defining general principles for the design of robots, based on the novel concept of the embodied-simplicity of movement. This aim means designing robots around a unified reference frame given by their vestibular system and providing them with predictive behaviour (i.e. the capability to predict the effect of their own movements, in terms of perception). The other basic idea of the present project is to analyse the characteristics of the anticipation of movement of the taekwondo's athletes in order to define a novel dynamic balance model and control in symbiosis with the embodied robotic platform. Taekwondo is a modern sport that offers a relatively unique environment where the limits of human achievement are challenged continually. A study focused on taekwondo athletes will help to develop new neurophysiological models that will be implemented and validated in the robotic platform. The robot control will be based on networks of non-linear oscillator and feedback strategies [3] in order to describe omnidirectional movements and an on-line feedback mechanism that uses multi-sensory information. The modelling of the robot kinematics and dynamics will use a screw-theory based method, with the platform seen as a multiloop, redundantly actuated mechanism with flexibility in the joints. A complex vision system will be also developed: it will be able to perform image stabilization and fast target tracking and recognition of specific close objects and to implement an anticipation of movement loop in the visual field. Finally, the humanoid robots will have the maximum height of 1200 mm and the maximum weight of 50 kg., a maximum of 7 DOF in each leg and arm, 2 DOF in the waist, 4 DOF in the trunk, 5 DOF in each hand and foot, and many passive DOFs in the body. Soft-intelligent materials will be used in order to reproduce the cartilage and other connective tissues associated with the joints in the final structure. Ranges of motion will be defined in reference to the taekwondo's athlete motion measurements (Fig. 1).

Fig. 1. Taekwondo martial art and Olympic sport.

References

1. Pfeifer, R., Bongard, J.: How the Body Shapes the Way We Think: A New View of Intelligence. Bradford Books, Cambridge (2007)
2. Berthoz, A.: Simplexity: Simplifying Principles for a Complex World. Yale University Press, New Haven (2012)
3. Liu, C., Chen, Q.: Walking control strategy for biped robots based on central patterns generator. In: Proceedings of IEEE International Conference on Robotics and Automation, pp. 57–62 (2012)

Navigation

Navigation

A Robotic Geospacial Surveyor

Sam Wane[✉]

Staffordshire University, Staffordshire, UK
s.o.wane@staffs.ac.uk

Abstract. This paper discusses the prototyping and testing of a mobile robot system programmed to follow a defined set of tram-lines calculated from a surrounding geo-fence which can be entered using mapping software or from a GPS device. The tracking method uses an on-board compass and a GPS sensor and an application to archaeological scanning is trialled. A solution to controlling the trajectory between latitude and longitude lines is formulated and implemented.

Keywords: Mobile robot · GPS · Global positioning system · Automated scanning · Tram-lines · Geolocation · GNSS · Global navigation satellite systems

1 Introduction

There are many situations where an open area requires a surface scan, this can be when searching for buried treasure using a metal detector, surveying beneath the soil using ground penetrating RADAR, measuring the surface quality of plant growth, or preparing and planting for robotic farming. It is in the last application that most of the research was found, particularly from Japan in which robotic agriculture is being heavily invested. One issue with GPS is accuracy, with the cost increasing prohibitively if an accuracy greater than 2 m is needed. This paper covers the development and test of a mobile robot programmed to scan an area using the fused data of a GPS, compass and odometry, store the locations visited and match the scanning sensor values with the coordinates.

2 Current Application of GPS and Sensors as Applied to Tractor Control

The sensors and application techniques for trajectory guidance of outdoor robot systems was investigated. A majority of experiments involved the retro-fitting of control mechanisms and sensors to tractors.

In [1], the ideal situation of outdoor navigation are summarised in that there should be no clutter and thus the problem is simplified, however, disadvantages

This project was made possible using the Higher Education Innovation Fund (HEIF). Many thanks to Guy Peters and Caroline Sturdy-Colls.

A. Natraj et al. (Eds.): TAROS 2013, LNAI 8069, pp. 315–327, 2014.
DOI: 10.1007/978-3-662-43645-5_35, © Springer-Verlag Berlin Heidelberg 2014

include: large area, uneven ground, wheel slippage, interference from cultivation, soil changes, rain, fog, and interference from dust. Companies are unwilling to invest in this area as it is not perceived as worth-while, and farmers are not financially able to participate. In the experiment, machine vision was used on row crops for orientation. Differential GPS was used (RTK-real time kinematic) which is already available on some tractors, however, this needs a base station within 10 km. Latency problem with control was also discussed. The results report less than a 100 mm deviation using a Fibre Optic Gyroscope (FOG), Doppler shift sensors, inertia sensors, accelerometers and a gyro to contribute to an Inertial Navigation System (INS). A Geomagnetic Direction Sensor (GDS) is a magnetometer used as a heading sensor using magneto-meter for heading. Utilising GPS (with an accuracy of 200 mm) combined with GDS gives a 10 mm accuracy.

Stanford University [2], used a GPS on a John Deere 7800 Tractor resulting in 1° accuracy in heading, and line tracking accuracy with 25 mm deviation. They used Carrier Phase Differential GPS (CDGPS), Integrity Beacon Landing System at 10 Hz, 4 antennas.

The University of Florida used GPS and machine vision guidance [3] resulting in a tractor average error of 28 mm, this was improved to 25 mm using LASER RADAR.

The National Agriculture Research Center, Japan, used RTK GPS, FOG PH-6 on a Iseki Co., Ehime transplanter resulting in a deviation of less than 120 mm, and a yaw angle offset of about 55 mm at 2.52 km/h [4]. The calculations of the deviation from desired path 'd', heading angle error 'ω', steering angle 'δ_{aim}' are shown in Fig. 1 and defined as:

$$\delta_{aim} = K_{p1}d + K_{p2}\omega \qquad (1)$$

where the 'K' gains were determined experimentally:

$|d| < 0.06$	$K_{p1} = 0.07, K_{p2} = 0.4$
$0.06 < |d| < 0.12$	$K_{p1} = 0.1, K_{p2} = 0.3$
$|d| > 0.12$	$K_{p1} = 0.29, K_{p2} = 0.1$

GPS-based guidance technology has successfully been reported in use for many field operations such as sowing, tilling, planting, cultivating, weeding and harvesting [5, 6]. Stoll and Kutzbach [7] studied the use of the RTK GPS as the only positioning sensor for the automatic steering system of self-propelled forage harvesters. Kise et al. studied the use of an RTK GPS guidance system for control of a tractor as an autonomous vehicle travelling along a curved path [8, 9]. Ehsani et al. [10] evaluated

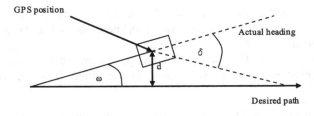

Fig. 1. The use of steering gains to allow trajectory following [4]

the dynamic accuracy of several low-cost GPS receivers with the position information from an RTK GPS as reference. Nagasaka et al. [4] used an RTK GPS for positioning, and fibre optic gyroscope (FOG) sensors to maintain heading. Despite these limitations, researchers use odometry as an important part of robot navigation systems [11, 12].

Gyroscopes measure the rate of rotation independent of the coordinate frame. They can also provide 3D position information and have the potential to detect wheel slippage. Unfortunately, these types of sensors are prone to positional drift [13]. In [15] GPS and a gyroscope are used to follow tram lines and the actual heading is compared to the calculated heading.

3   The Tracking Procedure

This project requires an area to be scanned in a linear fashion, the area is bounded by four geo-fence posts and the area covered by the linear tram-lines between these posts.

Outline of Tracking Procedure

1. Enter the geo-fence latitude and longitude points and the scan distance.
2. Calculate the bearing.
3. Head to the start of scan location.
4. Read the coordinates and convert these coordinates to a local coordinate frame.
5. Track along the scanning line until the scanning distance is reached.
6. Rotate in an arc equal to the width of the scanning area until heading in the opposite direction.
7. Aim for a point adjacent to the starting location offset by the scanning width.
8. Repeat from point two until the area is completely scanned.
9. Stop and download the scan data.

3.1   Calculate the Intermediate Points to Form a Series of Column Trajectories (Tram-Lines)

The input is four latitude and longitude points (termed the 'geo-fence'), and the scan pattern happens in the direction from geo-fence point 1 to geo-fence point 2. Assuming the geo-fence points make a rectangle, the intermediate points forming the tramlines between points '1'-'4', and '2'-'3' are calculated by calculating the bearing between these points, and using the equation to calculate position from a bearing and distance equation where the tram-line distance is used and iterating it until the end points '3' and '4' are reached respectively. This is shown in Fig. 2.

This is repeated for each of the tramline end points and alternated to define the direction of scan motion with even lines going parallel to the direction of point 1 and odd lines heading parallel to point 2.

The bearing, φ of the origin points (lat_0, $long_0$) to the displacement points (lat_1, $long_1$) are calculated using the Haversine formula as:

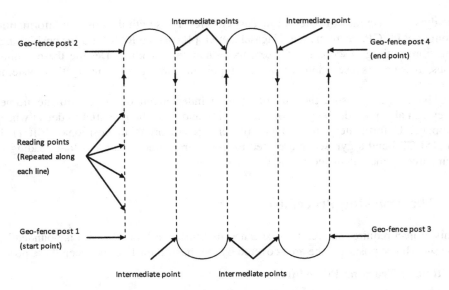

Fig. 2. The motion of the scanning robot following the desired scan lines

$$\varphi = a\tan 2(\sin(\Delta long)\cos(lat_1), \cos(lat_0)\sin(lat_1)$$
$$- \sin(lat_1)\cos(lat_1)\cos(\Delta long)) \tag{2}$$

where lat_0, $long_0$ are the coordinates of the lower left scan origin, lat_1, $long_1$ are the coordinates at the current point, Δlat, $\Delta long$ is the current displacement from the origin, and φ is the bearing from North (increasing opposite to the direction of compass rotation), all values are in radians.

This is measured in radians and is positive rotating clockwise from North. The system uses a local coordinate system which is rotated to the bearing of the scan lines (Fig. 3). The rotation is measured from the North axis and the angle is *positive* when rotating anti-clockwise (opposite to the direction of compass rotation). A bearing of 0° places the x-axis along the direction of increasing longitude facing East, and the y-axis along the direction of increasing latitude facing North. The GPS coordinates are read and converted to a local coordinate system x, y with a bearing φ.

The x and y values of the local coordinate system are calculated:

$$x = R\cos\varphi\cos(lat_0)[\Delta long] + R\sin\varphi[\Delta lat]$$
$$y = -R\sin\varphi\cos(lat_0)[\Delta long] + R\cos\varphi[\Delta lat] \tag{3}$$

where R is the mean radius of the earth (6378137 m).

3.2   Head to the Starting Location

The robot heads to the location in which it is to start scanning. It does this by fixing its translation velocity (large if the robot is a long distance from this point, and small as

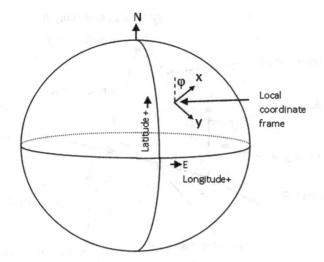

Fig. 3. The coordinate system

the robot approaches it). The bearing of the start location is calculated from reading the current robot GPS location and comparing it with the start location using Eq. (2). Once the start location is reached, the robot rotates on the spot to face the tracking direction and then commences to follow the first tram-line.

3.3 Trajectory Tracking-Heading Towards a Desired Location

In order for the mobile robot to follow a desired path, the robot bearing, path bearing, and the robot deviation from the path need to be considered. The robot bearing is sensed using an onboard compass, the path bearing is calculated from knowing the start and end latitude and longitude coordinates, the position of the robot is sensed using GPS, and the tangential distance from the robot to the desired trajectory line is calculated, referring to Fig. 4, and using vectors.

The path bearing 'δ_{aim}' is calculated from the latitude and longitude points (Eq. 2). Points A(xa,ya) – B(xb,yb) define the desired trajectory line. Point P(xp,yp) is the robot measured (GPS) position.

To calculate deviation from the desired trajectory line (FP):

$$AP^2 = FP^2 + AF^2$$
$$\therefore FP = \sqrt{AP^2 - AF^2}$$
$$AP^2 = (xp - xa)^2 + (yp - ya)^2$$

(4)

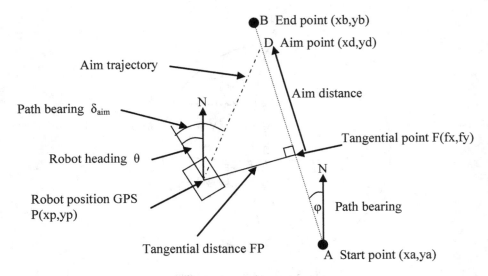

Fig. 4. Heading towards a desired location

Find AF:

$$AF = AP \cdot \frac{AB}{|AB|}$$

$$AF = AP. \frac{(xa - xb), (ya - yb)}{\sqrt{(xa - xb)^2 + (ya - yb)^2}} \qquad (5)$$

$$AF = \frac{(xa - xg) * (xa - xb) + (ya - yg) * (ya - yb)}{\sqrt{(xa - xb)^2 + (ya - yb)^2}}$$

To find the coordinates of the perpendicular point F(xf, yf), we know the equation of the line A-B, and the distance along it, AF, so:

$$AB = \sqrt{(xa - xb)^2 - (ya - yb)^2}$$

$$x_f = xa + (xb - xa) + \frac{AF}{AB} \qquad (6)$$

$$y_f = ya + (yb - ya) + \frac{AF}{AB}$$

Then, the displacement from the path (FP) as previously calculated along with knowing the perpendicular point on the path F(xf, yf) are use to find an 'aim point'. This is a point, some way ahead of the ideal point on the path where the robot should be D(xd,yd) and is given as a reasonable point on the ideal path for the robot to aim for. From this, δ_{aim} is calculated and used to control the rotation of the robot.

$$x_d = x_f + (xb - xa) + \frac{Aim\ dist}{AB}$$

$$y_d = y_f + (yb - ya) + \frac{Aim\ dist}{AB} \tag{7}$$

$$\delta_{aim} = atan2\left(\frac{yd - yp}{xd - xp}\right)$$

All coordinates are in terms of decimal latitude and longitude. In the area of testing the coordinates are (52.813632, −2.083800) and, since small distances are covered, an approximate conversion to 1 m distance is a change in the coordinates of 0.000023 depending on the bearing, and the 'Aim dist' is an approximation value of difference in latitude and longitude values from the distance in metres where 3 m was chosen through experimentation. The robot is thus steered by setting the velocity of rotation as:

$$\dot{\theta}_{des} = K_{rot}(\delta_{aim} - \theta) \tag{8}$$

k_{rot} is chosen experimentally to allow a fast response with a minimum of overshoot. This value was initially determined in simulation. This, along with the effects of tuning $Krot$ are in the next section.

The distance 'd' in kilometers between two latitude and longitude points can be used to determine the distance of the robot to the end point of the trajectory, and is also used in calculating the aim point:

$$a = \sin^2(\Delta lat/2) + \cos(lat_1)\cos(lat_2)\sin^2(\Delta long/2)$$

$$c = 2atan2\left(\sqrt{a}, \sqrt{(1-a)}\right) \tag{9}$$

$$d = R.c$$

where 'R' is the mean radius of the earth and a value of 6371 km is used.

4 Kinematics and Simulation

The robot was simulated in MATLAB in order to test the tracking procedure, verify and aim distance, and to tune the rotational gain.

A Pioneer 3AT [23] was to be used, the input to this is desired translational velocity(V), and desired rotational velocity ($\dot{\theta}$), and the corresponding wheel velocities are controlled in a local feedback loop:

$$\dot{L} = V - \dot{\theta}$$

$$\dot{R} = V + \dot{\theta} \tag{10}$$

Knowing the translational velocity of the robot and its current angle, the x,y components are calculated as:

Local velocity:

$$\dot{x} = V \cos \theta$$
$$\dot{y} = V \sin \theta$$
(11)

Position (simulation):

$$x = x_0 + \int_0^t \dot{x}\, dt$$
$$y = y_0 + \int_0^t \dot{y}\, dt$$
(12)

'θ' is determined by integrating the rotational velocity:

$$\theta = \theta_0 + \int_0^t \dot{\theta}\, dt$$
(13)

The velocity, 'V' is fixed at a rate suitable for ground scanning, the robot is kept on course by altering the desired steering velocity ($\dot{\theta}$). There are limits to the velocities of the real robot and rate limiters were implemented into the simulation.

4.1   Simulation to Find Steering Tuning Values

A simulation was run using MATLAB. The input was the trajectory line to follow. The equations to find the steering velocity was implemented from the equations of 'Trajectory Tracking' above and the simulation run with a variety of initial robot bearings and positions. A screen shot of the initial setup is shown in Fig. 5.

The simulation was run with varying values of steering rotational gain (*Krot*) and the results demonstrated that *Krot* values $\ll 1$ result in a sluggish tracking, whilst $1 < Krot < 10$ results in a good tracking with little difference between the values, this is due to the feedback nature of the continuous application of the tracking equations and the rotational velocity limits in the simulation.

Having *Krot* $= 1$ gave a good 'steering' response in that the robot turns towards the trajectory line with no overshoot, this is due to the steering angle being

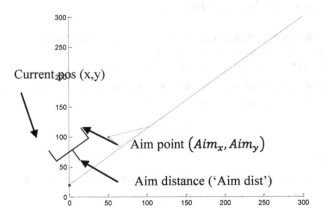

Fig. 5. How the current position, aim distance and aim point relate

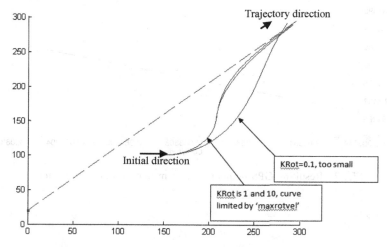

Fig. 6. Comparison of altering the 'Krot' value for steering

recalculated for each new position of the robot. The initial direction of the robot in Fig. 6 is an extreme example that shows recovery of the desired trajectory.

5   Application

The experiment was conducted on a Pioneer 3AT all-terrain robot [23] equipped with an onboard GPS and magnetic compass. The robot was programmed using C++ on a Linux Ubuntu platform and used ARIA [23] as the interface to control the robot.

Functions were written to drive the robot to a latitude and longitude location, follow a pair of latitude and longitude lines, store its GPS location, and turn in an arc at the end of each tramline section.

Markers were placed on an open field to define a 25 m^2, the latitude and longitude values were manually read using GPS and input as a geo-fence to the robot. The robot was commanded to move one metre, stop, read its magnetic bearing, read the GPS position, and store it to a.gpx file. The robot was stationary for GPS readings to allow a more accurate location and bearing to be read (both sensors had to be mounted on a plastic pole away from the metallic body of the robot). The one metre movement was measured using the robot's internal odometry using wheel encoders, since this was not cumulative and was only used to measure one metre intervals, it was deemed to have sufficient accuracy.

The robot successfully followed the lines and turned 180° to follow the next tramline even though the accuracy of the GPS receiver was two metres. The combination of multiple readings and the use of the compass resulted in very little deviation of the robot as long as a fix of five or more satellites was established.

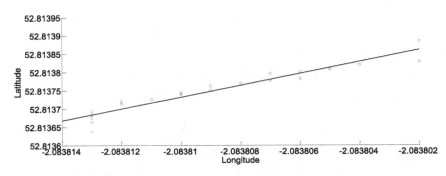

Fig. 7. Resulting GPS points read at 1 m intervals along trajectory

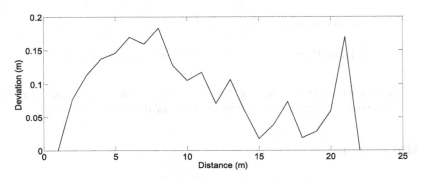

Fig. 8. Deviation from the trajectory line as measured by GPS in moving 0–25 m

6 Results

The .gpx file was uploaded into MATLAB for analysis. This showed small deviations from the trajectory line and this is shown in Fig. 7.

A deeper analysis of these points to calculate the perpendicular distance from the trajectory line using the Eq. 5 resulted in the graph in Fig. 8. This shows the measured deviation did not exceed 0.18 m throughout the 25 m trajectory.

7 Conclusion and Further Work

The robot followed the trajectory lines with a deviation of less than 0.18 m, which is better than the accuracy of the GPS alone. The sensors used were the compass and the internal odometry and for the application of scanning an area using e.g. a metal detector, this would be sufficient. A greater use of sensor fusion and including gyro and accelerometer data could improve the tracking accuracy.

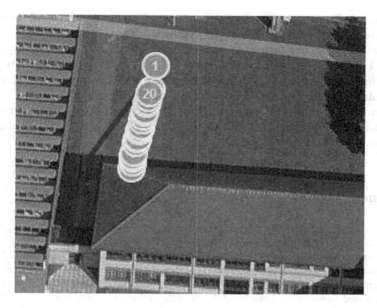

Fig. 9. The plotted .gpx file on Bing Maps

The points were stored as a.gpx file and analysed in MATLAB. The points were also imported into Bing Maps [22] to view the progress. Although this accurately confirms the location of the robot, it is inefficient for trajectory analysis as shown in Fig. 9.

Since the work began, considerable interest has come from the Geophysical side of the Geography department at Staffordshire University, particularly in the use of automating the collection of data in WWII concentration camp grave sites [18–20] and

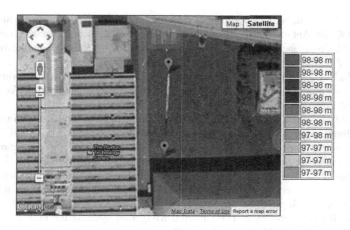

Fig. 10. Creating the geo-fence using Maplorer [21]

further work will interface the robot with the ground penetrating RADAR sensors that are used in this and fuse the data onto mapping software [24].

The experiment consisted of gathering the geo-fence points manually by using a hand-held GPS receiver, work has been investigated whereby these points can be selected in a map, sent to the robot, and the robot will scan in the highlighted area. A particular mapping tool is available at [21] and shown in Fig. 10.

Future work will involve using an airborne quadcopter to scan the geo-fence area and grab images and particular heights and times of the day, furthering the information available to archaeologists.

References

1. Li, M., Imou, K., Wakabayashi, K., Yokoyama, S.: Review of research on agricultural vehicle autonomous guidance. Int. J. Agric. Biol. Eng. 2(3), 1–16 (2009)
2. O'Connor, M., Bell, T.: Automatic steering of farm vehicles using GPS. In: 3rd International Conference on Precision Agriculture, Minneapolis, Minnesota, June 1996
3. Subramanian, V., Burks, T.F., Arroyo, A.A.: Development of machine vision and laser radar based autonomous vehicle guidance systems for citrus grove navigation. Comput. Electron. Agric. 53(2), 130–143 (2006)
4. Nagasaka, Y., Umeda, N., Kanetai, Y., Taniwaki, K., Sasaki, Y.: Autonomous guidance for rice transplanting using global positioning and gyroscopes. Comput. Electron. Agric. 43(3), 223–234 (2004)
5. Heidman, B.C., Abidine, A.Z., Upadhyaya, S.K., Hills, D.J.: Application of RTK GPS based auto-guidance system in agricultural production. In: Robert, P.C., et al. (eds.) Presented at the Proceedings of the 6th International Conference on Precision Agriculture and Other Precision Resources Management, Minneapolis, MN, USA, 14–17 July 2002, pp. 1205–1214 (2003)
6. Gan-Mor, S., Clark, R.L., Upchurch, B.L.: Implement lateral position accuracy under RTK-GPS tractor guidance. Comput. Electron. Agric. 59(1–2), 31–38 (2007)
7. Stoll, A., Kutzbach, H.D.: Guidance of a forage harvester with GPS. Precis. Agric. 2(3), 281–291 (2000)
8. Kise, M., Noguchi, N.: Development of agricultural autonomous tractor with an RTK-GPS and a FOG. In: Automation Technology for Off-Road Equipment, Proceedings of the 26-27 July 2002 Conference, Chicago, Illinois, USA, pp. 367–373. ASAE publication number 701P0502 (2002)
9. Kise, M., Noguchi, N.: The development of the autonomous tractor with steering controller applied by optimal control. In: Proceedings of Conference on Automation Technology for Off-Road Equipment, Chicago, USA, pp. 367–373 (2002)
10. Ehsani, M., Sullivan, M.: Evaluating the dynamic accuracy of low-cost GPS receivers. In: Proceedings of the ASAE Annual International Meeting, Nevada, USA (2003)
11. Borenstein, J.: Experimental results from internal odometry error correction with the OmniMate mobile robot. IEEE Trans. Robot. Autom. 14, 963–969 (1998)
12. Chenavier, F., Crowley, J.L.: Position estimation for a mobile robot using vision and odometry. In: Proceedings of the 1992 IEEE International Conference on Robotics and Automation, vol. 3, pp. 2588–2593 (1992)
13. Barshan, B., Durrant-Whyte, H.F.: Inertial navigation systems for mobile robots. IEEE Trans. Robot. Autom. 11(3), 328–342 (1995)

14. Noguchi, N., Terao, H.: Path planning of an agricultural mobile robot by neural network and genetic algorithm. Comput. Electron. Agric. **18**(2–3), 187–204 (1997)
15. Nagasaka, Y., Taniwaki, K., Otani, R., Shigeta, K.: An automated rice transplanter with RTKGPS and FOG (2002)
16. Haversine formula: http://www.movable-type.co.uk/scripts/latlong.html
17. Lat long of a google map point http://itouchmap.com/latlong.html. Inspired by http://members.tripod.com/vector_applications/distance_point_line/index.html
18. Sturdy Colls, C.: Holocaust archaeology: archaeological approaches to landscapes of Nazi genocide and persecution. J. Conflict Archaeol. **7**(2), 71–105 (2012)
19. Hunter, J., Sturdy, C.C.: Archeology. In: Siegel, J., Saukko, P. (eds.) Encyclopaedia of Forensic Sciences, vol. 1, 2nd edn, pp. 18–32. Academic Press, Waltham (2013)
20. Sturdy Colls, C.: The archaeology of the holocaust. British Archaeology **130**, 50–53 (2013)
21. GPX viewer: http://page42.ovh.org/maplorer/gpx_view.php
22. Bing Maps: http://www.bing.com/maps/
23. Mobile Robots: http://www.mobilerobots.com
24. ArcGIS: http://www.esri.com/software/arcgis

Motion Planning and Decision Making for Underwater Vehicles Operating in Constrained Environments in the Littoral

Erion Plaku[1]([✉]) and James McMahon[1,2]

[1] Department of Electrical Engineering and Computer Science,
Catholic University of America, Washington, DC 20064, USA
plaku@cua.edu
[2] U.S. Naval Research Laboratory, Washington, DC 20375, USA

Abstract. This paper seeks to enhance the mission and motion-planning capabilities of autonomous underwater vehicles (AUVs) operating in constrained environments in the littoral zone. The proposed approach automatically plans low-cost, collision-free, and dynamically-feasible motions that enable an AUV to carry out missions expressed as formulas in temporal logic. The key aspect of the proposed approach is its use of roadmap abstractions in configuration space to guide the expansion of a tree of feasible motions in the state space. This makes it possible to effectively deal with challenges imposed by the vehicle dynamics and the need to operate in the littoral zone, which is characterized by confined waterways, shallow water, complex ocean floor topography, varying currents, and miscellaneous obstacles. Experiments with accurate AUV models carrying out different missions show considerable improvements over related work in reducing both the running time and solution costs.

1 Introduction

As maritime operations shift from deep waters to the littoral, it becomes increasingly important to enhance the mission and motion-planning capabilities of AUVs. Operating in the littoral brings new challenges which are not adequately addressed by existing approaches. As the littoral is characterized by highly-trafficked zones, the AUV must operate near the ocean floor in order to avoid collisions with medium- to small-sized ships. Operations are further complicated when tankers are present, which often cause large drafts, or when fishing vessels are trawling, which could cause the AUV to get entangled. Additional complications arise from boulders, wreckage, and other objects found at the bottom of the ocean, sudden changes in elevation of the ocean floor, and varying ocean currents associated with tides and weather patterns. Dealing with these challenges requires efficient motion-planning approaches that take into account the vehicle dynamics and quickly plan feasible motions that enable the AUV to operate close to the ocean floor while avoiding collisions.

The demand to enhance the capabilities of AUVs operating in the littoral becomes more pressing when coupled with critical missions, such as monitoring

A. Natraj et al. (Eds.): TAROS 2013, LNAI 8069, pp. 328–339, 2014.
DOI: 10.1007/978-3-662-43645-5_36, © Springer-Verlag Berlin Heidelberg 2014

harbors or searching for mines and other objects of interest. AUV operations, however, are currently limited to simple missions given in terms of following a set of predefined waypoints or covering an area of interest by following specific motion patterns. Conventional approaches to motion planning for AUVs based on potential fields, numeric optimizations, A*, or genetic algorithms [1–5] have generally focused on largely unobstructed and flat environments where dynamics and collision avoidance do not present significant problems. As a result, while conventional approaches are able to plan optimal paths in 2D environments, they become inefficient when dealing with the challenges arising from dynamics and the need to operate close to the ocean floor in constrained 3D environments. The restrictions to follow predefined waypoints or motion patterns and to operate in essentially flat environments greatly limits the feasibility of such approaches being used for complex missions in constrained 3D environments.

To enhance the mission and motion-planning capabilities of AUVs, this paper develops an approach that makes it possible to express maritime operations in Linear Temporal Logic (LTL) and automatically plan collision-free and dynamically-feasible motions to carry out such missions. As missions are characterized by events occurring across a time span, LTL allows for the combination of these events with logical ($\wedge$ and, $\vee$ or, $\neg$ not) and temporal operators ($\square$ always, $\lozenge$ eventually, $\cup$ until, $\bigcirc$ next). As an illustration, the mission of inspecting several areas of interest while avoiding collisions can be expressed in LTL as

$$\square \pi_{safe} \wedge \lozenge \pi_{A_1} \wedge \ldots \wedge \lozenge \pi_{A_n}, \tag{1}$$

where π_{safe} denotes the proposition "no collisions" and π_{A_i} denotes "searched area A_i." As another example, searching A_1 or A_2 before A_3 or A_4 is written as

$$\square \pi_{safe} \wedge ((\neg \pi_{A_3} \wedge \neg \pi_{A_4}) \cup ((\pi_{A_1} \vee \pi_{A_2}) \wedge \bigcirc (\pi_{A_3} \vee \pi_{A_4}))). \tag{2}$$

The expressive power of LTL makes it possible to consider sophisticated missions, making planning even more challenging. Motions need to be coupled with decision making to determine the best course of action to accomplish a given mission. The discrete aspects, which relate to determining the propositions that need to be satisfied to obtain a discrete solution, are intertwined with the continuous aspects, which need to plan collision-free and dynamically-feasible motions in order to implement the discrete solutions.

As a result, approaches that compute discrete solutions without taking into account the intertwined dependencies with the continuous aspects have been limited in scope and capability [6, 7]. Such approaches rely on the limiting assumption that a motion controller would be available to implement *any* discrete solution. However, due to dynamics, obstacles, complex ocean floor topography, and drift caused by currents, only some discrete solutions could be feasible. Indeed, determining which discrete solutions are feasible is one of the major challenges when dealing with the combined planning problem. As a result of such strong requirement on the availability of motion controllers to implement any discrete solution the applicability of these approaches is limited to 2D environments, disk robot shapes, low-dimensional state spaces, and simplified dynamics.

To address the challenges imposed by the intertwined dependencies of the discrete and continuous aspects of the combined planning problem, the proposed approach simultaneously conducts the search in the discrete space of LTL sequences and the continuous space of feasible motions. The proposed approach builds on our recent framework [8,9], which combined discrete search with sampling-based motion planning to compute collision-free and dynamically-feasible trajectories that satisfy LTL specifications.

The proposed approach significantly enhances our previous framework [8,9] by (i) improving the computational efficiency; (ii) generating lower cost solutions; and (iii) enhancing the application from ground vehicles to AUVs, taking into account drift caused by ocean currents and complex ocean floor topography. A key aspect of the technical contribution is the introduction of roadmap abstractions in configuration space to guide a sampling-based exploration of the state space. The roadmap is constructed by sampling collision-free configurations and connecting neighboring configurations with simple collision-free paths. The roadmap represents solutions to a simplified motion-planning problem that does not take into account the vehicle dynamics. These solutions are converted into heuristic costs and heuristic paths to guide sampling-based motion planning as it expands a tree of feasible motions in the state space, taking into account the vehicle dynamics, drift caused by ocean currents, and collision avoidance. The roadmap abstraction also facilitates the decision-making mechanism by providing estimates on the feasibility of reaching intermediate goals as part of the overall mission specified by the LTL formula. Simulation experiments with accurate AUV models carrying out different missions show considerable improvements over related work in reducing both the running time and solution costs.

2   Problem Formulation

AUV Simulator: This paper uses the MOOS-IvP framework [10], which provides a 3D AUV simulator that accurately models the vehicle dynamics. The simulator propagates the dynamics based on a set of control inputs and external drift forces caused by the ocean currents, i.e., $s_{new} \leftarrow \text{SIMULATOR}(s, u, drift, dt)$, where s_{new} is the new state obtained by applying the control input u and the external forces $drift$ to the current state s and propagating the dynamics for dt time steps. The MOOS-IvP vehicle model has a state space which consists of the vehicles position and orientation, the actuator values (thrust, rudder, elevator), and the vehicle dynamics (speed, depth rate, buoyancy rate, turn rate, acceleration, etc.). The model represents a second-order dynamical system that has been tested in many applications and shown to be robust and accurate [10].

From Co-Safe LTL to Finite Automata: As mentioned, LTL formulas are composed by combining propositions with logical ($\wedge$ and, $\vee$ or, $\neg$ not) and temporal operators ($\square$ always, $\lozenge$ eventually, $\cup$ until, $\bigcirc$ next). The temporal semantics are as follows: $\square\phi$ indicates that ϕ is always true; $\lozenge\phi$ indicates that

ϕ will eventually be true; $\phi \cup \psi$ indicates that ϕ will be true until ψ becomes true; $\bigcirc\phi$ indicates that ϕ will become true in the next time step.

Since LTL planning is PSPACE-complete [11], as in prior work [8,9], this paper considers co-safe LTL, which is satisfied by finite sequences. Co-safe LTL formulas are converted into Deterministic Finite Automata (DFA) [12], which is more amenable for computation. Figure 1 shows an example.

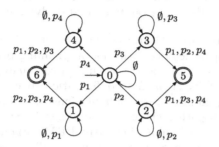

Fig. 1. The mission "visit any two of the areas p_1, p_2, p_3, p_4" encoded as a finite automaton and as a co-safe LTL formula $\bigvee_{i \neq j}(\Diamond p_i \wedge \Diamond p_j)$

Let Π denote the set of propositions. A DFA is a tuple $\mathcal{A} = (Z, Q, \delta, z_{init}, Accept)$, where Z is a finite set of states, $Q = 2^{\Pi}$ is the input alphabet corresponding to the discrete space, $\delta : Z \times Q \to Z$ is the transition function, $z_{init} \in Z$ is the initial state, and $Accept \subseteq Z$ is the set of accepting states.

To facilitate presentation, let $\mathcal{A}([q_i]_{i=1}^{n}, z)$ denote the state obtained by running $\mathcal{A}$ on $[q_i]_{i=1}^{n}$, $q_i \in Q$, starting from the state z. Then, $[q_i]_{i=1}^{n}$ is accepted iff $\mathcal{A}([q_i]_{i=1}^{n}, z_{init}) \in Accept$. Moreover, let $Reject$ denote the states that cannot reach an accepting state. Let $\delta(z)$ denote all the non-rejecting states connected by a single transition from z, i.e., $\delta(z) = \{\delta(z, q) : q \in Q\} - Reject$. Let $props(z_{from}, z_{to})$ denote all the propositions $\pi_{i_1}, \ldots, \pi_{i_k}$ labeling the transition from z_{from} to z_{to}. As an example, referring to Fig. 1, $props(4, 6) = \{p_1, p_2, p_3\}$.

Discrete Semantics in the Continuous State Space: The semantics of a proposition $\pi \in \Pi$ is defined over the continuous state space S by a function $\text{HOLDS}_\pi : S \to \{\texttt{true}, \texttt{false}\}$, which indicates if a continuous state satisfies π. This interpretation provides a mapping from S to Q, i.e.,

$$qstate(s) = \{\pi : \pi \in \Pi \wedge \text{HOLDS}_\pi(s) = \texttt{true}\}.$$

Moreover, as the continuous state changes according to a trajectory $\zeta : [0, T] \to S$, parametrized by time, the discrete state $qstate(\zeta(t))$ may also change. In this way, ζ maps to a sequence of discrete states, $qstates(\zeta) = [q_i]_{i=1}^{n}$, $q_i \neq q_{i+1}$. As a result, ζ satisfies a specification given by $\mathcal{A}$ iff $\mathcal{A}$ accepts $qstates(\zeta)$.

The objective is then to compute a dynamically-feasible trajectory $\zeta : [0, T] \to S$ that enables the AUV to carry out the mission, i.e., $\mathcal{A}$ accepts $qstates(\zeta)$, while avoiding collisions and operating close to the ocean floor.

3   Method

3.1   Roadmap Abstraction

The approach uses solutions to a simplified version of the problem which ignores the vehicle dynamics to guide the overall search. The vehicle motions in this simplified version are defined over the vehicle's configuration space, denoted as C, which accounts for translations and rotations, but not for velocities, accelerations, curvature, and other constraints related to dynamics.

Roadmap Construction: Motivated by the success of the Probabilistic RoadMap (PRM) [13] in dealing with motion-planning problems in configuration spaces, the proposed approach constructs a roadmap $RM = (V_{RM}, E_{RM})$ to capture the connectivity of the free configuration space. The roadmap is constructed by sampling collision-free configurations and connecting neighboring configurations via local paths, where a distance metric $\rho : C \times C \rightarrow R^{\geq 0}$ defines the distance between two configurations. Any PRM variant [13–15] can be used for the roadmap construction. Since the roadmap will be used to define heuristic costs based on shortest paths it is important to allow cycles during roadmap construction. Since C is of lower dimensionality than S and the motions are less constrained in C, the roadmap construction takes only a fraction of the overall running time.

Implicit Partition of the State Space Induced by the Roadmap: The roadmap and the distance metric ρ induce an implicit partition of S into equivalence classes where roadmap configurations act as centers of Voronoi sites. In particular, let $cfg(s) \in C$ denote the configuration corresponding to $s \in S$. For example, $cfg(s)$ could correspond to position and orientation. The state s is then associated with the configuration $c \in V_{RM}$ closest to $cfg(s)$ according to ρ, i.e.,

$$NearestCfgRM(s) = \arg \min_{c \in V_{RM}} \rho(cfg(s), c).$$

This partition, as described later in the section, is particularly useful when determining the region from which to expand the search in the state space S.

Mapping Roadmap Configurations to Propositions: Each configuration $c \in V_{RM}$ is mapped to a discrete state based on the propositions it satisfies, i.e.,

$$qstate(c) = \{\pi : \pi \in \Pi \land \text{HOLDS}_\pi(c) = \textbf{true}\}.$$

The inverse map from propositions to configurations is defined as

$$cfgs(\pi) = \{c : c \in V_{RM} \land \text{HOLDS}_\pi(c) = \textbf{true}\}.$$

Additional sampling may take place during roadmap construction to ensure $cfgs(\pi)$ is nonempty. Based on the problem under consideration, it may be easier to sample directly from the regions associated with the propositions. For example, propositions in the experiments in this paper correspond to areas of interest in the environment. As such, a configuration can be generated by sampling a position inside the area and then a random orientation.

Heuristic Costs Based on Shortest Paths: The roadmap is used to define heuristic costs of the form $hcost(c, \pi_{i_1}, \ldots, \pi_{i_k})$ as the length of the shortest path from c to a roadmap configuration c' satisfying one of the propositions $\pi_{i_1}, \ldots, \pi_{i_k}$, i.e., $c' \in \cup_{j=1}^{k} cfgs(\pi_{i_j})$. This heuristic cost serves as an estimate on the difficulty of expanding the motion tree T from tree vertices associated with the Voronoi site defined by c to reach states that satisfy one of the propositions $\pi_{i_1}, \ldots, \pi_{i_k}$. Dijkstra's shortest-path algorithm is used for the computation of the heuristic cost, where $\rho(c_i, c_j)$ defines the weight of an edge $(c_i, c_j) \in E_{RM}$.

3.2   Overall Search

The roadmap abstraction in conjunction with the automaton A guides the search in S. A tree data structure T is used as the basis for conducting the search in S. The tree starts at the initial state s_{init} and is incrementally expanded by adding new collision-free and dynamically-feasible trajectories as branches.

The tree vertices are partitioned according to their corresponding automaton states. More specifically, let $traj(v)$ denote the trajectory obtained by concatenating the trajectories associated with the edges connecting the root of T to v. The vertex v keeps track of the automaton state obtained by running $qstates(traj(v))$ on A, denoted as $astate(v)$. The computation of $astate(v)$ is done incrementally when checking for collisions the trajectory from the parent of v to v, as described in Sect. 3.2. The partition induced by A is then defined as

$$TreeVertices(z) = \{v : v \in T \wedge z \in Z \wedge z = astate(v)\}.$$

To speed up computation, the approach keeps track of the automaton states reached by T, referred to as active automaton states and defined as

$$ActiveAStates = \{z : z \in Z \wedge |TreeVertices(z)| > 0\}.$$

A solution is obtained if T reaches an accepting automaton states, i.e., $ActiveAStates \cap Accept \neq \emptyset$. The search proceeds incrementally at each iteration as follows:

1. an automaton state z_{from} is selected from $ActiveAStates$;
2. an automaton state z_{to} is selected from $\delta(z_{from})$;
3. attempts are made to expand T from $TreeVertices(z_{from})$ toward z_{to}.

These steps are repeated until a solution is obtained or an upper bound on computational time is exceeded. A description of each of these steps follows.

Selecting Automaton States: The selection of z_{from} from $ActiveAStates$ determines the tree vertices from which to expand the search. Since the overall objective is to compute a trajectory that reaches an accepting automaton state, the selection is biased toward states that are close to an accepting state, i.e., $prob(z_{from}) = 2^{-d(z_{from})} / \sum_{z' \in ActiveAStates} 2^{-d(z')}$, where $d(z)$ denotes the minimum number of transitions to reach an accepting state in A from z.

The selection of z_{to} aims to promote expansion toward unreached automaton states ($|\mathit{TreeVertices}(z_{to})| = 0$) in order to explore new regions. Similar to the bias in the selection of z_{from}, preference is given to those unreached automaton states that are close to accepting states. Taking these into account, z_{to} is selected from the neighbors of z_{from} according to the probability distribution $prob(z_{to}) = 2^{-d(z_{to})} / \sum_{z' \in \delta(z_{from}) - ActiveAStates} 2^{-d(z')}$.

Tree Expansion: The tree expansion from $\mathit{TreeVertices}(z_{from})$ adds new collision-free and dynamically-feasible trajectories aiming to reach z_{to}. Since $z_{to} \in \delta(z_{from})$ the automaton transition from z_{from} to z_{to} determines the propositions that need to be satisfied in order to reach z_{to}, which, as described in Sect. 2, are denoted as $props(z_{from}, z_{to})$. Recalling again the example in Fig. 1, the automaton state z_6 can be reached from z_4 by satisfying one of the propositions p_1, p_2, p_3. Then, the objective is to expand $\mathit{TreeVertices}(z_{from})$ towards states $s \in S$ such that $\text{HOLDS}_\pi(s) = \textbf{true}$ for some $\pi \in props(z_{from}, z_{to})$.

In this way, the expansion from $\mathit{TreeVertices}(z_{from})$ toward z_{to} gives rise to a motion-planning problem. The roadmap abstraction is used to effectively guide the tree expansion. To speed up computation, $\mathit{TreeVertices}(z)$ are further grouped according to the nearest configuration in the roadmap RM, i.e.,

$$\mathit{TreeVertices}(z, c) = \{v : v \in \mathit{TreeVertices}(z) \land c = NearestCfgRM(sstate(v))\},$$

where $sstate(v)$ denotes the state in S associated with v. The approach also keeps track of all the active roadmap configurations, i.e.,

$$ActiveCfgs(z) = \{c : c \in V_{RM} \land |\mathit{TreeVertices}(z, c)| > 0\}.$$

From an implementation perspective, each time a vertex v is added to $\mathcal{T}$, it is also added to the corresponding $\mathit{TreeVertices}(z, c)$. Moreover, c is added to $ActiveCfgs(z)$ if not already there.

The tree expansion proceeds incrementally for several iterations. During each iteration, a roadmap configuration c is selected from $ActiveCfgs(z_{from})$. The selection is based on $hcost(c, \pi_1, \ldots, \pi_k)$ where $\{\pi_1, \ldots, \pi_k\} = props(z_{from}, z_{to})$. In particular, the procedure selects the configuration c in $ActiveCfgs(z_{from})$ with the lowest $hcost(c, \pi_1, \ldots, \pi_k)$. As described in Sect. 3.1, $hcost(c, \pi_1, \ldots, \pi_k)$ is initially computed as the length of the shortest-path in the roadmap from c to a configuration satisfying one of the propositions $\pi_1, \ldots, \pi_k$. As such, the selection strategy provides a greedy component necessary to effectively expand the search toward z_{to}. To balance the greedy exploitation with methodical exploration, $hcost(c, \pi_1, \ldots, \pi_k)$ is increased after each selection. Increasing the heuristic cost, for example, by doubling it, promotes selection of other configurations in future iterations. In this way, the tree expansion has the flexibility to make rapid progress toward z_{to} while effectively discovering new ways to reach it. A hash map keep tracks of the heuristic costs using $\langle c, \pi_1, \ldots, \pi_k \rangle$ as keys.

After selecting c from $ActiveCfgs(z_{from})$, a vertex v is selected from vertices in $\mathit{TreeVertices}(z_{from}, c)$. A simple strategy that has worked well in practice is to select v at random. A collision-free and dynamically-feasible trajectory is

then obtained by applying control inputs for several time steps starting from
$sstate(v)$ and running the AUV simulator to obtain the new state. The approach
uses motion controllers to generate the control inputs. In particular, PID controllers are used to adjust the heading and steer the AUV toward selected target
positions. To expand the search toward z_{to}, the target is often selected near configurations associated with the roadmap path from c to configurations satisfying
propositions in $props(z_{from}, z_{to})$. At other times, the target is selected uniformly
at random from the entire workspace to expand the search along new directions.
After selecting the target position, the PID controllers are invoked until the trajectory comes close to the target, a collision is found, or a maximum number
of steps is exceeded. Intermediate states along the generated trajectory and the
final state are added to $\mathcal{T}$. Bookkeeping information is updated accordingly.

4   Experiments and Results

The approach is tested on a simulated environment using an accurate and robust
AUV simulator, as described in Sect. 2. Several LTL formulas are used to specify
different missions.

4.1   Experimental Setup

Ocean Floor: A simulated ocean floor map is created by adding random peaks.
A uniform grid is imposed over (x, y) and the center of each peak is sampled
at random. The height of each grid cell is determined based on the distance to
the closest peak. The height map is then converted to a triangular mesh. As
shown in Fig. 2, this setting provides a challenging environment due to changes
in topography. In the experiments, the dimensions of the grid are set to 1.5 km ×
1.5 km, which correspond to realistic settings for AUV missions.

The AUV is 4.93 m long and has a diameter of 0.53 m. The AUV moves at a
maximum speed of 1.5 m/s. The AUV is restricted to an altitude of 2 m to 8 m
from the ocean floor. Operation this close to the ocean floor is required in order
to provide high quality acoustic bathymetric data along with detecting objects
of interest. The environment is populated with 100 obstacles placed at random
positions to test the ability of the approach to avoid collisions.

Ocean Currents: The drift caused by ocean currents is modeled as vector
fields. The drift magnitude becomes smaller as the depth increases since the
ocean currents are stronger closer to the surface. For the experiments, we used
publicly-available data from the National Data Buoy Center [16] describing the
ocean currents for the Chesapeake Bay Channel (see Fig. 3(a)). From this data,
the vector field was extracted and then scaled to match the area dimensions used
in the experiments (see Fig. 3(b)).

LTL Mission Specifications: Each area of interest A_i defines a proposition
π_i. The function $\text{HOLDS}_{\pi_i}(s)$ is **true** iff the AUV is in area A_i when its state is s.

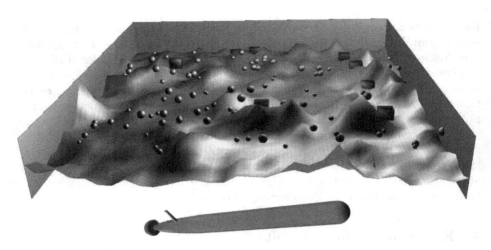

Fig. 2. One of the environments used in the experiments. The ocean floor and the obstacles are shown in gold. The areas of interest shown as red boxes. The AUV in its initial state is shown in red. A zoomed-in illustration is shown below the environment (Color figure online).

An area A_i is defined as a box, placed at random inside the test environment. The inspection mission, which requires the AUV to visit each area of interest, is defined by the following LTL formula:

$$\phi_1 = \bigwedge_{i=1}^{n} \Diamond \pi_{A_i}.$$

Note that the mission specification leaves it up to the approach to determine an appropriate order in which to visit the areas. Experiments were also conducted with a mission that requires visiting the areas of interest in a predefined order $A_1, \ldots, A_n$. This mission, referred to as "sequencing," is given by

$$\phi_2 = \beta \cup (\pi_{A_1} \wedge ((\pi_{A_1} \vee \beta) \cup (\pi_{A_2} \wedge (\ldots (\pi_{A_{n-1}} \vee \beta) \cup \pi_{A_n})))), \text{ where } \beta = \wedge_{i=1}^{n} \neg \pi_{A_i}.$$

Computational Time and Normalized Solution Cost: Due to the probabilistic nature of sampling-based motion planning, results are based on 30 different runs for each problem instance. The five worst and the five best runs are discarded to avoid the influence of outliers. Results report on average computational time and normalized solution costs. Standard deviations are also included as bars in the plots. To understand how good a trajectory cost is, we normalize it with respect to the cost of the shortest-path in the roadmap abstraction from $config(s_{init})$ that satisfies the LTL mission specification. Such shortest path is computed by running the graph search over the product graph formed by combining the roadmap with the automaton [9]. The roadmap cost provides a lower bound as it is over the configuration space and does not take dynamics into account. Experiments are run on an Intel Core i7 machine (CPU: 1.90 GHz, RAM: 4 GB) using Ubuntu 12.10. Code is compiled with GNU g++-4.7.2.

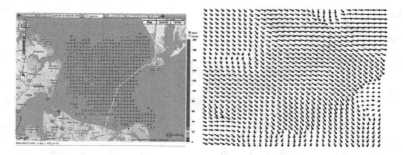

Fig. 3. (left) Ocean currents for the Chesapeake Bay Channel. Data obtained from National Data Buoy Center [16]. (right) Current field with the missing data filled in and scaled to fit each area used in the experiments. See Sect. 4.1 for more details.

4.2 Results

The proposed approach is compared to our prior framework [8]. Figure 4 summarizes the results when varying the number n of areas of interest.

Results show considerable improvements both in reducing the running time and the solution cost. The approach in prior work has difficulty solving the problem instances as n is increased. In contrast, the proposed approach effectively solves the problem instances. Prior work relies on a workspace decomposition to guide the tree expansion, which, as the problems become more challenging, can lead the exploration along infeasible discrete paths. The lead will eventually be corrected, but it may take the approach in considerable time to do so and find other more feasible leads. In contrast, the proposed approach relies on a roadmap abstraction to capture the connectivity of the free configuration space, which more effectively guides the tree expansion in the state space.

The results show that in addition to computational time the proposed approach effectively computes low-cost, dynamically-feasible, solution trajectories. The costs of these solutions are comparable to those obtained by the roadmap abstraction, which provides a lower bound as it is over the configuration space and does not account for the motion dynamics (recall that the plots show the normalized solution cost over the roadmap abstraction solution cost). By using the roadmap abstraction as a guide, and the expansion heuristics, the proposed approach is able to effectively find low-cost solutions.

5   Discussion

This paper focused on enhancing the mission and motion-planning capabilities of AUVs operating in the littoral zone, which is characterized by confined waterways, shallow waters, complex ocean floor topography, varying currents, and miscellaneous obstacles. The proposed approach can take into account maritime operations expressed in LTL and automatically plans collision-free and dynamically-feasible motions to carry out such missions. The key aspect of the

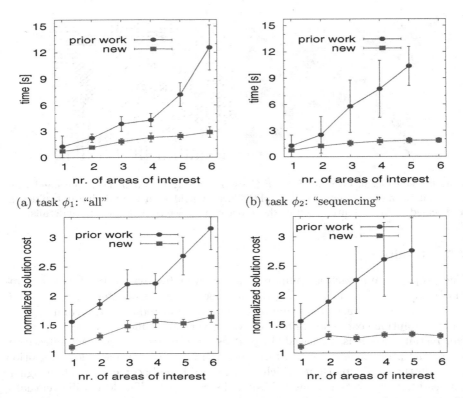

(a) task ϕ_1: "all"   (b) task ϕ_2: "sequencing"

Fig. 4. Results when comparing the proposed approach (labeled as new) to prior work [8]. Bars indicate one standard deviation. A missing data point in the graph indicates the method timed out before obtaining a solution. Results include the time to construct the DFAs from the LTL formulas (in the order of milliseconds). For the proposed approach, results also include the time to construct the roadmap abstraction, which took between 0.1 s–0.4 s. Solution costs are normalized with respect to those obtained by the roadmap abstraction, which provides a lower bound as it is over the configuration space and does not take the vehicle dynamics into account.

approach is the introduction of roadmap abstractions to capture the connectivity of the free configuration space and, in conjunction with an automaton representing the LTL formula, to effectively guide the expansion of a tree of feasible motions in the state space.

In future work, we plan to enhance the approach so that it can be applied to real AUVs operating in the littoral. This would require adaptations to fit a replanning framework, building the map from sensory information, and quickly responding to unanticipated obstacles, changes in elevation, or other events.

Acknowledgment. The work of J. McMahon is supported by the Office of Naval Research, code 32.

References

1. Petres, C., Pailhas, Y., Patron, P., Petillot, Y., Evans, J., Lane, D.: Path planning for autonomous underwater vehicles. IEEE Trans. Robot. **23**(2), 331–341 (2007)
2. Alvarez, A., Caiti, A., Onken, R.: Evolutionary path planning for autonomous underwater vehicles in a variable ocean. IEEE J. Oceanic Eng. **29**(2), 418–429 (2004)
3. Soulignac, M.: Feasible and optimal path planning in strong current fields. IEEE Trans. Robot. **27**(1), 89–98 (2011)
4. Yilmaz, N.K., Evangelinos, C., Lermusiaux, P., Patrikalakis, N.M.: Path planning of autonomous underwater vehicles for adaptive sampling using mixed integer linear programming. IEEE J. Oceanic Eng. **33**(4), 522–537 (2008)
5. Ho, C., Mora, A., Saripalli, S.: An evaluation of sampling path strategies for an autonomous underwater vehicle. In: IEEE International Conference on Robotics and Automation, pp. 5328–5333 (2012)
6. Fainekos, G.E., Girard, A., Kress-Gazit, H., Pappas, G.J.: Temporal logic motion planning for dynamic mobile robots. Automatica **45**(2), 343–352 (2009)
7. Kloetzer, M., Belta, C.: Automatic deployment of distributed teams of robots from temporal logic specifications. IEEE Trans. Robot. **26**(1), 48–61 (2010)
8. Plaku, E.: Planning in discrete and continuous spaces: from LTL tasks to robot motions. In: Herrmann, G., Studley, M., Pearson, M., Conn, A., Melhuish, C., Witkowski, M., Kim, J.-H., Vadakkepat, P. (eds.) TAROS-FIRA 2012. LNCS, vol. 7429, pp. 331–342. Springer, Heidelberg (2012)
9. Plaku, E.: Path planning with probabilistic roadmaps and linear temporal logic. In: IEEE/RSJ International Conference on Intelligent Robots and Systems, Vilamoura, Algarve, Portugal, pp. 2269–2275 (2012)
10. Benjamin, M.R., Schmidt, H., Newman, P.M., Leonard, J.J.: Nested autonomy for unmanned marine vehicles with MOOS-IvP. J. Field Robot. **27**(6), 834–875 (2010)
11. Sistla, A.: Safety, liveness and fairness in temporal logic. Form. Asp. Comput. **6**, 495–511 (1994)
12. Kupferman, O., Vardi, M.: Model checking of safety properties. Form. Methods Syst. Des. **19**(3), 291–314 (2001)
13. Kavraki, L.E., Švestka, P., Latombe, J.C., Overmars, M.H.: Probabilistic roadmaps for path planning in high-dimensional configuration spaces. IEEE Trans. Robot. Autom. **12**(4), 566–580 (1996)
14. Yeh, H.Y., Thomas, S., Eppstein, D., Amato, N.M.: UOBPRM: a uniformly distributed obstacle-based PRM. In: IEEE/RSJ International Conference on Intelligent Robots and Systems, pp. 2655–2662, Vilamoura, Algarve, Portugal (2012)
15. Hsu, D., Latombe, J.C., Kurniawati, H.: On the probabilistic foundations of probabilistic roadmap planning. Int. J. Robot. Res. **25**(7), 627–643 (2006)
16. National Data Buoy Center.: NOAA HF radar national server and architecture project. http://www.ndbc.noaa.gov/

Using Range and Bearing Observation in Stereo-Based EKF SLAM

Yao-Chang Chen[1]([⊠]), Tsung-Han Lin[2], and Ta-Ming Shih[3]

[1] School of Defense Science, Chung-Cheng Institute of Technology,
National Defense University, 33551 Tauyuan County, Taiwan
joseph.chen4210@gmail.com
[2] Department of Computer Science and Information Engineering,
National Taiwan University, 10617 Taipei, Taiwan
[3] Department of Avionics, China University Science and Technology,
31241 Hsinchu County, Taiwan

Abstract. In this work, we have developed a new observation model for a stereo-based simultaneous localization and mapping (SLAM) system within the standard Extended-Kalman filter (EKF) framework. The observation model was derived by using the inverse depth parameterization as the landmark model, and contributes to both bearing and range information into the EKF estimation. In this way the inherently non-linear problem cause by the camera projection equations is resolved and real depth uncertainty distribution of landmarks features can be accurately estimated. The system was tested by real-world large-scale outdoor data. Analysis results show that the landmark feature depth estimation is more stable and the uncertainty noise converges faster than the binocular stereo-based approach. We also found minor drift in the vehicle pose estimation even after extended periods demonstrating the effectiveness of the new model.

Keywords: Simultaneous localization and mapping (SLAM) · Stereo · Inverse depth parameterization · Stereo observation model · Extended Kalman filter (EKF)

1 Introduction

Over the past years, Simultaneous Localization and Mapping (SLAM) has received extensive research interest because it serves as a basic methodology for robots moving autonomously in an unknown environment. Current methods of SLAM have found to achieve accurate mapping for extended periods of time especially by using laser sensor with well bounded result for both indoor and outdoor environments [1].

Since vision systems have properties of being low cost, lightweight and contain rich information when compared to traditional robotic sensors, like laser scanners or sonars, vision-based systems are employed in a wide range of robotic applications. These applications include object recognition, obstacle avoidance, navigation, topological global localization and, more recently, in simultaneous localization and mapping, which in this case is the so-called visual SLAM.

A. Natraj et al. (Eds.): TAROS 2013, LNAI 8069, pp. 340–352, 2014.
DOI: 10.1007/978-3-662-43645-5_37, © Springer-Verlag Berlin Heidelberg 2014

One can distinguish visual SLAM as either monocular or binocular approaches. The first remarkable work in monocular visual SLAM was done by Davison et al. [2], in which a single camera is used under the Extended Kalman filtering (EKF) framework. Since camera is a bearing-only sensor, crucial limitation of monocular SLAM is the unobservability of the scale, and this cause the scale of the map to slowly drift in large environment. On the other hand, stereo vision systems are often applied in which the absolute measurement of 3D space, especially the feature depths, can be directly estimated to avoid the scale ambiguity.

In stereo systems, the observation pair (u_1, v_1, d) contains both bearing and range information, where (u_1, v_1) is the left image coordinate and d_i is the disparity. By projecting this observation pair through the pinhole model one can get the 3D Euclidean XYZ position of landmark relative to the camera [3]. Many stereo SLAM systems, including indoor and outdoor, build the map using the 3D Euclidean representation for landmark model [3–6].

Standard pinhole model projection equations used in the vision systems with EKF framework suffers from nonlinearity [7, 8]. Due to this nonlinearity, the true uncertainty of 3D Euclidean XYZ landmark can be modeled by Gaussian only for nearby features. However, true distributions of faraway features are non-Gaussian, which makes the EKF filter estimation inconsistent [9]. Other work also tried using UKF (Unscented Kalman Filter) for a stereo system on an unmanned aerial vehicle [10] to solve the nonlinearity issue, but a better way is to find a landmark model that has a high degree of linearity. Therefore, Montiel et al. [11] proposed an inverse depth parameterization to represent the landmark model. The key concept is to parameterize the inverse depth of features relative to the camera locations from which they were first viewed directly. This way of parameterization would achieve a high degree of linearity, and furthermore, the features are initialized with no delay and can successfully estimate for both near and distant features.

The drawback of inverse depth parameterization is that the 6-D state vector representation is computational intensive. Therefore, Civera et al. [12] proposed a linearity index, that inverse depth representation can be safely converted to Euclidean XYZ form; once the depth estimate of a feature has converged. The speed of convergence therefore is important.

The most closely related work is by Paz et al. [13], in which they proposed a binocular stereo-based EKF SLAM. They combine both the inverse depth parameterization and Euclidean XYZ parameterization in the map, which alleviate the nonlinearity issue effectively. The system can map both near and far features with proper uncertainty distribution, and it can be used in large-scale outdoor environment.

However, the observation model in previous studies is bearing-only. When a 3D feature is acquired simultaneously by left and right camera images, the stereo system are treated as two bearing-only observers. In each instance, EKF update is done once for left and right camera without using any range information. Thus, range information such as disparity does not directly contribution to camera poses and map spatial location estimation. In contrast we wish to incorporate not just bearing but also range information.

In this research, we want to focus on using EKF to solve the stereo-based SLAM problem. It is essential to find an appropriate probabilistic models for observations of a stereo camera that is still consistent to the linear property. Thus, our contribution is to derive a new stereo observation model that incorporates the inverse depth parameterization with observation pair (u_1, v_1, d). In this way, both range and bearing information can be directly injected into the EKF estimation process to handling the nonlinear projection issue. In order to develop the new observation model, two Jacobians needs to be derived for the EKF framework. The first instance is in the feature initialization step and the second instance is in the feature prediction step. Based on our knowledge this is the first time this observation model is proposed in the stereo-based EKF SLAM.

2 Stereo-Based EKF SLAM System Models

2.1 State Vector Definition

Following the standard EKF-based approach of SLAM, the system state vector x consists the current estimated pose of camera and physical location of features. x will change in size dynamically as features are added to or deleted from the map.

$$x = (x_C, y_1, \ldots y_i, \ldots y_n)^T \tag{1}$$

The camera state X_C is composed by the position r^{WC} with respect to a world reference frame W, and q^{WC} quaternion for orientation, and linear and angular velocity v^W and ω^C relative to world frame W and camera frame C, respectively.

$$x_C = (r^{WC} \quad q^{WC} \quad v^W \quad \omega^C)^T \tag{2}$$

The feature y_i is defined by the inverse depth parameterization [11] using a 6-D state vector:

$$y_i = (x_i y_i z_i \theta_i \phi_i \rho_i)^T \tag{3}$$

The y_i vector encodes the ray from the first camera position from feature observed by x_i, y_i, z_i, the camera optical center, and θ_i, ϕ_i azimuth and elevation (coded in the world frame) defining unit directional vector $m(\theta_i, \phi_i)$. The feature point's depth along the ray d_i is encoded by its inverse $\rho_i = 1/d_i$.

y_i models a three-dimensional point located at

$$\begin{pmatrix} X_i \\ Y_i \\ Z_i \end{pmatrix} = \begin{pmatrix} x_i \\ y_i \\ z_i \end{pmatrix} + \frac{1}{\rho_i} m(\theta_i, \phi_i) \tag{4}$$

$$m = (\cos \emptyset_i \sin \emptyset_i, \ -\sin \emptyset_i, \ \cos \emptyset_i \cos \theta_i)^T \tag{5}$$

2.2 Motion Model

The motion model in this work describes an ego motion with 6 DOF. The camera orientation is represented in terms of quaternions, which can deal with the issue of gimbal lock in Euler angles. It is assumed to be both in constant velocity and angular velocity with a zero-mean Gaussian acceleration noise $n = \begin{pmatrix} a^W & \alpha^C \end{pmatrix}^T$ uncertainty. At each step, there is an impulse of linear velocity $V^W = a^W \Delta t$ and angular velocity $\Omega^C = \alpha^C \Delta t$, with zero mean and known Gaussian distribution.

$$x_{C_{k+1}} = \begin{pmatrix} r_{k+1}^{WC} \\ q_{k+1}^{WC} \\ v_{k+1}^{W} \\ \omega_{k+1}^{C} \end{pmatrix} = f_v(x_{C_k}, n) = \begin{pmatrix} r_k^{WC} + (v_k^W + V_k^W)\Delta t \\ q_k^{WC} \times q((\omega_k^C + \Omega^C)\Delta t) \\ v_k^W + V^W \\ \omega_k^C + \Omega^C \end{pmatrix} \tag{6}$$

where $q((\omega_k^C + \Omega^C)\Delta t)$ is the quaternion defined by the rotation vector $(\omega_k^C + \Omega^C)\Delta t$.

2.3 Stereo Observation Model

In this work, we define a new nonlinear function $h(X_C, y_i)$, which allows the prediction of the value of observations measurement $\hat{z}_i$ given the current estimatiom camera pose x_C and the ith feature y_i in the map. The observation model can be written in a general form as:

$$\hat{z}_i = \begin{bmatrix} u_{li} \\ v_{li} \\ d_i \end{bmatrix} = h(x_C, y_i) + w_{u_i v_i d_i} \tag{7}$$

The vector $\hat{z}_i$ is a observation of the feature y_i relative to the camera pose x_C, where $w_{u_i v_i d_i}$ is a vector of uncorrelated observation errors with zero mean Gaussian noise and covariance matrix $R_{u_i v_i d_i}$.

The observation model can be divided in three steps. In step 1, through inverse depth parameterization, feature y_i is transformed to Euclidean XYZ landmark representation with respect to the world reference frame W:

$$\begin{pmatrix} X_i \\ Y_i \\ Z_i \end{pmatrix} = \begin{pmatrix} x_i \\ y_i \\ z_i \end{pmatrix} + \frac{1}{\rho_i} m(\theta_i, \phi_i) \tag{8}$$

In step 2, Euclidean XYZ is transformed into camera reference frame C, while plugging into x_C camera pose:

$$h^C = \begin{bmatrix} h_x^C \\ h_y^C \\ h_z^C \end{bmatrix} = (R^{WC})^T \left(\begin{pmatrix} X_i \\ Y_i \\ Z_i \end{pmatrix} - r^{WC} \right) \tag{9}$$

Next, using the typical pinhole camera model [3], we will project this 3D position h^C to its expected image coordinates and compute its expected disparity in the new view:

$$\widehat{z}_i = \begin{bmatrix} \widehat{u}_{li} \\ \widehat{v}_{li} \\ \widehat{d}_i \end{bmatrix} = \begin{bmatrix} \frac{h_x^C}{h_z^C} f + C_x \\ \frac{h_y^C}{h_z^C} f + C_y \\ \frac{fb}{h_z^C} \end{bmatrix} \tag{10}$$

where $(C_x C_Y)$ are the camera center in pixels, f is the focal length, b the baseline of stereo.

3   The Estimation Process of Stereo-Based EKF SLAM

3.1   The Prediction Step

In the prediction stage, the camera motion model Eq. (6) is used to produce a state prediction from the previous state. Since the camera motion model only propagates the pose of previous state, we leave the map states unchanged:

$$x(k+1|k) = f_v(x_C(k|k), n) \tag{11}$$

In the prediction of covariance, state-augmentation methods [14] is used, which results in an optimal SLAM estimate with reduced computation from cubic complexity to linear complexity, and has the form:

$$P(k+1|k) = FP_{xx}(k|k)F^T + GQ_kG^T \tag{12}$$

where $F = \frac{\partial f_v}{\partial x_C}$ is the Jacobian of $f_v()$ evaluated at the estimate $x_C(k|k)$, Q_k the Gaussian noise covariance and $G = \frac{\partial f_v}{\partial n}$ is the Jacobian of $f_v()$ evaluated with the noise n.

3.2   The Update Step

In the update step, an observation $z_i(k+1) = (u_{li}, v_{li}, d_i)$ of the ith feature will be available. Stereo observation model Eq. (7) is used to form an observation prediction $\widehat{z}_i(k+1|k)$ and innovation $v_i(k+1)$

$$\widehat{z}_i(k+1|k) = h(x_C(k+1|k), y_i(k|k)) \tag{13}$$

$$v_i(k+1) = z_i(k+1) - \widehat{z}_i(k+1|k) \tag{14}$$

and then, one can calculate the innovation covariance matrix:

$$S_i(k+1) = HP(k+1|k)H^T + R_{u_i v_i d_i}(k+1) \tag{15}$$

where H is the Jacobian of h(.) evaluated at $x_C(k+1|k)$ and $y_i(k|k)$, and the Kalman gain $K_i(k+1)$ can be obtained as

$$K_i(k+1) = P(k+1|k)H^T + S_i(k+1)^{-1} \qquad (16)$$

The observation matrix $z_i(k+1)$ is used to update the predictions and form a new estimation of the state by using the standard EKF update equations:

$$x(k+1|k+1) = x(k+1|k) + K_i(k+1)v_i(k+1) \qquad (17)$$

$$P(k+1|k+1) = P(k+1|k) - K_i(k+1)HP(k+1|k) \qquad (18)$$

The main focus here is the innovation $v_i(k+1)$ (14), which represents the difference between the actual sensor measurement $z_i(k+1)$ and the predicted measurement $z_i(k+1|k)$, both containing range and bearing information. It means that when multiplying innovation with Kalman gain $K_i(k+1)$, both bearing and range information optimize the state estimation directly

3.3 Landmark Initialization

The initialization process includes both the feature state initial values and the covariance assignment. Therefore we use inverse depth parameterization to represent features initial values, and derived the feature initialization model, $g\left(r^{WC}_{(k+1|k+1)}, q^{WC}_{(k+1|k+1)}, z_{i(k+1|k+1)}\right)$ to describe the initial values in terms of current camera pose $r^{WC}_{(k+1|k+1)}$, $q^{WC}_{(k+1|k+1)}$ and a new sensor observation pair $z_{i(k+1|k+1)} = (u_{li}, v_{li}, d_i)$

$$y_i = g\left(r^{WC}_{(k+1|k+1)}, q^{WC}_{(k+1|k+1)}, z_{i(k+1|k+1)}\right) = (x_i y_i z_i \theta_i \ \phi_i \ \rho_i)^T \qquad (19)$$

The end-point of the projection ray is taken from the camera location estimate:

$$(x_i y_i z_i)^T = r^{WC}_{(k+1|k+1)} \qquad (20)$$

The feature spatial location vector (in the camera reference frame) is computed from the observation pair $z_i = (u_{li} v_{li} d_i)^T$, by rearranging stereo observation model Eq. (10), we have

$$h^C = \begin{pmatrix} (u_{li} - C_x)\dfrac{b}{d_i} \\ (v_{li} - C_Y)\dfrac{b}{d_i} \\ \dfrac{fb}{d_i} \end{pmatrix} \qquad (21)$$

where $(u_{li} v_{li})$ are the pixels on the left image, and d_i is the horizontal disparity. The inverse depth prior ρ_i can be computed from h^C

$$\rho_i = \frac{1}{\|h^C\|} \tag{22}$$

Using the current camera orientation estimation from the state vector, h^C can be transformed to the world reference frame and the azimuth and elevation angles are extracted as:

$$h^W = R^{WC}(q^{WC}_{(k+1|k+1)})h^C \tag{23}$$

$$\begin{pmatrix} \theta_i \\ \emptyset_i \end{pmatrix} = \begin{pmatrix} \arctan(h^W_x, h^W_z) \\ \arctan\left(-h^W_y, \sqrt{h^W_x + h^{W^2}_z}\right) \end{pmatrix} \tag{24}$$

The newly initialized feature $y_i = (x_i y_i z_i \theta_i \phi_i \rho_i)^T$ is added to the state vector $x(k+1|k+1)$.

In order to model the uncertainty of the newly initialized feature, we derived the Jacobian matrix of the functions in (19), using a first-order error propagation to approximate the distribution of the variables in (19) as multivariate Gaussians. The covariance matrix of newly feature is:

$$P_{y_i y_i}(k+1|k+1) = J_{xc}P_{xx}J^T_{xc} + J_R R J^T_R \tag{25}$$

R includes σ_u, σ_v, σ_d, which represents the pixel uncertainties in image $(u_{li} v_{li})$ location and disparity d_i . In our experiments, we use $\sigma_u = 1$ pixel , $\sigma_v = 1$ pixel, $\sigma_d = 1$ pixel. Since R is the error covariance describing the noisy measurements of the stereo system, the uncertainty through J_R is propagated to the newly feature yi of landmark model space. J_R is the Jacobian of $g(.)$ which is derived by the observation pair z_i. P_{xx} is the camera pose covariance matrix, representing current pose estimation uncertainty. This uncertainty through J_{xc} is propagated to the newly feature y_i of landmark model space. J_{xc} is the Jacobian of $g(.)$ which is derivative by $r^{WC}_{(k+1|k+1)}$, $q^{WC}_{(k+1|k+1)}$.

4 Experimental Results

4.1 Analysis of Landmark Uncertainty

In order to validate that our proposed observation model describes the uncertainty of 3D points accurately, we have simulated an experiment where the true uncertainty of the landmark location (derived from a Monte Carlo simulation) is compared to the estimated uncertainty from Eq. (25) and the traditional Euclidean XYZ landmark model.

The actual intrinsic parameters of the stereo camera, such as the baseline, are accounted in the simulation. The origin of the left camera is set as the reference frame, with the principal axis pointing to Z and X axis pointing to the right.

Consider a landmark point in front of the left camera that is at 70 m distance along the Z axis, a Monte Carlo simulation has been performed by drawing a set of 10,000 samples from the Gaussians distributions of u_{li}, v_{li}, and d_i (assuming a standard deviation of $\sigma_u = \sigma_v = 1$ and $\sigma_d = 2$ pixels, respectively), and by projecting them through Eq. (21), yielding a set of 10,000 samples of the landmark 3D position (X Y Z). In Fig. 1 (Left), the black sample points show the true measurement uncertainty from stereo systems, green point shows the real position of the landmark point. Next, the estimated uncertainty is calculated using first-order error propagation based on our observation model Eq. (25), shown by the enclosing red lines. The traditional Euclidean XYZ model is shown by the enclosing blue lines. One can see that the red lines enclose the true uncertainty noise, while blue lines do not. In Fig. 1 (Right), histogram is used to show the uncertainty distribution. Gray rectangles show the true uncertainty of the landmark location, red rectangles show our proposed observation model uncertainty, and blue indicates the traditional Euclidean XYZ model. The red rectangles have more closely covered the true gray rectangles distribution.

In Fig. 2(a), (b), (c), black sample points indicates the real distributions with various distances, using 15 m, 30 m, 45 m respectively, and the red points shows the real position of the landmark point. The estimated uncertainty is calculated using first-order error propagation using our observation model, shown by blue enclosing lines. One can see that for any distance close or far, the uncertainty region estimated by our model accurately bounds the true uncertainty.

From the simulation result, we shown that the measure error can be accurately estimated basing on our proposed observation model, which will help the EKF filter estimation to be consistent and avoid filter divergence.

4.2   Real World Experiments

4.2.1   Dataset and Feature Points Matching

All experiments are verified by using the Karlsruhe dataset [15], a real-world, large-scale, grayscale stereo sequences. Odometry data is available from OXTS RT 3000 GPS/IMU system. An experimental vehicle is equipped with a stereo camera rig

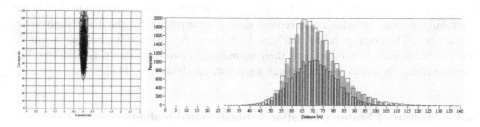

Fig. 1. (Left) The black point clouds represent the true uncertainty which are samples of distribution of a real landmark point position, given that the pixel noise in the images is Gaussian. Red line enclosed regions represent the estimated uncertainty using our proposed observation model. Blue line enclosed regions represent the estimated uncertainty using the traditional Euclidean XYZ model. (Right) The histogram is used to show the uncertainty distribution, gray rectangles show the true uncertainty of the landmark location, red rectangles show the estimated uncertainty using our proposed observation model, and blue indicates the traditional Euclidean XYZ model. Our model describes true noise distribution more accurately.

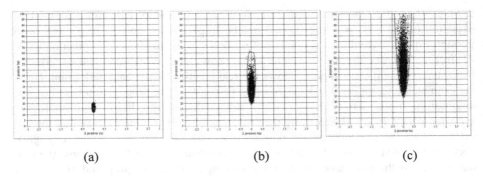

(a)   (b)   (c)

Fig. 2. Simulated experiment of a point reconstruction from a stereo pair observation for a point at (a) 15 m, (b) 30 m, (c) 45 m distance. The point clouds are samples of distribution of a real landmark point position, given that the pixel noise in the images is Gaussian. The black sample points show the distribution. Blue line enclosed regions represent the estimated uncertainty using our proposed observation model (color figure online).

Pointgrey Flea2 firewire cameras covering inner city traffic. All the frames are rectified with a resolution of 1344 × 391 pixels running at 10 fps. In our experiment we set the left image frame as the reference coordinate system. Our algorithm is implemented by LabVIEW on an Intel Core i5 with 1.7 GHz and 4 GB RAM computer.

In order to obtain the matching stereo observation pairs, stereo points matching based on LIBVISO2 [16], an open-source library is used to demonstrate real-time computation of point feature match of left and right images. The matching stereo pairs are done only in the first frame for initialization.

We use a different method to track the initialized points in the subsequent frames. When each stereo observation pair is initialized and saved in the map, it also save corresponding 11 × 11 surrounding patch, which serves as a photometric identifier. When the next image comes, the saved pairs in 3D space are projected back to image plane. The pairs are deleted if it is outside the visible field of view of left and right images. If they are within the field of view, we use active search concept to decide the searching region, in which the region size is determined by the EKF innovation covariance. The corresponding patch of the features first will be warped according to the predicted camera motion, and then normalized cross-correlation is performed in the searching region to find the matching point, for details see [17].

4.3 Analysis of Features Location Estimation and Stability

In this part of the experiment, we want to compare the two monocular observers based visual SLAM systems and our proposed stereo observer based visual SLAM systems with regard to the accuracy of landmark features spatial location estimation.

We used a subset of the Karlsruhe dataset, the 2009_09_08_drive_0010 from frame 61 to frame 72. The scenario is that the vehicle is moving forward on the road, and then turns right. The features are initialized first in frame 61 (Fig. 3 Left), and then were continuously tracked until frame 72 (Fig. 3 Right). Therefore the camera

Fig. 3. (Left) At frame 60 features are initialized. (Right) At frame 70, features tracking result.

poses and features locations were updated 10 times. We recorded every state vector and covariance matrix along the way, and select a near (number 7) and far (number 0) feature point for analysis.

In Fig. 4 (Left) and (Right) each time the feature updates its depth uncertainty, no matter the feature is either far or near, it can be seen that our proposed method depth and uncertainty converge faster.

Figure 5 (Left) and (Right) shows the raw measurement of each feature's depth (green points), and the depth estimation. The blue points shows the two monocular observers based visual SLAM systems estimation result, and the red points shows the stereo observer based visual SLAM systems estimation result.

From Fig. 5 it shows that our proposed method estimation is more stable, therefore the curve is smoother. Also after 10 updates, the estimation result is also closer to the mean values of raw measurement. After 10 updates, only the estimation of near feature is closer to the raw measurement mean values, Fig. 5 (Left), while the features far away cannot get close to the raw measurement mean values Fig. 5 (Right).

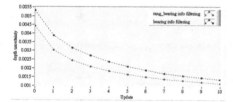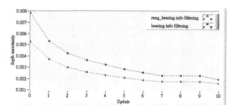

Fig. 4. (Left) Depth uncertainty of feature number 7, over 10 times EKF updates, (Right) depth uncertainty of feature number 0, over 10 times EKF updates.

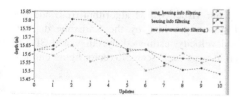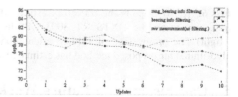

Fig. 5. (Left) The depth estimated value of feature number 7 (close by), over 10 times EKF updates, (Right) the depth estimated value of feature number 0 (far away), over 10 times EKF updates. The proposed model (red line) is more stable and closer to raw measurement mean (color figure online).

4.4   Motion Estimate Results

To evaluate how the capability of the proposed algorithm can correctly estimate the camera pose and velocity in a large-scale environment after extended periods of time, we use one of the Karlsruhe dataset, the 2009_09_08_drive_0015 images. Within the dataset we pick out 800 stereo frames, with a total length of the route approximately 350 m. These images are sequential scenes of inner city urban driving, including two wide angle turns.

Figure 6 depicts the trajectory estimated by our visual SLAM algorithm (in red) and 'groundtruth' output of a OXTS RT 3003 GPS/IMU system (in green). Note that the GPS/IMU system can only be considered as 'weak' groundtruth [16], because localization errors of up to two meters may occur in inner-city scenarios due to limited satellite availability. Thus instead, LIBVISO2 [16] visual odometry (in black) is argued to be a better groundthruth. As we can see, our proposed model estimated trajectory is quite close to the 'groundtruth'.

From Fig. 6, because the car at frame 250 made a negative x-directional turn, yaw angle started to increase drastically. This cause the forward velocity Vz to decrease drastically, while the negative x-direction Vx started to increase. Afterward at frame 600, both Vz and Vx started to decrease, and then at frame 650 the car turns toward the z-axis direction, making the yaw decreases. Finally the car continues toward the z-axis direction while Vz velocity started to increase.

From the experimental result, we can see that the proposed algorithm can be used in large-scale outdoor scenario, that the car pose and velocity can be decently estimated quite well, close to the groundtruth.

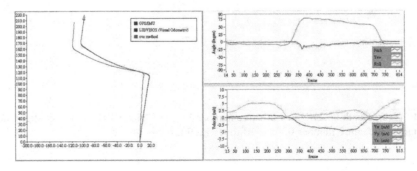

Fig. 6. (Left) Depicts the trajectory estimated by our visual SLAM algorithm (red) and the trajectory using the OXTS RT 3003 GPS/IMU system provided by the Karlsruhe dataset. (Right) The progression of vehicle orientation and velocity, estimated by our visual SLAM algorithm. Our proposed method has estimated well in large-scale scenario (color figure online).

5   Conclusions and Future Works

The focus of this work was to develop a new probabilistic observation model in EKF-based stereo SLAM. The statistical behavior in stereo vision is known to have inherently non-linear problem. Therefore, we use inverse depth parameterization as the landmark model to deal with the non-linear problem, and in contrast to the binocular stereo-based approach, we used stereo observation pair (u, v, d) to derive a new observation model, allowing both bearing and range information to be incorporated into the EKF estimation process. Furthermore, our new observation model is also computationally faster than the previously mentioned binocular stereo-based approach. In our approach, we only have to do projection and update in the EKF framework once, in contrast to the binocular approach which requires projection and update two times each. Moreover, because update step requires inverse of large innovation matrix, doing update step twice would be computationally intensive in large-scale SLAM. Based on our knowledge this is the first time this observation model with inverse depth parameterization is proposed in the stereo-based EKF SLAM.

From our experiments, it shows that even in large-scale outdoor environment, there is only a little 'scale drift', which means that our observation model together with inverse depth parameterization has kept true to the real noise distribution of the landmark feature. This also means the proposed observation models that is designed with the additional range information has helped and worked consistently under the strict requirement of EKF framework. We also demonstrated that the proposed system has converged faster and has more stable depth estimation from experiment data. These convergence and stableness characteristics will be critical to our future work. For future work we want to develop moving objects detection along with static map into a nice single EKF framework.

References

1. Cole, D.M., Newman, P.M.: Using laser range data for 3D SLAM in outdoor environments. In: Proceedings of IEEE International Conference on Robotics and Automation (ICRA), pp. 1556–1563 (2006)
2. Davison, A.J.: Real-time simultaneous localisation and mapping with a single camera. In: Proceedings of Ninth IEEE International Conference on Computer Vision, vol. 2, pp. 1403–1410. IEEE (2003)
3. Se, S., Lowe, D., Little, J.: Mobile robot localization and mapping with uncertainty using scale-invariant visual landmarks. Int. J. Robot. Res. 21(8), 735–758 (2002)
4. Herath, D., Kodagoda, S., Dissanayake, G.: Simultaneous localisation and mapping: a stereo vision based approach. In: IEEE/RSJ International Conference on Intelligent Robots and Systems, pp. 922–927 (2006)
5. Lemaire, T., Berger, C., Jung, I.-K., Lacroix, S.: Vision-based SLAM: stereo and monocular approaches. Int. J. Comput. Vision 74(3), 343–364 (2007)
6. Berger, C., Lacroix, S.: Using planar facets for stereovision SLAM, intelligent robots and systems. In: IEEE/RSJ International Conference on Intelligent Robots and Systems (IROS), vol. 22, no. 26, pp. 1606–1611 (2008)
7. Sibley, G., Matthies, L., Sukhatme, G.: Bias reduction and filter convergence for long range stereo. In: 12th International Symposium of Robotics Research (ISRR), San Francisco, CA, USA (2005)
8. Sibley, G., Sukhatme, G., Matthies, L.: The iterated sigma point filter with applications to long range stereo. In: Robotics: Science and Systems II, Cambridge, USA (2006)

9. Tim, B., Nieto, J., Guivant, J., Stevens, M., Nebot, E.: Consistency of the EKF-SLAM algorithm. In: International Conference on Intelligent Robots and Systems (IROS), Beijing, China (2006)
10. Li, X., Aouf, N., Nemra, A.: 3D mapping based VSLAM for UAVs. In: 2012 20th MediterraneanConference on Control & Automation (MED), pp. 348–352 (2012)
11. Montiel, J.M.M., Civera, J., Davison, A.J.: Unified inverse depth parametrization for monocular slam. In: Robotics: Science and Systems, Philadelphia, USA, August 2006
12. Civera, J., Davison, A.J., Montiel, J.M.: Inverse depth to depth conversion for monocular SLAM. In: Proceedings of IEEE International Conference on Robotics and Automation, pp. 2778–2783. IEEE (2007)
13. Paz, L., Piniés, P.: Large-scale 6-DOF SLAM with stereo-in-hand. IEEE Transactions on Robot. 24(5), 946–957 (2008)
14. Durrant-Whyte, H., Bailey, T.: Simultaneous localization and mapping: part I. IEEE Robot. Autom. Mag. 13(2), 99–110 (2006)
15. Karlsruhe Dataset. http://www.cvlibs.net/datasets/karlsruhe_sequences.html
16. Geiger, A., Ziegler, J., Stiller, C.: StereoScan: dense 3d reconstruction in real-time. In: IEEE Intelligent Vehicles Symposium (IV), (Iv), pp. 963–968 (2011)
17. Civera, J., Davison, A.J., Martínez, M.: Structure from motion using the extended Kalman filter. In: Civera, J., Davison, A.J., Montiel, J.M.M. (eds.) Springer Tracts in Advanced Robotics, vol. 75. Springer, Heidelberg (2012)

On New Algorithms for Path Planning and Control of Micro-rover Swarms

H.D. Ibrahim$^{(\boxtimes)}$ and C.M. Saaj$^{(\boxtimes)}$

Surrey Technology for Autonomous Systems and Robotics Lab,
Surrey Space Centre, University of Surrey, Guildford GU2 7XH, UK
{H.Ibrahim, C.Saaj}@surrey.ac.uk

Abstract. In this paper, a new approach to modelling the dynamics of the Pioneer-3AT robot on planetary soil is presented. This model is used to design a robust traction control algorithm that makes use of the Sliding Mode Control (SMC), the Artificial Potential Field (APF) and a Fethi Beckoche navigation function for collision avoidance with predefined obstacles. Simulations were carried out for swarms of micro-rovers using the parameters of an in-house planetary soil simulant. The controller was designed to control the wheel movements of the robot on an unstructured environment using the SMC and APF. Simulation results show that the controller and the navigation function effectively achieves the permitted slip rate, sinkage, angular and longitudinal velocities, sliding surface and torque for the wheels, thus offering efficient traction while avoiding the excessive wheel slip and sinkage.

Keywords: Dynamic model · Traction control · Navigation · Swarms · Sliding mode control · Artificial potential field

1 Introduction

There is an increasing interest to investigate the field of terramechanics and control of micro-rovers, following the success of the PathFinder Mission [1, 2]. The first robot to explore an extra-terrestrial body was Lunakhod, launched by the then Soviet Union in 1970 to explore the moon. Remotely operated from the earth, Lunakhod traversed 10.5 km of the moon surface, taking photos and analysing the soil. There are different planetary exploration missions that have brought autonomous vehicles to the surface of other planets, for in-stance Mars. There is the Sojourner of the Mars Pathfinder mission, followed by the Spirit and Opportunity of the Mars Exploration Rover (MER) missions. During the past decade, there were major breakthroughs in space robotics [3], notably by NASA's autonomous Mars rovers, Spirit and Opportunity, which landed on the Red planet in 2004. This has been followed closely by the Curiosity which was launched on 26 November 2011 and landed on Mars on 6th August, 2012. The MER rovers have sizes that allow them to venture into autonomous traverses on the planetary surface without support of an operator. They are capable of navigating rather rugged terrains.

Micro-rover is a small self-contained robot that can be capable of autonomous local navigation [4]. As research progresses in this area, there is need to continuously

A. Natraj et al. (Eds.): TAROS 2013, LNAI 8069, pp. 353–362, 2014.
DOI: 10.1007/978-3-662-43645-5_38, © Springer-Verlag Berlin Heidelberg 2014

keep abreast with new findings, without being limited by cost, hence the need for low-cost, repeatable missions and coordinated approach. With a swarm of coordinated micro-rovers, cooperative tasks could be achieved by the rovers during planetary explorations. Trajectory control of a four-wheeled skid-steering vehicle over soft terrain using physical interaction model was studied in [5]. The control strategy takes into account the slip and skid effects to extend the mobility over planar granular soils. The sliding mode controller for slip control of antilock brake systems has been carried out to regulate the slips in the front and rear wheels under some bounds of uncertainty [6]. According to [7], dynamic modeling and sliding mode control for lunar rover slip was investigated and the results shows that the controller robustly keeps the slip rate of the tracking wheel to the required value and avoids excessive spin and sink of the driving wheel. Sinkage has been a major problem of the Spirit Mars Exploration Rover which experienced difficulties due this factor and the mission, even though exceeded its design lifetime, came to an abrupt end. In path planning for planetary exploration rovers and its evaluation based on wheel slip dynamics [4], the difficulty in applying the wheel slip dynamics as a criterion into path planning algorithm was emphasized. These reasons spurn us to investigate further the slip rate and apply the outcome to our research on path planning and control of micro-rover swarms for planetary exploration. The steering characteristics of an exploration rover on loose soil was studied in [8]. The all wheel dynamic model developed was used to model the behavior of each wheel on the loose soil. This research is on developing innovative control and path planning algorithms for future micro-rover swarm missions. The goal of the research is to enhance a quick soil sample return using swarms of rovers for improved redundancy, efficiency, simplicity, timeliness and in a cost-effective, repeatable manner. The result of the developed algorithm using the dynamic model, artificial potential field, the sliding mode control and the navigation function has been successfully used for obstacle and collision avoidance for a swarm of Micro-rovers on the planetary soil.

2   Background on Dynamic Model of a Single Rigid Wheel on a Loose Soil

The force analysis of a single wheel was first carried out to determine the dynamic model of the single wheel and then derive the relationship between the wheel terrain contact forces [7]. The Bekker model [9–11] uses the relationship between certain physical soil characteristics and shearing strength to predict vehicle mobility. According to Bekker [12], the pressure sinkage relationship as illustrated in Fig. 1 can be expressed by Eq. (1),

$$p = \left(\frac{k_c}{w} + k_\phi\right) z^n \tag{1}$$

where p: is the pressure, w: width of the contact area with the soil, z: sinkage of the wheel, k_c: cohesive modulus of deformation, k_ϕ: frictional modulus of deformation, n: soil sinkage exponent.

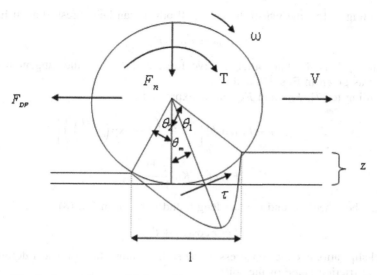

Fig. 1. Interaction model for a driven rigid wheel on loose soil [5, 12]

The free body diagram of a driven rigid wheel on a loose soil is shown in Fig. 1.

The vertical load F_n and a horizontal force F_{DP} are applied to the wheel by the rover suspension and the torque T is applied at the wheel rotation axis by the actuator [7]. The wheel has angular velocity of ω with a longitudinal velocity of v. The angle with the soil with which the wheel first makes contact with the soil is given by θ_1, while the angle at which the wheel leaves the soil is given as θ_2. The angle at which maximum shear stress τ_m occurs is θ_m. A stress region is created at the wheel–terrain interface with a normal component, known as the normal stress given as σ and the horizontal component known as shear stress given as, τ.

For the vertical load, the dynamic force equilibrium equations of a driven rigid wheel [5] can be expressed as given by Eq. (2),

$$F_n = K\left(1 - \frac{n}{3}\right)w\sqrt{2r}z^{\frac{2n+1}{2}} \tag{2}$$

Where

$K = \frac{k_c}{w} + k_\phi$, is relates the cohesive stress to the width of the wheel; r: radius of the wheel.

According to the Newton's second law, the equation of motion of the rover wheel is given by Eqs. (3–4)

$$M_w v = F_{DP} \tag{3}$$

and the torque can be expressed as

$$T - T_R = I_w \dot{\omega} \tag{4}$$

M_w: Mass of the wheel, I_w: Moment of inertial of the wheel, T_R: Reverse torque due to internal frictional and other disturbances, relates soil stiffness and the wheel width.

According to [5], the wheel drawbar pull power can be expressed as in Eq. (5)

$$F_{DP} = F_H - F_R \tag{5}$$

Where F_D : is the tangential tractive force and F_R : is the tangential motion resistance as given in Eqs. (6) and (7).

According to [7], F_{DP} and F_R can be expressed as

$$F_H = (cwl + F_n \tan \phi) \left[1 - \frac{j_0}{\lambda l} \left(1 - \exp \left(\frac{-\lambda l}{j_0} \right) \right) \right] \tag{6}$$

$$F_R(\lambda) = Kw \frac{z^{n+1}}{1 + n} \tag{7}$$

where l is the wheel-ground contact length and is given in Eq. (8)

$$l = r \cos(\theta_1 + \theta_2) \tag{8}$$

λ : wheel slip ratio, c: cohesion stress of the soil, j_0: tangential soil shear deformation, ϕ: internal friction angle of the soil.

According to [12], the sinkage z, can be expressed as $z = z_0 + z_d$, where z_0 is the static sinkage of the wheel resulting from the normal load, z_d is the dynamic sinkage due to the slip and $z_d = q_0 \lambda$, where q_0 is the coefficient of wheel dynamic sinkage. Assuming θ_2 is very negligible [10] and that the shear stress follows linear distribution laws,

$$T_R(\lambda) = \frac{1}{2} r^2 w \tau_m \theta_1,$$

where τ_m is the maximum shear stress of the soil [7] and is expressed as [7] in Eq. (10)

$$\tau_m = c + \frac{F_n}{wl} \tan(\phi) \tag{9}$$

with the variation of $\theta_1(\lambda)$ with the sinkage, z, given Eq. (10)

$$\theta_1(\lambda) = \cos^{-1} \left(1 - \frac{z}{r} \right). \tag{10}$$

The above equations were used for the simulations and Eqs. (2)–(4) represents the dynamic model of the single rigid wheel on the loose soil.

3   Control Torque Formulations for Pioneer 3AT Robot

The Pioneer 3AT robot consists of four wheels and is capable of reaching a maximum speed of 0.7 m/s. In this research, the dynamic equations used for the simulations, using the characteristics of the Pioneer 3AT was derived as follows. The dynamic force equilibrium of the robot in the x-direction can be represented as in Eq. (12)

$$M\ddot{x} = \sum_{i=1...n} (F_{Hi} - F_{Ri}),$$ (11)

and the torque is given in Eq. (13) as

$$I_{\omega i}\dot{\omega}_i = T_i - T_{Ri}$$ (12)

where M is the total mass of the robot, and T_i is the wheel tractive torque and T_{Ri} represents the resistance torque experienced by each wheel.

The longitudinal slip of the wheel on a loose soil is related to the longitudinal velocity and the angular velocity [9–12] and is given in Eq. (14)

$$\lambda_i = 1 - \frac{\dot{x}}{r\omega_i}$$ (13)

Differentiating the wheel slip and rearranging Eqs. (12)–(14) gives Eq. (15)

$$\dot{\lambda}_i = -(1 - \lambda_i) \left[\frac{\sum_{i=1...n} (F_{Hi} - F_{Ri})}{M\dot{x}} \right] - \frac{r(1 - \lambda_i)^2 T_{Ri}}{\dot{x} I_{\omega i}} + \frac{r(1 - \lambda_i)^2 T_i}{\dot{x} I_{\omega i}}$$ (14)

showing the rate of change of slip for multi-input non-linear system.

4   Design of Sliding Surface and Controller

The design of the sliding surface is based on [13–21], and this is given by:

$$s = \dot{x} + \nabla J(x)$$ (15)

However, Eq. (14) is valid for the point mass model only. Therefore, Eq. (14) is not valid for the four wheeled robots. Hence, a new approach to designing the controller is presented to account for the dynamics of the Pioneer-3AT robot. Rearranging Eq. (3), equating the sliding surface to zero, differentiating the resulting equation and substituting into Eq. (15), it can be shown that, the time rate of change of the gradient of the potential function is given Eq. (16)

$$\frac{\partial}{\partial t} \nabla J(x) = -\frac{\dot{\lambda} \nabla J(x)}{(1 - \lambda)} - r\dot{\omega}_i (1 - \lambda).$$ (16)

For satisfying the reaching condition, the Lyapunov criterion is applied. By differentiating the sliding surface in Eq. (15), and applying Lyapunov stability criterion as in Eqs. (17–18), for the reaching condition gives,

$$s\dot{s} \leq 0$$ (17)

$$s\dot{s} < -\varepsilon \|s\|.$$ (18)

The torque controller is designed as in Eq. (20), such that

$$\dot{s} = -\eta sign(s) \qquad (19)$$

Equating Eq. (19) and the differentiation of Eq. (15) and rearranging, gives the control torque as in Eq. (21):

$$T_i = \frac{I_{\omega i} \sum\limits_{i=1...n} (F_{Hi} - F_{Ri})}{rM(1-\lambda)} - \frac{\lambda I_{\omega i} \nabla J(x)}{r(1-\lambda)^2} + T_R + \frac{\eta I_{\omega i} sign(s)}{r(1-\lambda)}, \qquad (20)$$

This is equivalent to

$$T_i = T_{eq} + T_{sw}.$$

The artificial potential function $J(x)$ is given in Eq. (21) :

$$J(x) = -x\left(a - b\exp\left(-\frac{x^2}{c}\right)\right). \qquad (21)$$

where η is the control design parameter for the sliding mode controller design, a and b are design constants for the attraction and repulsion of the artificial potential field, while x represents the displacement variable, and $sgn(s)$ is a discontinuous function required for the sliding mode controller. This alternates between 1 and -1 depending on whether the sliding surface is greater or less than zero respectively and 0 when the sliding surface is zero.

5 Trajectory Tracking Navigation Function

A method of navigation of nonholomic wheeled, based on the kinematic model of the robot [22], was used for the trajectory tracking while avoiding collision with obstacles. This method was based on linear navigation functions, where the robot orientation angle is proportional to the visibility angle between the robot and its goal. This navigation function can be written as in Eq. (22)

$$\theta = B\delta + b_0 \exp(-at) + b_1 \qquad (22)$$

where θ: Orientation angle, B : constant real number greater or equal to 1, b_0 and b_1, angles representing the initial state and final state of the robot's orientation, δ : the visibility angle, given by the angle between the reference line parallel to the x-axis and the visibility line to the goal; and the constant a affects the shape of the robots' trajectory.

This navigation function was used to avoid collision by altering the values of the control parameters above or navigating towards an intermediary goal called the breakpoint. Navigating towards different breakpoints enables the robot to effectively avoid collision and also reach different sample points for their collection.

Table 1. Simulation parameters for Pioneer 3AT and in-house soil stimulant

Parameter	Values	Units
Mass, M	27	kg
Moment of Inertia, I_ω	0.413	kg.m^2
Wheel radius, r	0.11	m
Wheel width, w	0.86	m
Martian Gravitational force, g	3.7278	m/s^2
Coefficient of wheel dynamic sinkage, q_0 [7]	0.0375	
Tangential Shear Stress deformation, j_0 [7]	0.0778	m
Cohesion stress of the stimulant, c	1200	kg/m^2
Internal frictional angle of the soil, φ	40	Degrees
Soil stiffness related to the frictional angle for the stimulant, k_φ	2100	kg/m$^{(n+2)}$
Soil stiffness related to the cohesion for the stimulant, k_c	1100	kg/m$^{(n+1)}$
Soil Sinkage exponent, n	0.72	
Controller parameter, φ	0.001	

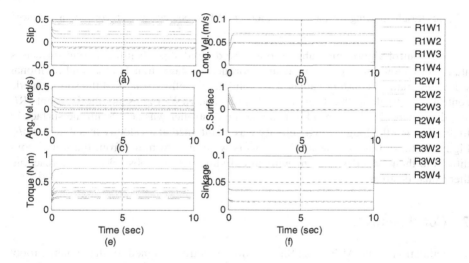

Fig. 2. Plots of (a) wheel slip, (b) longitudinal velocities (c) angular velocity (d) sliding surface (e) control torque (f) sinkage

6   Simulation Results

The simulation carried out assumes that the Pioneer 3AT robot follows a trajectory with static obstacles, which are captured as circular objects depending on the dimension as interpreted by the Laser Range Sensor. The parameters used for simulation is listed in Table 1. The robots move from the initial stationary position (initial velocity = 0 m/s), with different input wheel slip, sinkage, longitudinal and angular velocity, sliding surface and control torque, and attains final values as shown in Fig. 2, where three robots were used.

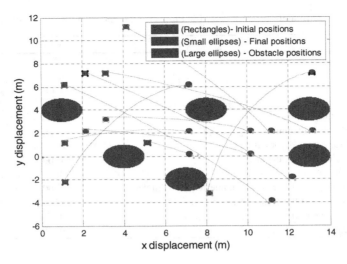

Fig. 3. Result showing the path of ten Pioneer 3AT Robot swarm with collision avoidance

The control algorithm is able to track the desired slip when the sliding surface is attained as shown in Fig. 2. The control torques, sinkage, translational and rotational velocities attained constant values. This validates the effectiveness of the sliding mode controller at tracking the desired slip. The legend $RiWi$ represents the ith robot (Ri) and the ith wheel (Wi). It can be seen that the controller was able to hit the surface in finite time and the longitudinal velocity was maintained at a velocity less than 1 cm/s. Figure 3 shows how a swarm of ten robots were able to move from the initial conditions to their final destinations without colliding with the obstacles before reaching there final destinations.

7   Conclusions

The validation of the APF based Sliding Mode Control designed for the dynamic robot model was carried out in simulation based on the characteristics of an in-house planetary soil stimulant. This algorithm can be applied for path planning and control of rover swarms. Simulation results show that the controller effectively keeps the slip rate of the wheel, tracking the desired value and avoids the excessive spin and sinkage of the driving wheels.

Tracking of the robots whilst avoiding obstacles to a predefined location was achieved using a navigation function proposed by Fethi Belkhouche. The result of the research will further advance the state-of-the-art of autonomous micro-rovers swarms operating on unstructured terrain, collaborating to achieve pre-defined goals, while maneuvering to avoid obstacles. The results presented in this paper demonstrate that APF together with SMC will not only enable collision–free navigation, but also improve mobility of the rover on unstructured terrains.

References

1. Pingshu, G., Rongben, W.: Terramechanics analysis and dynamics model for lunar rover on loose soil. In: 2010 3rd International Conference on Advanced Computer Theory and Engineering (ICACTE), pp. V6-418–V6-422 (2010)
2. Ishigami, G.: Path planning for planetary exploration rovers and its evaluation based on wheel slip dynamics. In: IEEE International Conference on Robotics and Automation, pp. 2361–2366 (2010)
3. Bogdanowicz, A.: Roving space explorers: next -generation robots are stepping in for humans. ieee.org/theinstitute (2010). Accessed June 2010
4. Miller, D.P.: Multiple behavior-controlled micro-robots for planetary surface missions. In: Proceedings of IEEE International Conference on Systems, Man and Cybernetics, pp. 289–292 (1990)
5. Lhomme-Desages, D., et al.: Trajectory control of a four-wheel skid-steering vehicle over soft terrain using a physical interaction model. In: IEEE International Conference on Robotics and Automation, pp. 1164–1169 (2007)
6. Harifi, A., et al.: Designing a sliding mode controller for slip control of antilock brake systems. Trans. Res. Part C Emerg.Technol. **16**, 731–741 (2008)
7. Kanfeng, G., et al.: Dynamic modeling and sliding mode driving control for lunar rover slip. In: IEEE International Conference on Integration Technology, ICIT '07, pp. 36–41 (2007)
8. Yoshida, K., Ishigami, G.: Steering characteristics of an exploration rover on loose soil based on all-wheel dynamic model. In: IEEE/RSJ International Conference on Intelligent Robots and Systems, pp. 3099–3104 (2005)
9. Wong, J.Y., (ed.): Theory of Ground Vehicles. Wiley, New York (2001)
10. Iagnemma, K., Dubowsky, S. (ed.): Mobile Robots in Rough Terrain: Estimation, Motion Planning, and Control with Application to Planetary Rovers. Spinger, Berlin (2004)
11. Gillespie, T.D.: Fundamentals of Vehicle Dynamics. Society of Automotive Engineers, PA, USA (1992)
12. Bekker, M.G.: Theory of Land Locomotion, pp. 167–207. University of Michigan Press, USA (1956)
13. Gazi, V.: Swarm aggregations using artificial potentials and sliding-mode control. IEEE Trans. Robot. **21**, 1208–1214 (2005)
14. Gazi, V., Passino, K.M.: Stability analysis of swarms. IEEE Trans. Autom. Control **48**, 692–697 (2003)
15. Gazi, V., Passino, K.M.: Stability analysis of social foraging swarms: combined effects of attractant/repellent profiles. In: Proceedings of the 41st IEEE Conference on Decision and Control, vol. 3, pp. 2848–2853 (2002)
16. Gazi, V., Passino, K.M.: Stability analysis of social foraging swarms. IEEE Trans. Syst. Man Cybern. Part B Cybern. **34**, 539–557 (2004)
17. Gazi, V., Passino, K.M.: Stability analysis of swarms in an environment with an attractant/repellent profile. In: American Control Conference, 2002. Proceedings of the 2002, vol. 3, pp. 1819–1824 (2002)
18. Koksal, M.I., et al.: Tracking a maneuvering target with a non-holonomic agent using artificial potentials and sliding mode control. In: 16th Mediterranean Conference on Control and Automation, pp. 1174–1179 (2008)
19. Passino, K.M.: Biomimicry of bacterial foraging for distributed optimization and control. IEEE Control Syst. Mag. **22**, 52–67 (2002)

20. Saaj, C.M. et al.: Spacecraft swarm navigation and control using artificial potential field and sliding mode control. In: IEEE International Conference on Industrial Technology, ICIT 2006. pp. 2646–2651 (2006)
21. Gazi, V., Passino, K.M.: A class of attractions/repulsion functions for stable swarm aggregations. Int. J. Control **77**, pp. 1567–1579, 2004
22. Belkhouche, F.: Nonholonomic robots navigation using linear navigation functions. In: Proceedings of the 2007 American Control Conference (2007)

Expecting the Unexpected: Measure the Uncertainties for Mobile Robot Path Planning in Dynamic Environment

Yan Li[✉], Brian Mac Namee[✉], and John Kelleher[✉]

School of Computing, Applied Intelligence Research Center,
Dublin Institute of Technology, Dublin, Ireland
yan.li.eric@gmail.com,
{brian.macnamee,john.d.kelleher}@dit.ie
http://www.comp.dit.ie/aigroup/

Abstract. Unexpected obstacles pose significant challenges to mobile robot navigation. In this paper we investigate how, based on the assumption that *unexpected* obstacles really follow patterns that can be exploited, a mobile robot can learn the locations within an environment that are likely to contain obstacles, and so plan optimal paths by avoiding these locations in subsequent navigation tasks. We propose the DUNC (Dynamically Updating Navigational Confidence) method to do this. We evaluate the performance of the DUNC method by comparing it with existing methods in a large number of randomly generated simulated test environments. Our evaluations show that, by learning the likely locations of *unexpected* obstacles, the DUNC method can plan more efficient paths than existing approaches to this problem.

Keywords: Mobile robot navigation · Dynamic environments · Learning

1 Introduction

A basic requirement for a mobile robot is the ability to plan and execute a path to a goal destination [1,2]. Unexpected obstacles, however, pose significant challenges to autonomous mobile robot navigation. For the purposes of this work we define unexpected obstacles as obstacles that may block corridors during robot navigation, but are not permanent features of the environment and hence do not appear on maps of the environment. For example, in a school building a robot may be trying to navigate a corridor and find that the corridor is blocked by a group of students who have just spilled out of a classroom. In situations where an unexpected obstacle blocks a path a robot may be forced to replan the route to the goal and take a detour, resulting in extra navigational costs in terms of time and energy.

Often, however, there is a regularity to the appearance of *unexpected* obstacles. For example, students may block a path at regular intervals (before and after

A. Natraj et al. (Eds.): TAROS 2013, LNAI 8069, pp. 363–374, 2014.
DOI: 10.1007/978-3-662-43645-5_39, © Springer-Verlag Berlin Heidelberg 2014

their scheduled classes) or particular doors may be closed at particular times for security reasons. If a robot could learn the pattern of unexpected obstacles and annotate the map used for planning with this information then more efficient paths could be planned.

Contributions: The main goal of this paper is to investigate how a mobile robot can learn features of unexpected obstacles in dynamic environments, so that optimal shortest paths can be planned. In particular, we will develop and evaluate a method for annotating topological maps with information regarding the probability of a corridor being blocked based on the robot's prior experiences in the environment.

Overview: This paper is organised as follows: In Sect. 2, we review background work. In Sect. 3, we describe our method for dynamically updating and maintaining the navigational confidence associated with an edge in a topological map, which we name DUNC. In Sect. 4, we present experiments and results on evaluating DUNC method. Finally, in Sect. 5, we discuss the performance of the DUNC method and suggest directions for future work.

2   Background

Path planning is a process of planning a collision-free path between a start position and a goal position [3]. It is a critical problem in mobile robot navigation. There are generally two types of robot path planning: static path planning and dynamic path planning [4,5]. Static path planning finds optimal passable paths based on a full knowledge of the environment [6–8]. However, static path planning cannot be used to deal with unexpected obstacles in dynamic environments.

Dynamic path planning generates passable paths based on environment changes (i.e., unexpected closed doors). There are existing methods that deal with unexpected obstacles locally. Examples include potential field methods [11], the dynamic window approach [12] and vector field histograms [13]. However, these local planning techniques focus on real time local efficiency instead of global optimal path planning [14] and suffer from the problem of local minima [15]. As a result they have no mechanism for retaining the information gleaned from one navigation (for example the fact that the robot encountered an unexpected obstacle at a particular location) and using it to improve subsequent navigations. Some other existing approaches deal with uncertainties based on probabilistic methods for improving robot localisation, but they are not proved to be applicable to dynamic path planning problems [9,10].

One existing dynamic global approach is proposed by Yamauchi and Beer [2]. They use the term *topological change* to describe variations in the environment that effect the decisions of a robot path planner. In their topological representation, a confidence value, defined as a *"robot's certainty that a topological link can be traversed"*, is associated with every link. In their approach, if a robot successfully navigates from a topological node A to a topological node B, the navigational confidence associated with the link between A and B, C_{AB}, in the map is increased. If, however, the robot fails to reach B from A then C_{AB} is decreased. During path planning the cost of each candidate path is the sum of the costs associated with each link in the plan and the cost of each link C is a function of the navigational confidence associated with that link. One of the strengths of Yamauchi and Beer's method is that successful navigation tasks are recorded. But their method suffers from the fact that path cost is solely based on confidence and neglects other criteria such as path length. As such, their approach would preference a very long path over a much shorter path if the longer path had a slightly higher confidence associated with it. In our approach, we take the length of path into account to measure the uncertainty cost of each link.

Stentz and Hebert [16] propose an autonomous navigation system that deals with the cost of potential obstacle positions based on a predefined coefficient value (a scalar value σ). In their system, potential obstacle positions are marked with high scalar values, so that the path planner will avoid planning paths across those positions. Then a graph search method is used to find an optimal path with the least cost. The scalar method suffers from the problem that since the cost of all the edges are updated based on the same σ, the cost update process has little effect on potential obstacle positions with low static cost and therefore a marked potential block position will be considered as unmarked by the path planner. In our approach, unexpected obstacles are measured separately for different edges and the measurements are normalised with maximum edge length.

Briggs et al. [17] propose a ratio method that deals with unexpected obstacles in dynamic environments based on a topological representation. In their approach, each topological edge is associated with a ratio value R. R is calculated as the ratio of the number of detections of unexpected obstacles and total number of traverses on the topological edge. After every edge traversal, the ratio value is updated based on whether an unexpected obstacle is detected or not. If the robot successfully traverses an edge, the ratio of the edge is decreased, otherwise R is increased. Hu and Brady [1] propose a similar method to Briggs et al.'s. They classify uncertainties as three types on a predefined path: obstacles that partially block the robot path, obstacles that fully block the robot path and moving obstacles that cross the robot path. They measure three types of uncertainties based on a ratio value R. Interestingly a decay process is used in Hu and Brady's approach to release edges with high R based on an exponential function for the subsequent navigation tasks. Essentially this decay process allows the system to forget that the robot encountered an obstacle on a particular edge and makes the edge viable during subsequent path planning. In contrast to Hu and Brady's approach, our approach uses a dynamic confidence scoring method with a confidence decay component to measure the uncertainties in dynamic environments.

3   The DUNC Method

In this section we present our method for dynamically updating the navigational confidence associated with edges in a topological map, which we will refer to as the DUNC method. The DUNC method is part of our Dynamic Confidence Topological Map Approach ($DCTMA$). $DCTMA$ is an approach for solving dynamic robot path planning problems based on a topological representation of the world. It is built on top of our landmark-based robot localisation and navigation system [18], with corridor following and obstacle avoidance behaviours.

A Dynamic Confidence Topological Map $DCTM = <N, E>$ is a topological map that can be maintained and updated dynamically based on the sensor data of the robot as it moves through the environment. A $DCTM$ contains a set of nodes $N = \{N_1, N_2, ..., N_n\}$, which represent landmarks (i.e., T-junctions, X-junctions) in the environment. And it contains a set of edges $E = \{E_1, E_2, ..., E_n\}$, which are passable edges between landmark positions. Each edge has a 3-tuple data structure $\{D, L, C\}$, where: D is the static cost of the edge (i.e., the length of an edge); L is running score that is increased by a fixed amount, δ_+, each time the edge is successfully navigated (see Eq. 1) and decreased by a fixed amount, δ_-, each time the robot encounters an obstacle on the edge (see Eq. 2); and C is the confidence that an edge can be traversed by the robot and is always calculated from L using Eq. 3. In the DCTM system L is initialised to 0 and C is consequently initialised to 0.5.

$$L = L + \delta_+ \tag{1}$$

$$L = L - \delta_- \tag{2}$$

$$C = 1 - \frac{1}{1 + e^L} \tag{3}$$

During path planning the cost of each edge E_x is calculated based on D and C via a weighted sum aggregation function given W_D and W_C as the corresponding weight for D and C. In this system the values of W_D and W_C are constrained such that the sum of the weights, $W_D + W_C$, always equals 1. Then E_x is used to feed the Dijkstra graph search method [19] to find optimal paths.

As stated in Eq. 3, the confidence score of an edge is a function of the L score of the edge which has the value range from $-\infty$ to ∞. This has the potential to cause problems as the L score for a particular edge may become so low that paths including the edge are never considered for navigation or so high that paths including the edge are always selected even if the robot encountered an obstacle on that edge very recently.

One way of addressing this issue is to impose a bounds on the range of values that L can take. We have found that bound the range of L between -4 and $+4$ permits a confidence range (0.018 to 0.982) that is large enough to reflect the navigation results (encounter an obstacle or successful traverse) while at the same improving the chance that a low confidence edge will be reused at limit points.

Another extension that could be made to the basic confidence representation is inspired by Hu and Brady's decay process [1]. In the DUNC method the proposed memory decay process implements a regression towards the initial a

confidence value of 0.5 on the edges in the $DCTM$. The memory decay component makes the robot periodically regress the L score of an edge (and hence the
edge confidence) towards the initial value of 0; essentially, allowing the robot
to forget the obstacles it has encountered on the edge or its previous successful
navigations of the edge. The motivation for proposing this extension is similar
to the motivation for imposing a bounds on the range of L namely that as the
robot navigates through the environment some edges in the $DCTM$ may be
marked with a low confidence, due to the robot failing to navigate through them
successfully. The effect of marking an edge with a low confidence is that that
edge has a low probability of being used for subsequent navigation tasks. The
memory decay component releases these edges to make them accessible again to
the robot. Also, it decreases L on high confidence edges to avoid persistent usage
of high confidence edges. The memory decay component updates the log odds L
on each edge based on a parameter ν, which is the frequency of activating the
memory decay component. In our system, memory decay component is switched
on every ν robot navigation tasks. When the memory decay process is run it
increase the L value of every edge with a confidence value lower than 0.5 by a
constant value, ρ_+, in the range 0:1, (see Eq. 4) and decreases the L value of
every edge with a confidence value greater than 0.5 by a constant value, ρ_-, in
the range 0:1, (see Eq. 5).

$$L = L + \rho_+ \qquad\qquad (4)$$

$$L = L - \rho_- \qquad\qquad (5)$$

In summary, the DUNC method dynamically updates the navigation confidence associated with an edge in a topological map representation based on
the robots experiences in navigating along the edge. The basic component of
the method is a logistic confidence model. There are two proposed extensions to
this model and bounding of the L values and a memory decay process. In the
next section we present a series of experiments that evaluate the contribution of
each of these components to the method and compare the method to the current
state-of-the-art methods.

4   Evaluation

This section describes experiments performed in order to evaluate the performance of the DUNC method. First the test environment used and the procedure
used to generate *unexpected* obstacles within this environment will be described.
Then we will describe 3 different experiments that evaluate the DUNC method
and compare it to existing methods discussed in Sect. 2.

4.1   Generating Test Environments

All of the experiments that we describe take place in simulated environments
with simulated robots (any type of robot can be used). In order to run experiments evaluating the performance of the DUNC method we generate random

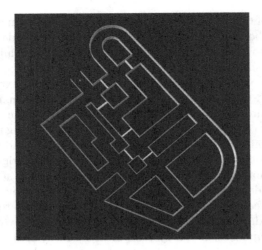

Fig. 1. An example of a randomly generated simulated test environment. Potential obstacle positions are marked with yellow bars (Color figure online).

indoor environments (represented as topological maps) in which robot navigations are simulated. The simulated robot actually jumps between topological nodes and the navigation distance is measured as the actual distance between the start node and the end node along topological edges. Each random environment is composed of 20 to 40 nodes connected by 40 to 80 edges. The nodes to be connected to each other are randomly selected. Figure 1 shows a representation of one such environment.

To determine which edges in a map will periodically become blocked by obstacles, we first simulate 1,000 journeys from random start nodes to random end nodes within this map without any unexpected obstacles being present. For each journey, the Dijkstra algorithm is applied given its start node and end node. The result of the Dijkstra algorithm is a path consist of a set of topological nodes. Navigations are simulated based on the results of Dijkstra algorithm to count the number of times each edge in the map is used in these journeys. Then, the edges that are used most frequently (the top 25 % in our experiments) are selected as potential obstacle positions. In this way we introduced obstacles into the most important edges within the environment. In Fig. 1 the potential blockage positions are marked as yellow rectangles. Each of the potential blockage positions has a randomly selected *obstacle appearance rate* associated with it. This is the probability (from 0 to 1) that an obstacle will appear on an edge when a robot tries to navigate it. Obstacle appearance rates remained constant throughout each experiment in the corresponding environmental unit.

4.2 Experimental Method

Each experimental unit consisted of a robot performing 1,000 journeys between randomly selected start and end nodes in a test environment. Based on the obstacle appearance rates associated with each edge in the map different edges would become blocked by obstacles during each of these journeys. When multiple techniques are compared, the same 1,000 journeys are always performed, during which the same edges are blocked. In this way different techniques are fairly compared.

When an unexpected obstacle is encountered during a navigation task, the robot will be forced to re-plan the route to the goal and detour to avoid the obstacle. As a result, each unexpected obstacle encountered will result in extra travel cost (extra travel distance in our experiments). For each journey a robot takes from a start node to an end node we record the length of the *actual path* it navigates, which is the summation of lengths of edges the robot traversed. We also record the length the *optimal path* would have been if the robot had known in advance where all the unexpected obstacles would occur. The key metric used in our experiments is the difference between the actual path length and the optimal path length. The average difference between these two measures is calculated across 1000 journeys for each experimental unit.

In the remainder of this section, we present 3 experiments that (1) fine tune the DUNC method, (2) find the best parameter values to use for the existing methods described in Sect. 2, and (3) compare the performance of the DUNC method with these existing methods.

4.3   Experiment 1: Fine Tuning the DUNC Method

The DUNC method is composed of three components: (1) the confidence representation itself, (2) the bounds component, and (3) the memory decay component. Thus, there are four candidate combinations of these components:

- confidence representation (*confidence*)
- confidence representation + bounds component (*confidence + B*)
- confidence representation + memory decay component (*confidence + D*)
- confidence representation + bounds component + memory decay component (*confidence + B + D*)

Each of the component parts of the DUNC method uses an associated set of parameters. Table 1 lists these parameters and, for each, the set of potential values they can assume. Table 2 shows parameter values that are used by each of the candidate combinations of the components of the DUNC method.

The first experiment sought to determine which was the most effective out of the four combinations of DUNC components, i.e. *confidence*, *confidence + B*, *confidence+D* and *confidence+B+D*. In these experiments the *best* parameter values shown in Table 2 were used. Robots using each of the four combinations of DUNC components were simulated performing 1,000 navigations each across 50 randomly generated environments (in all each robot performed 50,000 navigations). For each of the 50 environments the average difference between actual and optimal path lengths was recorded for each combination of DUNC components. The DUNC component combinations were then ranked based on their average

Table 1. The parameters used by the DUNC method and the possible values they can assume

Parameter	Description	Values
ω	The weight of the distance component used by the weighted sum cost function	10, 20, 30, 40, 50, 60, 70, 80, 90
ν	The frequency at which the memory decay component is invoked	10, 20, 30, 40, 50, 60, 70, 80, 90, 100
$\delta+$	The amount by which the running score is increased after a successful navigation	0.5, 1
$\delta-$	The amount by which the running score value is decreased after an unsuccessful navigation	0.5, 1
$\rho+$	The amount by which to update the running score value by the memory decay component on low log odds edges	0, 0.5, 1.0
$\rho-$	The amount by which to update the running score value by the memory decay component on high log odds edges	0.5, 1

Table 2. Best parameter values for DUNC component combinations

Combinations	ω	ν	$\delta+$	$\delta-$	$\rho+$	$\rho-$
confidence	50		0.5	1		
confidence + B	40		0.5	1		
confidence + D	50	50	0.5	1	0	0.5
confidence + B + D	40	100	0.5	1	0.5	0.5

difference between actual and optimal path lengths on each environment (ranks ranged from 1 (*best*) to 4 (*worst*)). Table 3 shows the average rank achieved across the 50 test environments by each DUNC component combination. From this it is clear that the *confidence* + B + D combination appears to perform the best, closely followed by *confidence* + B combination.

To test these results for statistical significance, we applied the Friedman test [20] followed by the Holm step-down procedure [21] as a post-hoc test on the average ranks. The results of the Friedman test show that there is a statistically significant difference in robot performance depending on which combination of DUNC components was used, $X_F^2(3) = 24.1446$, $p = <0.0001$.

In order to investigate significant differences between the best DUNC component combination, *confidence* + B + D, and the others we applied the Holm's step-down procedure as a post-hoc test to the Friedman test. Table 4 shows the result of this test. By comparing the p values in the second column to the critical values in the third column[1] we can see a statistically significant difference

---

[1] In a Holms step-down procedure a statistically significant difference exists when the p value is less than the critical value.

Table 3. Average ranks of candidate DUNC component combinations calculated for the Friedman test.

	$confidence$	$confidence + B$	$confidence + D$	$confidence + B + D$
Avg rank	2.84	2.12	3.04	2

Table 4. Results of the Holm step-down procedure.

Approaches	z	p	$\alpha/(k-i)$
$confidence + D$	4.0279	0.00056	0.017
$confidence$	3.2533	0.001	0.025
$confidence + B$	0.4648	0.642	0.05

exists between the $confidence + B + D$ and the $confidence$ and $confidence + D$ combinations but not the $confidence + B$ combination. This is not surprising given that the average ranks of the $confidence + B + D$ and $confidence + B$ combinations were so close.

Based on these experiment results, we see that the memory decay component does not really help the performance of the DUNC method. We believe that the reason for this is that, as long as the likelihood of an edge being blocked by an obstacle remains constant (as is the case in our experiments), *forgetting* about the existence of obstacles lends advantage. This is also evident in the very poor performance of the $confidence + D$ combination. The bounds component, however, has a significant impact on performance. We surmise that this is because it allows the update procedure remain agile and respond to successful and unsuccessful edge navigations quickly. In the remaining experiments when we refer to the DUNC method we refer to the $confidence + B$ component combination as, adopting the principle of *parsimony*, it makes sense to choose a slightly less complex method given almost equal performance.

4.4 Experiment 2: Fine Tuning Existing Methods

So as the comparison between existing methods and the DUNC method will be fair, we perform the same fine tuning experiment for the existing methods described in Sect. 2: the ratio method (*Ratio*), the ratio method with decay process (*RatioDecay*), the scalar method (*Scalar*) and the Yamauchi method (*Yamauchi*). Each of these methods (excluding the ratio method) has a set of parameters for which optimal values must be discovered. Table 5 shows the candidate methods and their corresponding parameters and possible values that are considered. Table 6 shows the resulting *best* parameter values and these are used in the final experiment described in this section.

Table 5. Existing methods and their corresponding parameter and possible values.

Method name	Parameters	Possible values
Ratio	No parameter used	
RatioDecay	Decay rate μ	0.005, 0.025, 0.05, 0.1, 0.5, 1, 2, 5
	Decay frequency Δ	1, 2, 3, 4, 5, 10, 20, 50, 100
Scalar	Scalar value σ	2, 3, 4, 5, 10, 15, 20, 50, 100
Yamauchi	Link learning rate λ	0.1, 0.2, 0.3, 0.4, 0.5, 0.6, 0.7, 0.8, 0.9

Table 6. Best parameter values for the *RatioDecay*, *Scalar* and *Yamauchi* methods.

Method name	Parameter values
RatioDecay	Decay rate $\mu = 0.005$
	Decay frequency $\Delta = 100$
Scalar	Scalar value $\sigma = 5$
Yamauchi	Link learning rate $\lambda = 0.1$

4.5 Experiment 3: Comparing the DUNC Method to Existing Approaches

The aim of this experiment is to compare the DUNC method to existing approaches that deal with unexpected obstacles in robot navigation across dynamic indoor environments. In this experiment we compare the performance of the DUNC method, the ratio method, the ratio method with decay process, the scalar method and the Yamauchi method. In this experiment robots using each of these methods were simulated performing the same 1,000 journeys across 50 randomly generated environments (50,000 journeys in total per robot). The average differences between actual and optimal paths for each of these sets of journeys were recorded. the performance of the 5 different methods were ranked (from 1 (*best*) to 5 (*worst*)) across the different random environments and the average ranks are shown in Table 7. It is clear from these ranks that the DUNC approach appears to choose more efficient routes than any of the other approaches.

In order to test for statistically significant differences in these results a Freidman test was used again. The results of the Friedman test show that there is a statistically significant difference in robot performance depending on which approach is used, $X_F^2(4) = 120.88$, $p = {<}0.0001$.

In order to further investigate this result we applied the Holms step-down procedure again. Table 8 shows the result of this analysis comparing the DUNC method with the other approaches. Again, by comparing the p-values in the second column with the critical values in the third column we can see that a statistically significant difference exists between the performance of the DUNC method and all of the other approaches. This result shows that, under our test conditions, a robot using the DUNC method to navigate within an indoor environment selects more efficient paths than a robot navigating using any of the other methods under test.

Table 7. Average rank values of candidate methods calculated for the Friedman test.

	$DUNC$	$Scalar$	$Ratio$	$RatioDecay$	$Yamauchi$
Avg rank	1.18	2.4	3.66	4.32	3.44

Table 8. Results of the Holm step-down procedure.

Methods	z	p	$\alpha/(k-i)$
$RatioDecay$	9.93	<0.0001	0.125
$Ratio$	7.842	<0.0001	0.0167
$Scalar$	7.148	<0.0001	0.025
$Yamauchi$	3.858	0.00011	0.05

5   Conclusion

In this paper we presented the DUNC method that a mobile robot can use to learn likely positions of unexpected obstacles in dynamic environments, so that optimal shortest paths can be planned. This is built on the assumption that most *unexpected* obstacles are not really unexpected at all and follow a pattern occurring in similar places over time and that such patterns can be exploited. The DUNC method is composed of three key components: the confidence component and a set of bounds on running outcome scores (a forgetting mechanism was considered for inclusions but tests showed that it did not improve overall performance).

The DUNC method was compared against the most important existing methods for addressing this problem - the ratio method, the ratio method with decay process, the scalar method and the Yamauchi method. The DUNC method was found to consistently plan more efficient paths than all of these other approaches across a large number of test runs in different environments.

There are a range of ways in which we could improve the DUNC method. In particular, in the future we plan to examine how the approach can be modified to handle partial blockages of routes (currently all obstacles completely block an edge on the map), different ways the decay mechanism could be adjusted to make it more effective, ways in which the approach could be modified to handle moving obstacles and the introduction of a time dimension in which different routes are assumed to be more likely to be blocked at different times of the day.

References

1. Hu, H., Brady, M.: Dynamic global path planning with uncertainty for mobile robots in manufacturing. In: IEEE International Conference on Robotics and Automation, pp. 760–767 (1997)
2. Yamauchi, B., Beer, R.: Spatial learning for navigation in dynamic environments. IEEE Trans. Syst. Man Cybern. B Cybern. **26**, 496–505 (1996)

3. Hsu, D., Latombe, J.C., Motwani, R.: Path planning in expansive configuration spaces. In: IEEE International Conference on Robotics and Automation (1997)
4. Phillips, B., Likhachev, M.: SIPP: safe interval path planning for dynamic environments. In: IEEE International Conference on Robotics and Automation, Shanghai, pp. 9–13 (2011)
5. Wang, T., Dang, Q., Pan, P.: Path planning approach in unknown environment. Int. J. Autom. Comput. **7**, 310 (2010)
6. Hao, G., Zhang, D., Feng, X.: Model and algorithm for shortest path of multiple objectives. J. Southwest Jiaotong Univ. **42**, 641–646 (2007)
7. Kulyukin, V., Kutiyanawala, A., LoPresti, E., Matthews, J., Simpson, R.: iWalker: toward a rollator-mounted wayfinding system for the elderly. In: IEEE International Conference on RFID (2008)
8. Murphy, L., Newman, P.: Planning most-likely paths from overhead imagery. In: IEEE International Conference on Robotics and Automation (ICRA) (2010)
9. Mathibela, B., Osborne, M.A., Posner, I., Newman, P.: Can priors be trusted? learning to anticipate roadworks. In: 15th International IEEE Conference on Intelligent Transportation Systems (ITSC) (2012)
10. Wolf, D.F., Sukhatme, G.S.: Mobile robot simultaneous localization and mapping in dynamic environments. J. Auton. Robot. **19**, 53–65 (2005)
11. Koren, Y., Borenstein, J.: Potential field methods and their inherent limitations for mobile robot navigation. In: IEEE International Conference on Robotics and Automation (1991)
12. Baehoon, C., Beomseong, K., Euntai, K., Kwang-Woong, Y.: A modified dynamic window approach in crowded indoor environment for intelligent transport robot. In: International Conference on Control, Automation and Systems (2012)
13. Borenstein, J., Koren, Y.: The vector field histogram-fast obstacle avoidance for mobile robots. In: IEEE International Conference on Robotics and Automation, vol. 7, pp. 278–288 (1991)
14. Ding, F., Jiao, P., Bai, X., Wang, H.: AUV local path planning based on virtual potential field. In: IEEE International Conference on Mechatronics and Automation (2005)
15. Majdi, M., Anvar, H.S., Barzamini, R., Soleimanpour, S.: Multi AGV path planning in unknown environment using fuzzy inference systems. In: International Symposium on Communications, Control and Signal Processing (2008)
16. Stentz, A., Hebert, M.: A complete navigation system for goal acquisition in unknown environments. Auton. Robot. **2**, 127–145 (1995)
17. Briggs, A.J., Detweiler, C., Scharstein, D., Vandenberg-rodes, A.: Expected shortest paths for landmark-based robot navigation. Int. J. Robot. Res. **23**, 717–728 (2004)
18. Li, Y., Mac Namee, B., Kelleher, J.D.: Navigating the corridors of power: using RFID and compass sensors for robot localisation and navigation. In: The 11th Towards Autonomous Robotic Systems (TAROS 2010) (2010)
19. Edsger, W.D.: A note on two problems in connexion with graphs. Numer. Math. **1**, 269–271 (1959)
20. Friedman, M.: A comparison of alternative tests of significance for the problem of m rankings. Ann. Math. Stat. **11**, 86–92 (1940)
21. Holm, S.: A simple sequentially rejective multiple test procedure. Scand. J. Stat. **6**, 65–70 (1979)

Swarm Robotics

Novel Method of Communication in Swarm Robotics Based on the NFC Technology

Ulf Witkowski[✉] and Reza Zandian[✉]

Electronics and Circuit Technology Department,
South Westphalia University of Applied Science, Soest, Germany
{witkowski.ulf,zandian.reza}@fh-swf.de

Abstract. Swarm robotics has pulled the attention of scientists for its flexibility, redundancy, stability, emergent intelligence and many other features. One major issue in design of swarm robots is the communication system. In spite of existence of currently employed communication systems in swarm robots such as IR, Bluetooth, ZigBee, etc. still it has not been concluded which communication system is the best fit in swarm robotics. In this paper an alternative communication system based on the NFC technology is proposed and its advantages over other available systems are evaluated. In order to prove the concept, a simple food scavenging scenario is implemented using three robots which communicate together based on the NFC technology. At the end of this paper, it has been deduced that advantageous features of the NFC communication systems seem to satisfy the requirements of the swarms up to some extents.

Keywords: NFC (Near Field Communication) · Swarm robotics · Wireless communication techniques · Food scavenging robots · NFC antenna design

1 Introduction

Reliability, robustness and flexibility are the features of swarm robots which spur scientists to do more research in this area [1]. One of the main elements that make these features possible, is the ability of the communication in between the swarm members. Jacques Penders [6] has compared the swarm robots which has communication capability with the ones without communication or limited level of communication and resulted that communication system improves the efficiency of the swarm and is able to multiply their capabilities. Coordination information, tasks distribution, navigation data and cooperation requests are typical types of messages that can be transferred in between the robots in swarms.

Y. Mohan et al. [3] have categorized the swarm robotics literature into two parts, namely implicit/indirect and explicit/direct based on their communication systems. The first part addresses the communication systems which are conveyed through the environment. The pheromone communication system is one example of that, which is based on odorimetry science. In this paper, this type of communication is not considered rather direct form of communication is the focus of the discussion. Although the communication systems of swarms are rather application dependant and

A. Natraj et al. (Eds.): TAROS 2013, LNAI 8069, pp. 377–389, 2014.
DOI: 10.1007/978-3-662-43645-5_40, © Springer-Verlag Berlin Heidelberg 2014

S. Kernbach et al. [5] introduced swarms with no communication and Ming Li et al. [4] introduced mesh networks in swarms, the swarm robots with the local communication capability are at the center of attention in this paper. Based on a research which is conducted by Manuele Brambilla et al. [1] in the swarm robotics literature, these systems have to be autonomous, be able to sense the environment and communicate with the other robots locally, have no access to a centralized control unit or a knowledge center and finally follow the scheme of cooperation for overcoming their duties. Considering these characteristics, it can be noticed that the robots have only the chance to communicate locally and just with the other members of the swarm [2]. In such cases each member of the swarm has to communicate with another member in its neighborhood to transmit and receive data. If special care has not been taken in selection of proper communication method, jamming of the communication network, interferences on the other robot's communication, noise and corruption of the data can be the major consequences. Some of the parameters that must be considered in the swarm designs are the radius of the communication, using of low level signals, possibility of omni-directional communication, energy consumption, scalability and routing issues [7].

Several different types of communication systems are implemented in the swarm robots by swarm engineers. S. Kornienko et al. [7] have proposed using of IR sensors for communication between the robots while using the sensors for perception and distance measurement at the same time. Ming Li et al. [4] have used Wi-Fi modules to form a mesh network with the routers installed in the environment. P. Benavidez et al. [8] have proposed swarm robots which use ZigBee to establish communication in between the robots. Hawick et al. [9] have a system which uses Wi-Fi and Bluetooth as the mean of communication between the robots and control center as well. S. Bhandari et al. [10] introduced an RF based communication system which suits swarm structure. Each of these systems has its own advantages and disadvantages. Brian Simms et al. [11] have discussed the pros and cons of these systems in their literature. In order to find out which of these systems better fits the application of swarm robotics, a comparison is done with our proposed method in Sect. 2.4.

In this paper, the NFC communication system is used as a mean for local data transfer in between two robots in the swarm. The NFC technology is the successor of the RFID concept which enables two systems to communicate within a short distance (less than 10 cm) and using RF frequency tuned on 13.56 MHz with the baud-rates of 212 Kbps, 424 Kbps and 848 Kbps (in some modes) [13]. Features such as security as a result of short distance of communication and internally implemented encryption protocols, simplicity of use, simple initialization process which just needs a short touch of two systems and easy data sharing, make NFC devices a good alternative for other opponents such as Bluetooth, Wi-Fi and ZigBee. The major improvement of NFC in comparison to RFID is an extra feature known as Peer to Peer (P2P) mode which enables two active devices to communicate together as soon as they are in the access range of each other. Use of this feature in the swarm robots makes the process of communication much simpler and more effective. At the end of this paper a platform is presented which has been made based on NFC device with the aim of evaluating the performances and efficiencies of the communication system.

2   NFC Characteristics

2.1   Modes of Operation

NFC technology has two concepts of communications namely active mode and passive mode. In passive mode one NFC device is using its own power supply and providing the power over its antenna circuit to the other device. In this case the generator of the RF signal is the initiator of the connection. Examples of such a connection are tags or Mifare cards which have not an internal power supply and receive the power from its initiator device. In the other concept (active mode), each NFC device uses its own power supply to generate RF signals. This is possible by using a collision detection method which provides an opportunity for NFC devices to sense the existence of the RF signal before turning on the antenna circuit. Therefore each device turns on the antenna circuit whenever a data transmission is needed and as soon as the transmission is completed it has to be turned off to let other devices start data transmission. Based on these two transmission concepts, three modes of operation are defined namely Read/Write mode, Peer to Peer mode (P2P) and Card emulation mode [13]. In swarm robotics' structure, P2P mode in active scheme is used which is suitable for local communication of the robots with their neighbors. In Fig. 1 comparison of the active mode operation vs. passive mode is shown graphically.

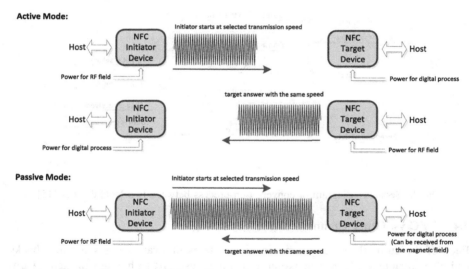

Fig. 1. Comparison of the active mode operation vs. passive mode [14]

2.2   Speed of the Connection

The speed of data transmission in the P2P mode based on the definition of the NFC forum [13] is 424 Kbps. This speed is fast enough for transmitting the small amount of data such as position information, coordination, tasks status, collaboration data, sensor data or even new updates of the firmware. In addition to that, the recognition of two devices takes place in only 0.1 s which is suitable for the swarms. However for transmission of a larger amount of data such as image data streams or analysis results of a complex process, other methods of communication can be used in parallel.

2.3   Initialization Process

In order to establish the connection between two devices in the P2P mode certain procedure has to be followed. The main requirement is one device has to be set in target mode and the other one as initiator of the connection. Right after initialization command, the initiator device turns the antenna circuit on and searches for the target device in its range. As soon as a target device is found by the initiator, the host of initiator can select the target device and transmit the data and receive data from the target device. The sequence diagram of this process is depicted in Fig. 2.

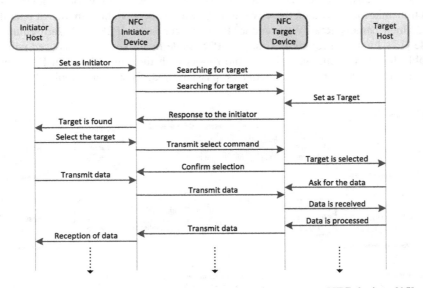

Fig. 2. Sequence diagram of conversation process between two NFC devices [15]

2.4   NFC vs. Other Communication Technologies

Each of the currently used communication systems in swarms has some drawbacks and also advantages. S. Kornienko et al. [7] have used IR techniques in their swarm robots. They claim that IR is easy to drive, no initialization is needed between the devices and it can also be used for the environmental perception. Although on the other side they had problems with the light in the environment and saturation problems as a result of sunlight. These problems do not exist in case of NFC and since almost not any other device is using the same operating frequency as NFC, there would be no interference problem. Ming Li et al. [4] have used Wi-Fi modules in their design but the problem of Wi-Fi is mostly seen in its structure. Basically Wi-Fi is designed to work in star topology as well as point to point ad hoc network. However setup of an ad hoc network is not suitable for swarms which have characteristics of rapid change of nodes and dynamic movements. As well as that, setup time of Wi-Fi is relatively long and cause delays in the communication of swarm robots. Another problem is the interference of the communication signals on the other nodes as the number of nodes increases. Comparing this technology with NFC, it can be realized that NFC has much faster initialization time and has no problem regarding the signal interference. Hawick et al. [9] used Wi-Fi but with the help of Bluetooth devices. Bluetooth can support point to point connection but the pairing issue is an obstacle. Bluetooth low energy can support a tree structure for communication. In this context the swarm robots should be divided into zones and each zone follows a leader in that zone. The leader of each zone can contact a master robot which makes the decision for all. On the other hand NFC is suitable for leader-less connection with individual decision making possibility. It has to be noticed that the main obstacle of Bluetooth can be solved using the pairing capability of the NFC devices in parallel. An overview of possible communication methods in swarm robotics is provided in the Table 1.

Table 1. Comparison of NFC specifications with other common technologies [12]

Technology	NFC	IrDa	Bluetooth	Bluetooth (low energy)	ZigBee
Selection	Touch	P2P	Scan	Scan	Scan
Setup time	<0.1 s	~<0.1 s	~3 s	<6 ms	<0.1 s
Range	<10 cm	<1 m	<100 m	<100 m	~10–1000 m
Data rate	424 Kbit/s	115.2 Kbit/s	2.1 Mbit/s	~1 Mbit/s	250 Kbit/s
Frequency	13.56 MHz	~120 KHz	2.4–2.5 GHz	2.4–2.5 GHz	2.4–2.5 GHz
Cost	Low	Low	Medium-High	Low-Medium	Low
Required power	Medium	Low	Medium	Low	Low

3 NFC Challenges in Swarm Robotics

Due to the topology of the connection in the NFC which is based on master-slave style and also the shape of NFC antenna and its performance, several issues have to be tackled in order to fit this method in swarm robotics. These issues are evaluated in this section and in case of each, proper solutions are provided.

3.1 Implementation of Master-Slave Topology in Swarm

As it was discussed in the Sect. 2, in order to establish a connection between two NFC devices, one has to be setup as an initiator and the other as a target device. This will be a problem in swarms which any robot should be able to communicate with another one. In order to solve this problem, a time division algorithm is used which setup the device as initiator for a short time (~ 100 ms) and if no target is found during this time the device will setup as target for the rest of the period (~ 400 ms). In this case each device will be initiator for 25 % of the time and will wait for an initiator for the rest of the period. These times are chosen considering the fact that at least 100 ms is required for establishing the connection by the hardware. The rest of the time should be small enough not to let the robots pass without recognizing each other. Based on this concept, the probability of overlapping can be calculated as follows:

$$\text{Probability of overlapping} \overset{\text{def}}{=} \frac{100}{400} * \frac{100}{400} = \frac{1}{16} \tag{1}$$

Considering the speed of each robot and their sizes in the performed project, feature of the NFC device which detects the existence of the field before turning on the antenna and finally the calculated probability in Eq. 1, it can be resulted that each robot has a low chance of overlapping the initial time with another robot.

3.2 Speed of Data Transmission

The speed of data transmission in the NFC is comparably high (424 kbps in P2P mode) for transmission of most formats of data in swarm robotics. In case a higher data rate is needed for transmitting larger amount of data in a shorter time, another radio type such as Bluetooth can be used in parallel. Steffen et al. [16] has used NFC in this context. In this case NFC can simplify and facilitate the pairing process which is a major obstacle in Bluetooth connection. The procedure is in a way that two devices transfer their pairing data through NFC devices and after reception of these data, Bluetooth connection can be setup in less than a second.

3.3 Security of the Connection

The physical structure of the NFC technology which is based on short distance communication (less than 10 cm) makes this technology a secure method. This means eavesdropping system should be close enough to locate in the antenna field otherwise it would be impossible to receive any data. This feature is suitable for swarm robotics because only the robots which are close enough can communicate and the farther robots do not receive any signal. Haselsteiner et al. [17] have evaluated the security issues in the NFC technology. They resulted that the NFC technology has a good level of security which can also be improved by applying a secure channel using encryption between the devices. As well as that, parity checking and BPSK modulation are implemented in the transceivers which simplify the processing tasks of the robots.

3.4 Multiple Targets Issue

One of the main issues that almost all communication systems have to deal with is the existence of several devices in the range. It has to be evaluated what will happen if two or more devices are in the range of the initiator device. In case of NFC systems, based on the protocol manual released by NXP [15, p. 115] two devices are supported simultaneously which means initiator can decide which of these two devices should be selected and therefore no overlap or interference would occur. Once one device is selected the other one does not receive any signal and cannot cause problems for the connection of the other pair. In case more than two targets are available the first two will be detected randomly and the initiator will not search for more devices until the initialization process of the current selected target device is finished.

3.5 Energy Consumption

Knowing the fact that swarm robots are running on batteries, the electronic parts should be optimized for using minimum possible energy. According to the specifications released by NXP [14] the transceiver consumes 30 mA when it is in normal mode and setup as target. In case the device is initialized as initiator, this value will be increased up to 60–150 mA depends on the antenna tuning circuit and its shape. This amount of power is relatively high compared to the other wireless communication options and has to be improved. According to the NXP document [14] it is possible to put the device in sleep mode which consumes 45 μA but has the possibility to wake up by sensing the existence of the initiator in the range. Keeping this in mind and applying the time modulation method which was explained in Sect. 3.1 the power consumption will drastically decrease and the major amount of power will be used only when the device is in initiator mode and just for a short time (100 ms). Overall power consumption in certain period of time can be calculated as follows:

$$0.1 \, \text{ms} * 100 \, \text{mA} + 400 \, \text{ms} * 45 \, \mu\text{A} = 10.02 \, \text{mAs} \tag{2}$$

This feature has been automatized in the transceiver released by NXP (PN532) in a way that the device goes to sleep mode if no external RF field is detected and stay in this mode until either an initiator comes in the field or the host sends a command [15].

4 Platform Design

4.1 Antenna Circuit

Based on the antenna design theory, for a device that needs to work in the frequency of 13.56 MHz a very large antenna is required. But the NFC has benefited a different method for transmission of the data which is using a magnetic field that enables the NFC to send the data and energy simultaneously. The antenna circuit is constructed of two loops which are mounted over each other in parallel [19]. The parallel equivalent circuit of the antenna and its circuit form is depicted in Fig. 3.

The typical range for C_{pa} is 3–30 pF, for L_{pa} is 0.3–3 μH and for R_{pa} is 0.1–4 Ω. The center of the antenna will be grounded and the signals will be receiving in the differential forms from each loop of the circuit. The final design has 4 loops in total which has 2 loops on each side. The distance that can be covered is highly dependent on the antenna size and quality of its matching circuit. In order to evaluate different forms of antenna, several shapes have been designed for this project. One issue that has been taken into account is the possibility of fitting the antenna in the robot's body.

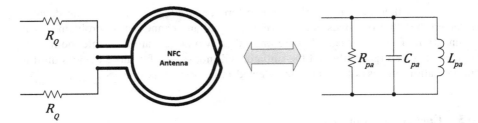

Fig. 3. Parallel equivalent circuit for the antenna circuit [19]

Therefore a flexible material has been used which let the antenna circuit fit in a curved space. It has to be noticed that in all of the antenna shapes, the length of the signal tracks stays the same to avoid having different matching circuit. The practical results that have been achieved prove that a target can be reached within the distance less than 8 cm. The final manufactured antenna circuit is shown in the Fig. 4.

Fig. 4. Different shapes of the NFC antenna based on the flexible material FPC

One interesting feature of the NFC technology is the possibility of energy transfer through its antenna circuit. Strömmer et al. [18] have proposed a method for wireless charging of devices such as mobile phones using NFC technology. This feature is not used in the current project but the possibility of getting the advantages of this technology for transferring the energy through the swarm members exists.

4.2 Hardware Platform

In order to equip a device with the NFC technology an NFC transceiver is required. This type of chip is provided by a few manufacturers. In this project a transceiver from NXP company is used (PN532) which has all required physical and low level software implemented in the chip. This feature simplifies the complex process of implementing the NFC protocols. The users need only to design a matching circuit for the antenna. The final NFC module designed for this project is shown in the Fig. 5.

Fig. 5. The NFC drive circuit used for equipping the swarm robots with the NFC technology

The user has several options to communicate with the transceiver chip such as SPI, I^2C or High Speed Uart (HSU). In order to have a modular robot with the possibility of using the NFC system for other devices all required components for driving the NFC chip is implemented on a separate PCB. The required parts are: the transceiver chip, antenna matching and tuning circuit and finally a microcontroller for simplifying the required software tasks and communicating with the host processor.

The robot platform that is used for evaluating the functionality and the performances of the designed module is the AMiREsot robot platform presented in [20]. The AMiREsot platform has been designed to be used for experiments with minirobots following the idea of the AMiRE symposium [21]. This platform is a modular robot system that provides a microcontroller (STM32F103), a powerful processor (TI OMAP, ARM Cortex A8 as part of the Gumstix Overo Fire board) running Linux OS, an FPGA (Xilinx Spartan-6 XC6SLX75) mainly used for image pre-processing, IR sensors for environmental perceptions, all required mechanical elements for driving the robots and a Bluetooth module for long range communication. The final platform equipped with the NFC module is shown in the Fig. 6.

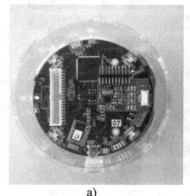
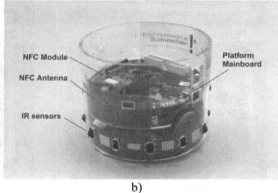

a)   b)

Fig. 6. NFC module installed on the AMiREsot platform (a) top view, (b) front view

5   Results and Discussions

In order to demonstrate the performance and efficiency of the NFC approach, a case study scenario has been defined in a way that three robots are used in swarm which scavenge for the food and distribute their findings using the NFC system. The food in this case is a Mifare NFC card which holds a piece of information containing the color code. This color code will be shown on the LEDs of the robots in case they find the source. At the beginning of the experiment, all the robots are programmed to show blue color and their task is to search for the source. The color code stored in the Mifare card is red. The first phase is the robots start searching the environment in order to find the source (Mifare Card) using Tag read mode (Fig. 7a). The second phase starts at the moment that the first robot finds the source. At this point, it shows the received color from the source on its LEDs and changes its task to a new one which is distributing the message to the others (Fig. 7b). The next phase is to detect another robot and transfer the message (Fig. 7c). A version control methodology is used to recognize which robot has the latest message. In the phase four, two robots exist which have the latest message; therefore three possibilities for the third robot exist namely robot A will find robot C, robot B will find robot C or finally the robot C will find the source on its own. In Fig. 7d the second possibility is shown.

Another experiment is performed to identify the area that an NFC antenna can discover another robot. In other words, it is necessary to know the effects of the angle and the distance between two robots communicating using their NFC system.

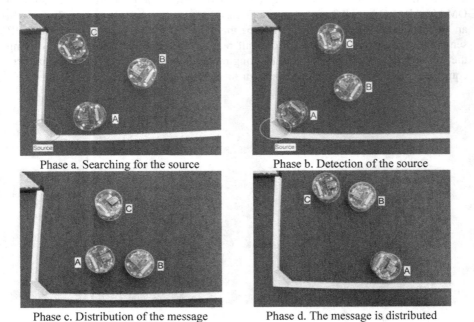

Phase a. Searching for the source   Phase b. Detection of the source

Phase c. Distribution of the message   Phase d. The message is distributed

Fig. 7. Implementation of the food scavenging scenario and data distribution using NFC

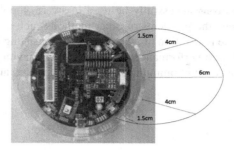

Fig. 8. The area that each NFC antenna can cover for detection of another NFC field

The result of this experiment is shown in the Fig. 8. As it can be seen in this figure an oval shape area is the range that a robot will be able to detect another NFC field.

6 Conclusion and Future Works

In this paper the NFC technology is introduced as a new method of communication between robot members of a swarm. The objective was to integrate a wireless communication system into robots that can form a swarm. For realizing behaviors in a swarm usually local interaction and communication is required. We have shown that a local communication based on the NFC system can meet these demands without causing problems like jamming, interference and packet loss as it usually occurs if large amount of communication nodes are used within a common area using techniques such as Bluetooth, Wi-Fi or ZigBee. The benefits of using such a system in swarm can be mentioned as follow:

- Low cost of the final system compared to the other options such as Bluetooth
- Very short time of initialization (<0.1 s) and automatic target recognition
- Simple structure of the antenna and possibility to have flexible antenna
- Simplified protocols and minimum software requirements for setup of the module
- Short range of communication which brings security and no interfering signals
- Possibility to setup a leader-less swarm structure with satisfactory efficiency
- Can be used in parallel with Bluetooth for solving pairing problem of Bluetooth
- Can be used for transferring energy in between swarm members

In spite of the above mentioned advantages some drawbacks still exist for NFC devices. These are namely inability to cover all areas around the robot; in case the robot's body is bigger than the covering range; and comparably large size of antenna which can be compromised sacrificing the maximum detection distance. Another option is to use two or three antennas including transceivers in parallel to cover the whole contour of a robot. In this case the hardware complexity is slightly increased but the robot will be able to communicate independently in different directions.

Considering the above mentioned advantages and disadvantages of NFC as well as the achieved results in the implemented scenario, it can be concluded that the NFC communication system can successfully be used in swarm topology and up to some extents even compensate the problems of other methods such as Bluetooth. The next task in future work is the realization of a larger swarm consisting of about 10 robots to measure the overall system performance, i.e. the swarm behavior as a global indicator but also number of successful communication events, failures and packet loss.

References

1. Brambilla, M., Ferrante, E., Birattari, M., Dorigo, M.: Swarm robotics: a review from the swarm engineering perspective. Swarm Intell. J. **7**(1), 1–41 (2013). Online ISSN: 1935-3820, Springer, New York
2. Kornienko, S., Kornienko, O., Levi, P.: Collective AI: context awareness via communication. In: IJCAI-05, Edinburgh, Scotland (2005)
3. Mohan, Y., Ponnambalam, S.G.: An extensive review of research in swarm robotics. In: World Congress on Nature & Biologically Inspired Computing, NaBIC (2009)
4. Li, M., Lu, K., Zhu, H., Chen, M., Mao, S., Prabhakaran, B.: Robot swarm communication networks: architectures, protocols, and applications. In: Proceedings of Third International Conference on Communications and Networking, China, pp. 162–166 (2008)
5. Kernbach, S., Häbe, D., Kernbach, O., Thenius, R., Radspieler, G., Kimura, T., Schmickl, T.: Adaptive collective decision-making in limited robot swarms without communication. Int. J. Robot. Res. **32**(1), 35–55 (2012)
6. Penders, J.: Robot swarming applications. In: Donkers, J. (ed.) Liber Amicorum ter gelegenheid van de 60e verjaardag van Prof. Dr. H. Jaap van den Herik. Maastricht, Netherlands, Maastricht ICT Competence Center, pp. 227–234 (2007)
7. Kornienko, S., Kornienko, O.: IR-based communication and perception in microrobotic swarms. In: 7th Workshop on Collective & Swarm Robotics, University of Stuttgart, Germany (2010)
8. Benavidez, P., Nagothu, K., Ray, A.K., Shaneyfelt, T., Kota, S., Behera, L., Jamshidi, M.: Multi-domain robotic swarm communication system. In: IEEE International Conference on System of Systems Engineering (2008)
9. Hawick, K.A., James, H.A., Story, J.E., Shepherd, R.G.: An architecture for swarm robots. Technical report DHPC121, School of Informatics, University of Wales, Bangor, North Wales, UK (2009)
10. Bhandari, S., Gautam, P., Chowdhury, M., Powers, M., Lin, S.: Implementation of RF communication with TDMA algorithm in swarm robots. In: International Conference on Technologies for Practical Robot Applications TePRA (2008)
11. Simms, B., Ng, E., Nguyen, D., Salomon, J., Woo, B.: Autonomous swarming robots. Interim report 423, Stevens Institute of Technology (2008)
12. Tuikka, T., Isomursu, M.: Touch the future with a smart touch. VTT Tiedotteita – Research notes 2492, pp. 4–7 (2009)
13. NFC-Forum. http://www.nfc-forum.org
14. NXP: PN532/C1 - Near Field Communication (NFC) controller product data sheet, 115434, Rev. 3.4. NXP (2011)
15. NXP Co: UM0701-02 PN532 User manual, Rev. 2, p. 192. NXP (2007)
16. Steffen, R., Preißinger, J., Schöllermann, T., Müller, A., Schnabel, I.: Near Field Communication (NFC) in an automotive environment. In: Proceedings of 2nd International Workshop on Near Field Communication, Monaco, France, pp. 15–20 (2010)

17. Haselsteiner, E., Breitfuß, K.: Security in near field communication (NFC), Philips Semiconductors, Printed Handout of Workshop on RFID Security (RFIDSec 06), July 2006
18. Strömmer, E., Jurvansuu, M., Tuikka, T., Ylisaukko-Oja, A., Rapakko, H., Vesterinen, J.: NFC-enabled wireless charging. In: 4th International Research Workshop on Near Field Communication, Published in: Proceedings of NFC 2012, Helsinki, pp. 36–41. IEEE (2012)
19. NXP: N1445 - Antenna circuit design for MFRC52x, PN51x, PN53x. NXP (2010)
20. Tetzlaff, T., Wagner, F., Witkowski, U.: Modular mobile robot platform for research and academic applications in embedded systems. In: Advances in Autonomous Robotics - Joint Proceedings of the 13th Annual TAROS Conference and the 15th Annual FIRA RoboWorld Congress, Bristol, U.K. (2012)
21. Tetzlaff, T., Witkowski, U.: Modular robot platform for teaching digital hardware engineering and for playing Robot Soccer in the AMiREsot league. In: Advances in Autonomous Mini Robots - Proceedings of the 6th International Symposium on Autonomous Mini-Robots for Research and Edutainment (AMIRE 2011), Bielefeld, Germany (2011)

Gesturing at Subswarms: Towards Direct Human Control of Robot Swarms

Gaëtan Podevijn[1]([✉]), Rehan O'Grady[1], Youssef S.G. Nashed[2],
and Marco Dorigo[1]

[1] IRIDIA, CoDE, Université Libre de Bruxelles, Brussels, Belgium
{gpodevij,rogrady,mdorigo}@ulb.ac.be
[2] Department of Information Engineering, University of Parma, Parma, Italy
nashed@ce.unipr.it

Abstract. The term *human-swarm interaction* (HSI) refers to the interaction between a human operator and a swarm of robots. In this paper,
we investigate HSI in the context of a resource allocation and guidance
scenario. We present a framework that enables direct communication
between human beings and real robot swarms, without relying on a secondary display. We provide the user with a gesture-based interface that
allows him to issue commands to the robots. In addition, we develop
algorithms that allow robots receiving the commands to display appropriate feedback to the user. We evaluate our framework both in simulation and with real-world experiments. We conduct a summative usability
study based on experiments in which participants must guide multiple
subswarms to different task locations.

1 Introduction

To date in the field of human-robot interaction, a great deal of effort has been
devoted to the study of the interaction between human beings and single agents
but little effort has been dedicated to human-swarm interaction (HSI) — the
interaction between human beings and robot swarms.

Swarm robotics systems are made up of a large number of relatively simple
and cheap robots that carry out complex tasks by interacting and cooperating
with each other. The distributed nature of such systems makes them robust (the
loss of an agent does not change the collective behavior), scalable (the same
control algorithms work with different swarm sizes) and flexible (the system
adapts to different types of environments). These characteristics make swarm
robotics systems potentially well suited for deployment in dynamic and a priori
unknown environments.

However, the large number of robots and the distributed nature of swarm
robotics systems also make them much harder to interact with. As the number of
robots increases, it becomes increasingly impractical for a human operator to give
instructions to or receive feedback from individual robots. Nor is it necessarily
easy to broadcast commands to the entire swarm. The distributed control used in

A. Natraj et al. (Eds.): TAROS 2013, LNAI 8069, pp. 390–403, 2014.
DOI: 10.1007/978-3-662-43645-5_41, © Springer-Verlag Berlin Heidelberg 2014

swarm robotics systems implies that each robot has a different frame of reference and is therefore liable to interpret the broadcast command differently.

In HSI literature, the interaction systems developed usually rely on a secondary display that provides a human operator with a real-time representation of both the environment and the robot swarms. In such approaches, therefore, the human operator does not interact with the real robots in their real environment, but with a modelled representation of both the robots and the environment. In order to create a modelling layer, it is necessary to collect telemetry data about the robots (i.e., their position and orientation) and data about the environment (i.e., size and obstacles). And to be useful for HSI purposes, such data must be collected and modelled in real-time. Simulated HSI approaches have used the omniscience afforded by robotic simulators to collect all of the relevant data. However, in the real-world, external tracking infrastructure would be required (e.g., GPS or external cameras). Such tracking infrastructure is often infeasible in the dynamic, a priori unknown environments for which swarm robotics systems are best suited.

In this paper, we present an approach to HSI that does not involve any modelling layer, and instead allows a human operator to interact directly with real-robots. We design, implement and validate our approach in the context of a resource allocation and guidance scenario. Our scenario involves a human operator selecting particular groups of robots from the swarm (henceforth referred to as *subswarms*) and then guiding them to specific locations in their environment. The key philosophy underlying our approach is that a human operator should be able to interact with a subswarm as if it were a single entity, issuing a single command to and receiving coherent feedback from the subswarm "entity". The challenge is to enable group-level responses in robot swarms that have fully distributed control.

We present a gesture-based interface that allows the operator to interact with a swarm. With a gesture-based interface, the operator can devote his full attention to the robots. In contrast, with a secondary display, the operator must divide his attention between both the robots and the display. In Fig. 1(a–c), we show a human operator using our gesture-based interface in order to interact with a swarm of 8 real robots. The robotic platform used in this study is the wheeled *foot-bot* robot (see Fig. 1(d)) [7]. We conduct experiments using both simulated and real foot-bots. The simulations of the foot-bots are provided by the swarm robotics simulator ARGoS [8].

We demonstrate our approach with proof-of-concept experiments on real robots. We also perform an analysis of the usefulness of our gesture-based interface through simulation based experiments with human operators. To evaluate the experiments, we use a summative approach that allows us to quantify the overall usability of the system.

1.1   Related Work

The most common approaches used in the current studies of HSI tend to rely on an intermediate modelling layer. In these studies, an abstract representation

(a) *Left:* The operator faces a group of 8 robots. *Right* The operator gestures initiate the interaction with the group (the *select* command).

(b) *Left:* The operator splits the group into two subgroups (the *split* command). *Right:* The operator now has two groups he can interact with independently.

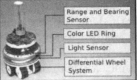

(c) *Left:* The operator holds a virtual steering wheel in the air to guide the robots. *Right:* To make the robots go right, the operator turns his virtual steering wheel to the right (the *steer* command).

(d) *Left:* Real foot-bot used in our real-world experiments. *Right:* Simulated foot-bot in ARGoS.

Fig. 1. Gesture-based interaction with a swarm of robots.

of both the robots and the environment is shown to the operator in a graphical user interface (GUI). McLurkin *et al.* [1] propose a centralized GUI based on real-time strategy video games where the user controls an army of hundreds of individuals. In addition to displaying modelled robots in a modelled environment, their GUI provides the user with extra debugging information (e.g., waypoints for individual robots, global positioning). The authors note that it can be difficult to display such a large amount of data while ensuring that the user still has a clear understanding of what is going on in the swarm. Kolling *et al.* [2] propose an approach based on so-called *selection* and *beacon* control. To select robots, the user draws a rectangular zone in a GUI. Robots inside this zone are considered selected. Once a subset of robots is selected, the user can send the selected robots different commands. With beacon control, the user can exert an indirect influence on the robots by adding virtual beacons in the GUI. Bashyal *et al.* [3] propose a GUI in which the operator takes the control of a single robot (an *avatar*) in the swarm. Because the avatar is perceived by the other robots as just another robot of the swam, the operator has the same limited influence as any single robot in the swarm. Bruemmer *et al.* [4] present a hierarchical communication architecture. They developed a GUI that allows the operator to send orders to a specific robot called "the sergeant".

Instead of GUIs, some modelling-layer based approaches propose an augmented reality view embedded in a dedicated head-mounted display. In Daily *et al.* [5], users wear an optical see-through head-worn device which receives simple robots' messages. When the device receives these messages, it analyzes them and augments the environment with a visual representation of these messages. A similar system is used in [6] where firefighters are helped in their mission by a robot swarm. The firefighters' helmets are augmented by a visual device, giving them direction information.

To the best of our knowledge, there is only one existing HSI approach that does not rely on an extra modelling layer: Giusti *et al.*'s [9] hand gesture-based interaction system. The goal of their work is to allow robots to decode hand gestures. Their method requires the robots to be placed in a particular spatial arrangement. In real-world situations, this requirement is not practical. Our research has a different focus. In our work, gestures are decoded into their corresponding commands by a central unit, which broadcasts the commands to the robots. The focus of our work is not on the gestures themselves, but on how single commands can be interpreted by decentralised robot swarms, and how decentralised swarms can provide composite feedback.

Usability studies are largely absent from the existing body of research in HSI (with the exception of [2,3]). We believe that this is a major omission, and that both objective and subjective usability results should be an essential part of any HSI research. In the research presented in this paper, we establish objective usability results via time-on-task based statistics and subjective usability results via a usability questionnaire.

2   Resource Allocation and Guidance Scenario

We base our work in this paper around a resource allocation and guidance scenario. In this scenario, a human operator moves selected robot groups of different sizes to specific locations in the environment. These different locations represent sites at which the robots would be required to carry out tasks. In this paper, robots do not actually carry out tasks – we represent task execution by pausing a group of robots for the amount of time that it takes them to "carry out" their hypothetical task. We do, however, let the human operator modify the size of the groups he selects and then guides, corresponding to the scale of different tasks the robot groups are required to "carry out".

This scenario allows us to exemplify, in the context of human-swarm interaction, the two major challenges inherent to any system that deals with bi-directional interaction. Firstly, an operator must be able to give commands to the system concerned. Secondly, the system must be able to provide the operator with appropriate feedback.

Giving commands to robots is challenging in HSI because each robot has its own reference frame. These different reference frames can lead to an operator's commands being interpreted differently by different robots. In the context of our guidance scenario, a command such as "turn left" would be meaningless as robots would interpret this command with respect to their individual reference frame (see Fig. 2). Meanwhile, understanding feedback provided by the robots in a robotic swarm is challenging because it is difficult to avoid overwhelming

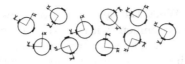

Fig. 2. Each robot in a swarm robotics system has its own local frame of reference. A command like *turn left* would not mean anything for the whole group since each robot would interpret the command differently.

the operator with a flood of data. If each robot provides individual feedback, the operator has too many data-points to process.

These issues could be solved if the operator were able to interact with groups of robots, rather than with individual robots. However, the distributed nature of control in swarm robotics systems makes such group-level interaction difficult to achieve. The challenge is to write distributed control code for a group of robots that lets each robot in the group interpret a single command meaningfully. In addition, this distributed control must provide the operator with group-level feedback, whereby the group of robots together provide a single data-point of feedback for the operator to process.

3   Our Approach: Interacting with Subswarms

We define a subswarm as a distinct group of robots within a swarm, that are identifiable both to a human operator, and to themselves. For a subswarm to be meaningful in the context of human-swarm interaction, a human operator must be able to visually distinguish a subswarm from other nearby robots. And robots in a subswarm must know that they belong to that particular subswarm, while robots outside the subswarm must know that they do not belong to that subswarm.

The first step towards implementing *subswarm-based interaction* is to define technically what the notion of subswarm means to both a human and a robot. In our subswarm-based interaction approach, a human perceives a subswarm as a set of robots that are close to each other and that are lit up in the same color. At the level of robot control code, subswarms are defined using *subswarm identifiers*. Every robot belonging to a subswarm has the same integer identifier and every subswarm identifier is unique.

We implement our subswarm-based interaction approach by first defining what commands and feedback make sense to a human operator in the context of our scenario. We then develop the distributed control code that allows groups of robots to process those commands and provide appropriate feedback.

3.1   Commands Available to Human Operator

In this section we present the list of commands that allow the operator to carry out our resource allocation and guidance scenario.

Steer The steer command is issued to guide the selected subswarm in the environment. When the selected subswarm receives the command, it starts moving straight. Subsequently, the operator can turn the subswarm left or right. In our gesture-based interface (see Sect. 4), the operator moves his hands just as he would turn the steering wheel of a car to change the subswarm's direction (see Fig. 1(c)).

Stop The stop command is issued to bring the selected subswarm to a halt.

Split The split command is issued to create new subswarms. When the selected subswarm receives the command, it splits into two independent subswarms of approximately the same size.

Merge The merge command is issued to reassemble two subswarms. When two selected subswarms receive the command, they move towards each other and unify into a single subswarm. The user can arrive at groups of required sizes by repeatedly splitting and merging subswarms.

Select The select command is issued by the operator in order to choose which subswarm to interact with. Once the select command is sent, one subswarm at random gets selected. All robots of the selected subswarm illuminate their yellow LEDs. If the selected subswarm is not the one the operator wants to interact with, he re-issues the command, and another subswarm becomes selected. He continues to issue the select command until the desired subswarm is highlighted. Note that before issuing the merge command, the operator must select two subswarms to merge. The gesture-based interface allows the operator to select a second subswarm. Robots belonging to the second selected subswarm illuminate their LEDs in red.

3.2   Distributed Robotic Control

In this section, we present the distributed behavioural control algorithms that implement the above commands. As is standard in swarm robotics systems, the same control code is run independently on each of the robots. The gesture-based interface broadcasts commands to the robots. Once the robots receive a command, they execute it only if they belong to the currently selected subswarm (see the Select algorithm below to know how a robot can determine if it belongs to the selected subswarm).

Our implementation assumes the existence of a fixed light source that defines a common point of reference in the environment. It also assumes the existence of a means for the robots to exchange short messages, calculate the distance and bearing between themselves and sense their own orientation. The short messages the robots have to exchange are their subswarm identifier. It is important for the subsequent algorithms that the robots all know in which subswarm they belong to. On the foot-bot platform, the range and bearing (R&B) module is used by the robots to communicate their subswarm ID and to know the position and bearing of their neighbors. The light sensor is used by a robot to measure its orientation with respect to a fixed light source.

For a subswarm to be clearly identifiable to a human operator, it is important that the robots of a subswarm remain close to each other and do not disperse.

To give subswarms this cohesive quality, we use a mechanism known as *flocking*. This flocking mechanism is implemented by having each robot use the distance and bearing information given by the R&B. The distance information allows each robot to adjust its position by placing itself at a constant distance from its neighbors. The bearing information allows each robot to adjust its orientation according to the average orientation of its neighbors. We based this flocking mechanism on [10], where robots are considered as particles that can exert virtual attractive and repulsive forces on one another. These forces are said virtual because the robots calculate them. At each time unit, each robot calculates a vector $\mathbf{f} = \mathbf{p} + \mathbf{h}$, which incorporates position information (encapsulated in vector $\mathbf{p}$) and orientation information (encapsulated in vector $\mathbf{h}$) of its neighbors. This vector $\mathbf{f}$ must then be converted into wheel actuation values.

In order for the robots to convert their vector $\mathbf{f}$ into wheel actuation values, each of them calculates its forward speed u and its angular speed ω. Robots set their forward speed to a constant value U, and their angular speed to a value proportional to the angle of vector $\mathbf{f}$ ($\theta = \angle \mathbf{f}$):

$$\omega = K\theta, \tag{1}$$

where K is a proportionality constant. Finally, robots convert their forward and angular speed into linear speed of their left (v_l) and right (v_r) wheel to $v_l = (u + \frac{\omega}{2}d)$ and to $v_r = (u - \frac{\omega}{2}d)$, where d is the distance between the wheels. The resulting behaviour of the robots is to place themselves at a constant distance of each other, to take the same orientation, then to move coherently in the same direction.

Steer. When the steer command is issued by the operator, robots of the current selected subswarm all compute vector $\mathbf{f}$. Then, they transform this vector into the relevant wheel actuation values. As a result, the robots start moving in the same (initially random) direction. When the operator decides to change the selected subswarm direction, each robot of the selected subswarm receives an angle of turn β from the gesture-based interface, corresponding to the angle at which the human operator has made his steer gesture. To turn β radians, each robot computes Eq. 1 by replacing the angle of vector $\mathbf{f}$, θ, by the angle of turn β.

Stop. When the stop command is issued, the robots of a selected subswarm stop moving by setting their linear wheel speeds to zero ($v_l = v_r = 0$).

Split. When the split command is issued to the robots of a selected subswarm, robots from this subswarm choose a new subswarm ID A or B with probability 0.5. Immediately after the robots chose their new subswarm ID, no robots from subswarm A and B have moved yet — they are still not spatially separated into two clearly distinct subswarms. To separate the newly distinct subswarms, we modify the cohesion behaviour of the constituent robots, so that robots with the same subswarm ID are attracted to each other, while robots with different subswarm IDs repel each other.

Merge. To reassemble two subswarms A and B, we assume that each robot r_a of subswarm A is able to compute its average distance to subswarm B. Every robot r_a calculates this average distance by averaging its distance (given by the R&B) to every robot r_b. Robots from subswarm B perform the same calculation with respect to the robots of subswarm A. Robots of the two subswarms then calculate the number t of time units necessary to travel half the average distance (assuming constant velocity) between the two subswarms. After moving for t time units in the direction of the subswarm they merge with, the two subswarms consider themselves joined. Robots from subswarms A and B then all adopt whichever of the two existing identifiers of A and B is smaller.

Select. In order for the robots to know if they belong to the subswarm that is currently selected by the operator, each robot of each subswarm maintains a variable `subswarm_selected` that contains either the selected subswarm ID or a sentinel value ϵ (if no subswarm is selected). Our distributed algorithms ensure that at any given moment, every robot across all the different subswarms has the `subswarm_selected` variable set to the same value. By comparing their own subswarm ID to the variable `subswarm_selected`, robots know if they belong to the selected subswarm.

Each time the select command is issued by the operator, every robot of every subswarm updates its `subswarm_selected` variable. To update the variable, every robot maintains a list of all subswarm IDs in the swarm. The update rules for this variable change based on context. There are three possible situations. In the first situation, the select command is issued while no subswarm is selected (`subswarm_selected` $= \epsilon$). In this case, the variable takes the lowest subswarm ID in the ID list. In the second situation, one subswarm is already selected. The selection must move to another subswarm. The variable is updated by taking the lowest ID in the list greater than `subswarm_selected`. In the third situation, every subswarm already has been selected once (there are no subswarm IDs greater than `subswarm_selected`). The variable is set to ϵ and no subswarm is selected anymore. Note that in case of a merge, the user must select two subswarms. The algorithm of the second selection command available to the user works as explained above with a minor modification: if the subswarm ID that is supposed to be selected (a second variable containing the second selected subswarm is maintained by the robots) is already selected with the first select command, then this ID is skipped and the next ID in the ID list is taken.

4   Gesture-Based Interface

Our interface allows a human operator to give commands to the robots by performing gestures. We use the Kinect system from Microsoft using the OpenNI library.[1] In our gesture-based interface, each command is associated with a specific gesture (see Fig. 3). When the interface recognizes[2] a gesture, it sends the corresponding command to every robot via a client-server mechanism.

---

[1] http://www.openni.org

[2] As it goes beyond the scope of this paper, we do not discuss the gesture recognition algorithm.

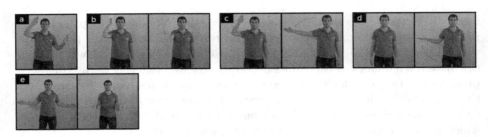

Fig. 3. Gestures associated with the commands (a) Steer (b) Split (c) Select (right arm) – The operator uses his left hand to select a second subswarm. (d) Stop (e) Merge

We discovered that the recognition was more accurate if we recorded gestures separately for each user, since different body shapes of the different users (e.g. short or tall) reduced the efficiency of our recognition algorithms. We recorded user gestures with a dedicated tool that we developed. Further work will focus on removing this constraint in order to recognize gestures with different types of body shapes without having the users to record the gestures [11].

5   Usability Study

The objective of our usability study is twofold. Firstly, we want to test if our participants can understand the concept of issuing commands to a robot swarm. Secondly, we want to study if they are able to carry out the test scenario with our gesture-based interface.

5.1   Experimental Test Scenario

We designed a specific instance of our resource allocation and guidance scenario, that would allow us to measure the performance of our interface. In this scenario, the participant has to use a swarm of 30 robots to carry out three tasks. The participant has to create three separate subswarms of robots by splitting the swarm, then guide each subswarm to one of the three task locations. Afterwards, the participant has to re-merge all of the robots back into a single swarm.

The simulated environment used in our scenario is depicted in Fig. 4. In Fig. 4 (Left), we show an initial swarm of 30 robots in the environment at the beginning of the experiment. In Fig. 4 (Right), the participant has split the swarm into three subswarms and has moved two of these subswarms to task locations.

Each participant had to perform the experiment with two interfaces: our gesture-based interface and a graphical user interface (GUI). We developed this simple button-based GUI as an alternative way for human operators to issue commands to the swarm. The GUI functionality is similar to that of our gesture-based interface, with one GUI button corresponding to each recognized gesture.

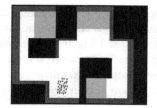 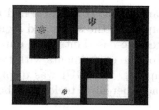

Fig. 4. (Left) A swarm waiting for orders. (Right) The initial swarm has been split into three subswarms. Two of them have been moved to task locations (the green/light gray areas) (Color figure online).

The one exception is the *steer* command. In the gesture-based interface, the operator can change the subswarm's direction while the subswarm is moving (the interface provides the operator with a steering wheel mechanism). In the GUI however, the operator cannot turn the subswarm while it is moving. The operator must first stop the subswarm moving, rotate it (right or left), then move the subswarm straight.[3]

5.2   Experimental Setup

We recruited 18 participants for this study: 6 of them were PhD students in robotics, 6 were Master students in Computer Science and 6 were non-technical people recruited from the entourage of the authors. Our participants were between 23 and 33 years old with an average age of 27.7 years ($SD = 2.9$). We started each experiment with a five minute presentation. The purpose of this presentation was to explain the resource allocation and guidance scenario and to present the simulated environment (see Fig. 4), the available commands and the two interfaces (the gesture-based interface and the GUI).

By having each participant perform experiments with both types of interfaces (the gesture-based interface and the GUI), we introduce the risk of a carryover effect. That is, the order in which the interfaces are tested by a participant might affect that participant's results. We prevent any possible carryover effect by alternating which interface the participants encountered first. We randomly divided our participants into two groups. The first group started the experiment with the gesture-based interface and finished with the GUI while the second group started with the GUI and finished with the gesture-based interface. Immediately before starting the experiment with the gesture-based interface, the participants underwent a brief preparation session in which they recorded the gestures and performed them several times in order to memorize them.

During the experiments with both the gesture-based interface and the GUI, the simulated environment was displayed on a large projection screen. By using a projection screen, we tried to approximate as closely as possible the experience

---

[3] Ideally, the GUI should also have had steering functionality. However, GUI design was not the purpose of this research, and for time constraints we kept the GUI as simple as possible and restricted to button-based functionality.

that a participant would have had with real robots — e.g. percentage of visual field occupied by the arena and by individual robots. For the experiment involving the gesture-based interface, our participants stood 1.5 m from the projection screen. For the experiments with the GUI, our participants were seated in front of a computer. The GUI was displayed on the computer monitor. The participants had, therefore, to look at both the projection screen and the computer monitor.

5.3 Usability Metrics

We evaluated the usability of both the gesture-based interface and the GUI with two kinds of metrics. The first metric is an objective metric: a time-on-task based statistic. We kept track of the amount of time taken by each participant to carry out their task. The counter was launched as soon as the first subswarm was selected and stopped exactly when the two last subswarms finished merging. The second metric is a subjective metric that allows us to measure the participants' evaluation of both the gesture-based interface and the GUI. After the experiment, each participant was asked to fill out two System Usability Scale questionnaires (SUS) [12] (one for each of the two interfaces). SUS is a reliable (i.e., consistent in its measure) and validated (i.e., it measures what it intends to measure) questionnaire with a scale of five options ranging from *strongly agree* to *strongly disagree*. The resulting score takes into account the efficiency (i.e., amount of effort needed to finish the task), effectiveness (i.e., ability to successfully complete the task) and satisfaction of participants. The score is a number that varies from 0 (low usability) to 100 (high usability) giving a global evaluation of the interface's usability.

5.4 Experimental Results

In Table 1 we present the time-on-task (ToT) results of both the gesture-based interface and the GUI. Results show that on average, our participants were slightly slower (+10.4 %) with the gesture-based interface. In Table 2 we present results regarding participants' subjective evaluation of the interfaces' usability. We can see that our participants evaluated the gesture-based interface with a mean score of 75.8 while they evaluated the GUI with a mean score of 78.5 (+3.4 %).

Results reveal that our participants managed to use both the gesture-based interface and the GUI effectively. Although our participants seemed to achieve marginally better results with the simpler GUI, we were quite satisfied by the overall usability of our gesture-based interface, especially given that all of our

Table 1. ToT statistics in minutes

$N = 18$	Mean	SD	C.I. (95 %)
Gesture-based	15.4	3.4	(13.7, 17.3)
GUI	13.8	3.3	(12.2, 15.5)

Table 2. SUS questionnaire results

$N = 18$	Mean	SD	C.I. (95 %)
Gesture-based	75.8	13.7	(69, 82.7)
GUI	78.5	12.4	(72.2, 84.7)

participants had prior experience in using GUI-based systems, and few had prior experience in gesture control systems.[4] We believe that the minor ToT superiority of the GUI is anyway counterbalanced by the numerous advantages of a gesture-based interface. With the emergence of wireless video cameras such as the Mobile Kinect Project[5], video camera deployment will get easier and easier. Furthermore, a single camera can be used to recognize the gestures of multiple users. On the other hand, a GUI requires that each user has his own device (e.g., personal computer or tablet). Moreover, with a gesture-based interface the operator can keep his attention wholly focused on the robots and their task, while with a GUI, the operator must concentrate on both the robots and the device displaying the GUI.

6   Real Robot Validation

We validated our approach with proof-of-concept experiments on real robots. We conducted experiments with groups of up to 8 real foot-bots. We do not report quantitative data as we did not run any usability experiments on the real robots. However, with only minor modifications to our distributed algorithms (e.g., parameters in the flocking algorithm), robots were able to receive and perform all the commands issued by the operator. In Fig. 1, we show an operator selecting a subswarm, sending the *split* command to the selected subswarm and then guiding a selected subswarm with the *steer* command.

7   Conclusions and Future Work

In this paper, we presented a gesture-based human-swarm interaction framework. We designed a resource allocation and guidance scenario in which a human operator is asked to move different robot subswarms to different locations in a physically detailed simulation environment with the help of five commands (steer, split, merge, stop and select). Instead of interacting with a representation of the robots, the operator interacts directly with the robots. We conducted a summative usability study and we collected both objective and subjective results. The results show that our participants (i) successfully managed to interact with a swarm of robots and (ii) were satisfied with using the gesture-based interface to carry out our scenario. Finally, we ran experiments on real robots in order to validate the technical feasibility of our approach in the real world.

---

[4] We did not succeed in establishing statistically significant difference between the gesture-based interface and the GUI ($p_{ToT} = 0.07$, $p_{SUS} = 0.55$). However, as the goal of this paper was not to design a GUI, such precise comparison would be fairly meaningless.

[5] http://big.cs.bris.ac.uk/projects/mobile-kinect

Future research will focus on improving feedback provided by the subswarms. Our goal is to leverage the same self-organised mechanisms that govern the robots' behaviour to generate the feedback to send to a human operator [13].

Acknowledgements. This work was partially supported by the European Research Council through the ERC Advanced Grant "E-SWARM: Engineering Swarm Intelligence Systems" (contract 246939). Rehan O'Grady and Marco Dorigo acknowledge support from the Belgian F.R.S.-FNRS.

References

1. McLurkin, J., Smith, J., Frankel, J., Sotkowitz, D., Blau, D., Schmidt, B.: Speaking swarmish: human-robot interface design for large swarms of autonomous mobile robots. In: Proceedings of the AAAI Spring Symposium, pp. 72–75. AAAI Press, Menlo Park (2006)
2. Kolling, A., Nunnally, S., Lewis, L.: Towards human control of robot swarms. In: Proceedings of the 7th Annual International Conference on H, pp. 89–96. ACM, New York (2012)
3. Bashyal, S., Venayagamoorthy, G.K.: Human swarm interaction for radiation source search and localization. In: Proceedings of Swarm Intelligence Symposium, pp. 1–8. IEEE Press (2008)
4. Bruemmer, D.J., Dudenhoeffer, D.D., Marble, J.L.: Mixed-initiative remote characterization using a distributed team of small robots. In: AAAI Mobile Robot Workshop. AAAI Press, Menlo Park (2001)
5. Daily, M., Cho, Y., Martin, K., Payton, D.: World embedded interfaces for human-robot interaction. In: Proceedings of the 36th Annual Hawaii International Conference on System Sciences, Big Island, pp. 125–130. IEEE Computer Society (2003)
6. Naghsh, A., Gancet, J., Tanoto, A., Roast, C.: Analysis and design of human-robot swarm interaction in firefighting. In: Proceedings of the 17th IEEE International Symposium on Robot and Human Interactive Communication, pp. 255–260. IEEE Press (2008)
7. Dorigo, M., Floreano, D., Gambardella, L.M., Mondada, F., Nolfi, S., Baaboura, T., Birattari, M., Bonani, M., Brambilla, M., Brutschy, A., Burnier, D., Campo, A., Christensen, A.L., Decugnière, A., Di Caro, G., Ducatelle, F., Ferrante, E., Förster, A., Martinez Gonzalez, J., Guzzi, J., Longchamp, V., Magnenat, S., Mathews, N., Montes de Oca, M., O'Grady, R., Pinciroli, C., Pini, G., Rétornaz, P., Roberts, J., Sperati, V., Stirling, T., Stranieri, A., Stützle, T., Trianni, V., Tuci, E., Turgut, A.E., Vaussard, F.: Swarmanoid: a novel concept for the study of heterogeneous robotic swarms. IEEE Rob. Autom. Mag. **20**(4), 60–71 (2013)
8. Pinciroli, C., Trianni, V., O'Grady, R., Pini, G., Brutschy, A., Brambilla, M., Mathews, N., Ferrante, E., Di Caro, G., Ducatelle, F., Gambardella, L.M., Birattari, M., Dorigo, M.: ARGoS: a modular, parallel, multi-engine simulator for multi-robot systems. Swarm Intell. **6**(4), 271–295 (2012)
9. Giusti, A., Nagi, J., Gambardella, L., Bonardi, S., Di Caro, G.A.: Human-swarm interaction through distributed cooperative gesture recognition. In: Proceedings of the 7th International Conference on HRI, pp. 401–402. ACM, New York (2012)
10. Ferrante, E., Turgut, A.E., Huepe, C., Stranieri, A., Pinciroli, C., Dorigo, M.: Self-organized flocking with a mobile robot swarm: a novel motion control method. Adapt. Behav. **20**(6), 460–477 (2012)

11. Nashed, Y.S.G.: GPU hierarchical quilted self organizing maps for multimedia understanding. In: Proceedings of International Symposium on Multimedia, pp. 491–492. IEEE (2012)
12. Brooke, J.: SUS: a 'quick and dirty' usability scale. In: Jordan, P.W., Thomas, B., Weerdmeester, B.A., McClelland, A.L. (eds.) Usability Evaluation in Industry, pp. 189–194. Taylor & Francis, London (1996)
13. Podevijn, G., O'Grady, R., Dorigo, M.: Self-organised Feedback in human swarm interaction. In: Workshop on Robot Feedback in Human-Robot Interaction: How to Make a Robot Readable for a Human Interaction Partner (RO-MAN'12), France, Paris (2012)

Profiling Underwater Swarm Robotic Shoaling Performance Using Simulation

Mark Read[1]([✉]), Christoph Möslinger[2], Tobias Dipper[3], Daniela Kengyel[2],
James Hilder[1], Ronald Thenius[2], Andy Tyrrell[1], Jon Timmis[1],
and Thomas Schmickl[2]

[1] Department of Electronics, University of York, York, UK
mark.read@york.ac.uk
[2] Artificial Life Laboratory, University of Graz, Graz, Austria
[3] Institut für Parallele und Verteilte Systeme, Universität Stuttgart,
Stuttgart, Germany

Abstract. Underwater exploration is important for mapping out the oceans, environmental monitoring, and search and rescue, yet water represents one of the most challenging of operational environments. The CoCoRo project proposes to address these challenges using cognitive swarm intelligent systems. We present here CoCoRoSim, an underwater swarm robotics simulation used in designing underwater swarm robotic systems. Collective coordination of robots represents principle challenge here, and use simulation in evaluating shoaling algorithm performance given the communication, localization and orientation challenges of underwater environments. We find communication to be essential for well-coordinated shoals, and provided communication is possible, inexact localization does not significantly impact performance. As a proof of concept simulation is employed in evaluating shoaling performance in turbulent waters.

1 Introduction

The ocean remains the least explored habitat on earth, hosting undiscovered organisms and resources of interest and value. The importance of addressing underwater search and environmental monitoring is exemplified by events such as the 2010 BP Deepwater Horizon oil spill and the 2009 Air France Flight 447 crash in the Atlantic ocean, where it took nearly 2 years to recover the black boxes from the ocean floor. Water is an extremely challenging environment to operate within, visibility is poor, and electromagnetic signals are heavily attenuated, complicating communication and GPS-based localization. The CoCoRo project[1] seeks to advance underwater exploration capability through use of swarm intelligent systems endowed with collective cognitive decision making abilities [11]. Collective cognition is intended to assist swarms in coping with a noisy and heterogeneous environment, identifying and discriminating between

---

[1] The EC funded CoCoRo Project, GA 270382; http://cocoro.uni-graz.at/.

A. Natraj et al. (Eds.): TAROS 2013, LNAI 8069, pp. 404–416, 2014.
DOI: 10.1007/978-3-662-43645-5_42, © Springer-Verlag Berlin Heidelberg 2014

multiple underwater targets, dynamically reallocating robots between tasks to meet requirements, compensating for failed or lost swarm members, and maintaining a communication network of robots between an exploratory swarm and the water surface.

Engineering swarm robotic systems is an inherently challenging field: the robotic platforms are complex, as are the environments in which they are deployed, and group behaviours must be engineered through the manipulation of interactions between individuals. As such computational simulation is frequently employed to aid in research, development and evaluation; the joint SYMBRION-REPLICATOR projects have developed a sophisticated 3D robotics simulation, Symbricator3D [13]. Symbricator3D employs highly realistic sensor and actuator models, and robotic controllers developed on the simulation should migrate directly onto real platforms. However it is a highly complex piece of software, and its documentation is lacking in comparison to other simulations reviewed here. The Jasmine swarm robotic platform is accompanied by a simulation of the platforms[2]; this software simulates robots on a 2D plane. Simbad is a recently developed 3D robotics simulation written in Java with a slant towards evolutionary robotics research [6]. Documentation is of a high quality, with a javadoc API published online along with details of the simulation's architecture and some tutorials. The Stage simulation, developed under the Player/Stage project, has been widely adopted for swarm robotic research and educational purposes [4,5]. Stage is however restricted to simulating 2D environments, and in addressing 3D environments the Gazebo simulation was developed [8]. Gazebo encompasses a realistic physics engine, where simulated bodies have properties such as mass, friction and bounce factors. However, due to its computational load Gazebo can simulate at most 10 robots, limiting its application in swarm robotic systems.

In this paper we present the CoCoRoSim underwater robotics simulation, developed by the CoCoRo project to facilitate controller conceptualization, development and evaluation. CoCoRoSim is implemented in NetLogo 3D, and as such can simulate large numbers of robots. Its Newtonian mechanics physics engine simulates water drag forces, buoyancy, translational and rotational motion, a variety of robotic sensor systems, and forces exerted on simulated bodies by water currents. We present CoCoRoSim in Sect. 2, and its calibration against a CoCoRo robotic platform in Sect. 4. CoCoRoSim's use in adapting a generic shoaling algorithm to the constraints of underwater environments and evaluating its performance is detailed in Sect. 5. This section also explores how CoCoRoSim can be used to evaluate controller performance in the presence of water currents and turbulence. Lastly, Sect. 6 concludes this paper.

2   The CoCoRoSim Simulation

The CoCoRoSim simulation is implemented in NetLogo 3D[3], a three dimensional multi-agent modelling environment. The role of CoCoRoSim in the CoCoRo

---

[2] http://www.swarmrobot.org/Simulation.html
[3] http://ccl.northwestern.edu/netlogo/

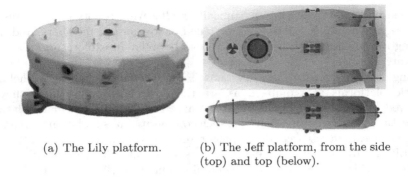

(a) The Lily platform.   (b) The Jeff platform, from the side (top) and top (below).

Fig. 1. Two underwater robotic platforms developed in the CoCoRo project.

project is as a fast means of prototyping algorithms and AUV deployment scenarios; the CoCoRo project's ethos is to migrate ideas to real-world robots as quickly as possible following their validation in simulation. As such, NetLogo 3D was selected as an implementation platform: it provides a powerful integrated 3D GUI interface, is well supported and documented, and using behaviorspace can be executed on a computation cluster to facilitate large-scale experimentation. The language is well suited for non-programmers on the project to pick up quickly with minimal effort, and its features have aided in minimising development time, hence allowing a greater focus on algorithmic development and real-world deployment. CoCoRoSim can be downloaded from http://cocoro. uni-graz.at/drupal/media under 'Software'.

2.1 Sensors, Actuators, Operating System and Robots

CoCoRoSim provides representations of the sensors and actuators developed and used within CoCoRo, and allows these to be configured and calibrated to reflect the autonomous underwater vehicles (AUVs) developed in the project (Fig. 1).

Lily is a modified toy submarine, used to facilitate fast hardware and controller prototyping. It has a relatively short operational life-span. In contrast, Jeff is developed and manufactured over a longer term; it is more robust, maneuverable and has a greater number of more powerful sensor and actuator systems. Both Lily and Jeff platforms are equipped with blue light sensor systems (BLSS), each of which comprises multiple LEDs and a receiver. Lily has 5 BLSS systems whereas Jeff has 6. BLSS receivers observe blue light within a cone of 120°. They can provide communication through the pulsing of LEDs, can be used for distance sensing, and are capable of detecting obstacles by observing LED reflections. BLSS are implemented in CoCoRoSim by registering the nearest AUVs within a cone of observation, centred according to the BLSS configurations on the AUVs. Simulated BLSS detect proximity to obstacles such as tank walls and floor by tracing rays projected along the centre and limits of the cone of observation. This implementation represents a compromise between computational efficiency and accuracy. BLSS communication is implemented through direct message passing to the nearest AUV within a sensor's cone.

CoCoRo AUVs are equipped with radio transmitters and receivers to provide omni-directional communication. These are implemented by direct message passing between AUVs of sufficient proximities. Pressure sensor, compass, accelerometer and gyroscope implementations are also provided.

Lily and Jeff platforms manoeuvre with propellor-based thrusters. These are simulated as simply providing a force of movement on the AUV along the axis in which they are oriented. Vertical manoeuvrability is provided by a buoyancy pump which changes the volume of the AUV. In CoCoRoSim such actuators are implemented as providing a vertical (in relation to the environment) force on the AUV based on its density in relation to water.

Controller interaction with CoCoRo AUV platforms is provided through an operating system based on FreeRTOS [1]. This OS's functionality is reflected in CoCoRoSim, with semantically similar functions being provided to ease migration of simulation algorithms to real-world platforms. The OS provides functions that allow, for example, controllers to specify a particular depth or heading and speed, and through the use of simple PD controllers the platform's actuators are manipulated to this end. The PD controller for heading adjusts the differential in thruster settings around a specified percent of backwards or forwards maximum thrust to maintain some specified forward speed and a particular heading based on compass readings.

3   Physics Engine

The CoCoRoSim physics engine discretizes time; the states of all simulated bodies are updated in each time step. The physics engine implements Newtonian mechanics with simulated bodies modelled as particles. The particles' translational and rotational velocities and accelerations in the x, y and z axes are influenced by forces acting on the body. There is currently no provision for simulated bodies to change their pitch or roll.

The translational drag force of water F_{dt} on a simulated body moving at velocity v is modelled using Eq. 1.

$$F_{dt} = 0.5 \cdot \rho \cdot v^2 \cdot A \cdot C_d \qquad (1)$$

where ρ represents the density of water, $1000\,\text{kg/m}^3$. A represents the cross sectional area of body in the direction of its movement through the water, and C_d represents the drag coefficient of the body. $A \cdot C_d$ is specific to the body, and these values are calibrated using the procedure outlined in Sect. 4.

Rotational acceleration and rotational velocity are modelled as:

$$a_r = \tau/I \quad \text{and} \quad \omega = a_r \cdot \delta \qquad (2)$$

where a_r is rotational acceleration in radians/s^2, τ is the net torque acting on the AUV, and I is the mass moment of inertia. τ is the sum of torque delivered by the thrusters, and rotational drag forces. ω represents rotational velocity in radians/second, and δ is the length of time represented by a simulation time-step. Rotational movement is countered by a drag torque τ_d arising from form drag and skin frictions which have been abstracted into a single parameter C_r that can be calibrated for a particular AUV:

$$\tau_d = \omega^2 \cdot C_r \tag{3}$$

Vertical movement of the AUV is dictated by the net forces of the buoyancy pump and gravity, and are subject to translational drag as in Eq. 1. A submerged AUV that is stationary in the vertical plane exhibits an equilibrium between gravity acting on the AUV's mass and the upwards force resulting from its buoyancy. Hence, the net vertical force, F_v, is:

$$F_v = (m \cdot g) - (\rho \cdot V \cdot g) \tag{4}$$

where ρ is the density of water, 1000, and m and V represent the AUV's mass and volume respectively. g represents acceleration under gravity. CoCoRoSim models AUV buoyancy pumps as changing an AUV's volume.

3.1   Simulation of Water Currents

CoCoRoSim can simulate water currents, represented as forces along each axis at points in discretised space (patches), that influence AUV motion. They are generated by AUV thrusters, and by 'cold-spots': low pressure patches that generate current forces towards them. Currents are subject to diffusion and to decay.

AUV thrusters suck water into them and propel it out the back, as depicted in Fig. 2a. The current force being pulled into the thrusters is equal to that pushed out, and is equivalent to the force the thrusters apply to the AUV. Each thruster's forces are considered individually. The left thruster creates a current towards the AUV in the patch to the left of it. An equal force is created behind the AUV: one fourth of this is created in each of the patches directly behind, behind and up, behind and down, and behind and left the AUV. The same applies, vice versa, to the right thruster.

A cold spot creates forces directed towards it in the patches surrounding it. They persist for a period of time, with the current forces they generate rising and falling in magnitude over this time to prevent any excessive force differentials being generated from seemingly nowhere. The periodic turnover of cold spots in randomly selected patches throughout the simulation creates dynamic convections in the water.

Current forces are subject to logarithmic decay and to diffusion; each time step the change in current component of each patch, F_c, is:

$$\Delta F_c = \Big(\big((1 - \kappa) \cdot \mathrm{avg_N}(F_c)\big) - (\kappa \cdot F_c)\Big) \cdot \gamma \tag{5}$$

where κ holds a value between 0 and 1 and represents the rate at which forces diffuse; $\mathrm{avg_N}$ is a function returning the mean of that current component amongst the neighbouring 26 patches[4], and γ is the decay rate.

---

[4] Or less if the patch being updated lies on the environment boundary.

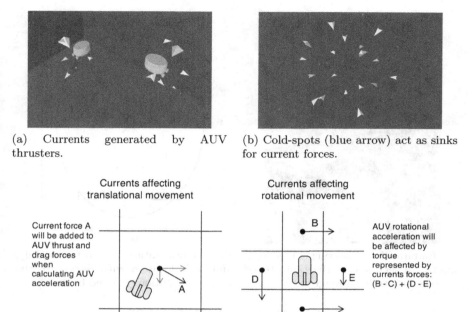

(a) Currents generated by AUV thrusters.

(b) Cold-spots (blue arrow) act as sinks for current forces.

(c) Current forces in the same patch as an AUV, and on the 4-neighbourhood patches around it affect its translational and rotational accelerations.

Fig. 2. Currents are represented as forces in three dimensions in discretised space (Color figure online).

Current forces affect both translational and rotational AUV movements (Fig. 2c). When calculating the net translational forces that dictate AUV acceleration, the forces on the patch that the AUV occupies are considered. The differential between opposing patches in the 4-patch neighbourhood are used to calculate a torque when calculating AUV rotational acceleration.

4   Calibration

This section reports the calibration of CoCoRoSim's representation of the Lily platform (Fig. 1a). The Jeff platform is still under development, and has not yet been calibrated in CoCoRoSim; when it is complete, it will undergo a similar calibration process in CoCoRoSim.

The Lily AUV weighs 0.44 kg. It has two thrusters capable of propelling the AUV forward, and has a buoyancy pump that provides vertical motion. Lily is unable to change its pitch or roll. The two thrusters have been empirically measured as able to deliver a combined maximum force of 0.01 N, through their ability to lift a weight (shown in Fig. 3a). The magnitude of the friction force

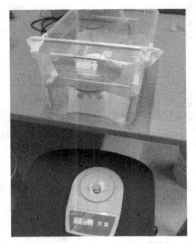

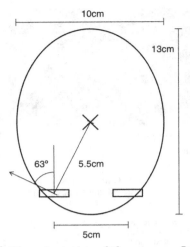

(a) Empirically measuring Lily maximum thruster force in the tank, based on its ability to lift a weight.

(b) The orientation of thrusters on Lily, and their location with respect to the centre of rotation (marked by an X).

Fig. 3. The calibration of Lily in CoCoRoSim.

between the string and the pipes over which it runs was not known, and as such two variations of the experiment were performed. Firstly the AUV is held by hand, with the string taught, and then released. Secondly, the string is slack, and the AUV accelerates to pull it taught. The direction of the frictional force on the string is opposite in the two experiments, which gave readings of the AUV lifting 0.8 g and 1.2 g respectively. The AUV was assumed to have lifted 1 g when calculating AUV maximum thrust force as being 0.01 N. Hence, each individual thruster can deliver 0.005 N of thrust.

Lily's translational cross-section and drag coefficient, $A \cdot C_d$ in Eq. 1, are difficult to calculate. Instead, they have been deduced based on maximum thrust and terminal velocity, given that:

$$v_t = \sqrt{\frac{2F_{dt}}{\rho \cdot A \cdot C_d}} \quad \text{rearranges to:} \quad A \cdot C_d = \frac{2F_{dt}}{\rho \cdot v_t^2} \quad (6)$$

Lily's terminal velocity has been empirically measured at 7.5 cm/s, and given its maximum thruster force (F_{dt}) of 0.01 N, $A \cdot C_d$ is 1.78. This provides simulated behaviours corresponding with empirical measurements of Lily's acceleration: from a standing start at full thrust Lily moves 10 cm in 3.03 s.

Using the buoyancy pump Lily can change its volume between 430 and 450 cm^3. A value of 440 cm^3 delivers a net vertical force $F_v = 0$. When Lily has a volume of 450 cm^3, the net force $F_v = 0.1$ N. Hence, the buoyancy pump can deliver a vertical force of ± 0.1 N. The change in density of water over Lily's 3 m diving limit is deemed negligible. As such, the buoyancy pump is not used to set a desired depth directly, but the buoyancy force, and hence the speed at which an AUV descends or ascends.

Rotational torque is delivered through the differential between the AUV's thrusters. Empirical measurements taken of Lily in the water reveal that the centre of rotation is the centre of the AUV, despite the thrusters being offset from this point, as shown in Fig. 3b. The rotational torque resulting from the thrusters, τ_T, is calculated as:

$$\tau_T = (T_l - T_r) \cdot 0.055 \cdot \cos(63°) \tag{7}$$

where T_l and T_r represents the thruster force exerted by the left and right thrusters respectively, the thrusters are 0.055 m from the centre of rotation, and oriented 63° from the perpendicular through which torque is applied. In calculating rotational drag, C_r can be estimated given Lily's terminal rotational velocity, empirically measured as 1.48 rad/s. At rotational terminal velocity, ω_t, the torque provided by the thrusters is equal to the opposing torque originating from rotational drag forces. Using an equation of the form of translational terminal velocity (Eq. 6 above), the following equation describing terminal rotational velocity is formed:

$$\omega_t = \sqrt{\tau_{max}/C_r} \tag{8}$$

where τ_{max} represents the maximum torque the thrusters can deliver. Re-arranging to solve for C_r:

$$C_r = \left((0.005 - -0.005) \cdot 0.055 \cdot \cos(63°)\right)/\omega_t^2 \tag{9}$$

which solves to give $C_r = 0.000114$. Lily's mass moment of inertia, I, is not known. However, given that the coefficient of rotational drag and maximum thruster forces are known, it can be calibrated in simulation to deliver similar rotational accelerations to those empirically observed of Lily. A value of $I = 0.0005$ is used, as this matched observations that from a standing start Lily can rotate 90° in 2.4 s, 180° after 3.7 s, 270° after 4.8 s, and completes a full turn after 5.9 s.

5   Profiling Controller Behaviour

This section demonstrates CoCoRoSim's use in controller design and evaluation. A principle challenge in underwater swarm robotics is collective motion; AUVs must be coordinated to efficiently explore the environment and not get lost in the ocean. Reynold's Boids algorithm is popular in computational simulations requiring collective motion [10], for example coordinating dinosaurs and bats in Jurassic Park and Batman Returns movies. Swarm member (termed a 'boid') motion is dictated by three rules: *cohesion* attracts boids to their neighbours, *separation* prevents them from colliding, and *alignment* promotes common velocities within the group [10]. We report here preliminary investigations into the suitability of the Boids algorithm for deployment on CoCoRo AUVs.

The principle challenges in real-world deployment are localization and communication. Lily AUVs can localize one another through use of blue light systems, which have a range of around 50 cm, and can detect distances to other AUVs only within a cone of observation (120°); exact triangulation is not possible. Furthermore it is likely that only the nearest neighbour will be detected. Communication is provided through omni-directional radio-frequency (range 50 cm), or directional-blue light systems where the nearest neighbour is the only likely recipient of a message. We have performed a series of experiments to examine how Boids's performance in underwater shoaling is effected by these constraints (Fig. 4). The *Vanilla* (Van) experiment refers to boids employing exact triangulation, each boid knows the exact location of all neighbours within 50 cm. This is impossible in real AUVs, but is performed in simulation as a baseline against which to examine performance of various adaptations of Boids on CoCoRo AUVs. In the *blue light triangulation* (BLT) group boids can only detect whether or not an AUV lies within a 120° cone, and the distance within that cone. If detected a neighbour is assumed to lie in the centre of the cone. Boid velocities are communicated omni-directionally to all neighbours within 50 cm using radio frequency. *Blue light communication* (BLC) extends BLT by communicating velocities only with the nearest neighbours over blue light, this leaves radio frequency communication free for other tasks CoCoRo shoals will have to perform. Lastly, because communication underwater is problematic and it will likely be needed for other swarm functions, *no alignment* (NA) examines shoal performance in absence of any velocity communication, effectively nullifying the alignment rule. Both BLC and BLT simulated algorithms assume noise-less loss-less instantaneous communication, implemented through direct message passing. Given Lily's relatively short sensor range in contrast to its terminal velocity, these algorithms were limited to using only 10 % of maximum thruster force to prevent erratic shoaling behaviour.

Six metrics of shoaling performance are applied, and experiments are conducted with 11 AUVs in total. Polarisation measures the degree to which all boids are pointed in the same direction, calculated as in [7]. A polarisation of 1 indicates that all shoal members have the same orientations, whereas 0 indicates a uniform spread of orientations; higher values are desirable. Angular momentum measures the degree of shoal rotation around its centre, calculated as in [7]. This measure complements polarisation: a shoal of boids rotating clockwise around some point can have a very low polarisation, yet high angular velocities indicate a shoal that is still well organized. Shoal speed measures the movement of the shoal's centre, a highly motile shoal is desirable. The number of times that an AUV is lost from the shoal is counted, and the mean number of distinct shoals throughout simulation time is also recorded. Fewer lost AUVs and low numbers of shoals are desirable. Shoal separation represents the median separation between AUVs in the shoal over the entire simulation time. Algorithms that can provide both wide and narrow separations are desirable, provided no more AUVs are lost. These metrics are shown as box plots in Fig. 4. The magnitude of effect change in comparison with the Vanilla experiment is calculated using the Vargha-Delaney A test [12], a non-parametric effect magnitude test that calculates the probability that a randomly selected sample from population A is larger than a randomly selected sample from population B. Values of ≥ 0.71 or ≤ 0.29 are assumed 'large'. Each experimental group comprises 400 simulations/samples, the number required to reduce the effect of stochastic variation on results to a "small" effect (procedure described in [9]). The boid's cohesion, separation and alignment weights were assigned values of 1.0, 1.0 and 10.0 respectively, a separation threshold of 15 cm was used, and boids had a full 360° vision around them. These values were identified through preliminary experimentation to give stable shoaling behaviour. The simulation was executed for 25,000 time steps, which represents 12,500 s. The simulated environment was 6 m × 6 m with a 2 m depth.

A switch from the vanilla group's perfect triangulation to blue-light based triangulation (BLT and BLC groups) does not deteriorate shoal polarisation, in fact it is increased. This is reflected in a reduced angular velocity in the BLT and BLC groups. This result is unexpected, as blue light-based triangulation is less precise. It may be explained through blue light's consideration of only the nearest neighbours, and boids thus perceiving swarm centres to be closer than they in fact are. Hence the impetus to turn towards the centre is reduced in contrast to the influences of alignment and separation. This may also explain why BLT and BLC groups have significantly reduced separations in contrast to the vanilla group: as demonstrated by the no alignment (NA) group, a lack of alignment results in a more spread out shoal. The BLT group has a significant reduction in the number of AUVs that lose the shoal, and in the number of shoals that emerge. However this change is lost in the BLC group where velocities are shared with only the nearest neighbours. It is clear from these results that no significant detriment to shoal quality occurs when adapting the vanilla boids algorithm to the constraints of CoCoRo platforms. It is also clear that although communication on these platforms is a scarce resource, reserving it purely for higher shoal functions at the expense of communicating velocities has a significantly detrimental impact on performance: the NA shoals were less polarised, had lower velocities, and lost more AUVs.

5.1   Controller Performance in Turbulent Waters

CoCoRoSim's simulation of current forces in the water permits analysis of controller performance in turbulent waters. This is demonstrated by examining the deterioration of the BLT boids variant's shoaling performance in increasingly turbulent waters. These experiments are run for 5,000 simulated time steps, representing 2,500 s, in a tank of 5 × 5 × 2 m. A smaller simulated environment and shorter runtime were selected to address the considerable computational requirements of simulating currents. Three experiments are conducted, with turbulence (represented by the maximum force a cold spot can exert) set at 0 N (*T0* in Fig. 4), 0.0005 N (*T5e-4*) and 0.005 N (*T5e-3*). The probability of any patch of discretised space becoming a cold spot in a time step is 0.0015, and cold spots persist for 12.5 s. 60 % of the current force in a patch is diffused to surrounding

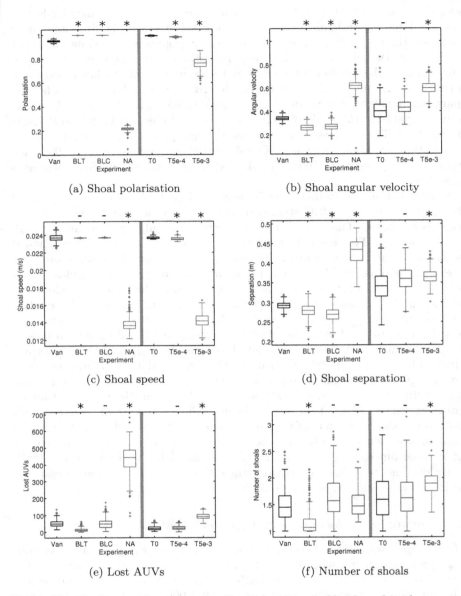

(a) Shoal polarisation

(b) Shoal angular velocity

(c) Shoal speed

(d) Shoal separation

(e) Lost AUVs

(f) Number of shoals

Fig. 4. The shoaling performance of various adaptations of boids and in the presence of various levels of water turbulence. Experiments left of grey vertical lines represent variations of boids, those to the right represent the BLT algorithm running in the presence of water currents and varying degrees of turbulent water. Effect magnitude is established through the A test [12], where red (∗) and blue (−) boxes indicate the presence or absence of large effect magnitudes with respect to the vanilla or T0 algorithm (depending on the experiment) (Color figure online).

patches every time step, and the decay rate is set to 0.65. These values have not been calibrated to any particular body of water, they were selected on the basis that a 0.5 kg floating simulated body exhibited visually appropriate trajectories. These experiments are to demonstrate that CoCoRoSim can be used to evaluate controller performance in the presence of water currents.

As shown in Fig. 4, shoaling performance is significantly altered in only 2 of the 6 metrics for current sink forces of 0.0005 N (*T5e-4*), and these changes are comparatively minor in contrast with forces of 0.005 N (*T5e-3*) where all metrics reveal significant changes. Although Boids could use at most 10 % of available thrust, the OS may make full use of the thrusters in attempting to maintain a particular heading. It is perhaps unsurprising that turbulence forces of 0.01 N, Lily's maximum thrust force, cannot be tolerated. However, it is notable that the shoaling algorithm tolerates, without significant deterioration of performance in many cases, turbulences of 10 % of the maximum available thrust.

6 Discussion and Further Work

This paper has presented the CoCoRoSim underwater swarm robotics simulation, detailed its calibration against empirical data taken from a real-world robot, and demonstrated its use in profiling robot controller behaviour. Collective coordination in underwater swarms is a challenging task owing to low visibility, the attenuation of electro-magnetic signals typically used for communication, and current forces inherent in moving water. We have made use of simulation in evaluating the quality of shoaling resulting from adapting Reynold's Boids algorithm to these underwater constraints. CoCoRo AUVs have limited communication bandwidth, and this might ideally be reserved for higher swarm functions when addressing a complex task of which shoaling is only one constituent activity. However, the presented results clearly indicate that without communication the coordination of shoals using Boids is problematic. Surprisingly, given the ability to communicate velocities amongst shoal members, the limited observational and triangulation capacity of CoCoRo AUVs does not have as detrimental affect on coordination as might be expected. There exist other flocking algorithms that are inherently communication-less, for example [2,3], however this algorithm has not been investigated in platforms with the present number of degrees of freedom. Investigating the suitability of such algorithms in controlling underwater shoals is left as further work, as is the deployment of the present shoaling algorithm on real-world hardware.

Simulation can substantially aid underwater systems research and development. The impracticalities of developing and calibrating algorithms on real-world robotic platforms that must be removed from the water, opened for reprogramming, charged, sealed, and redeployed prior to evaluation are prohibitive. CoCoRoSim facilitates must faster initial development of algorithms. Detailed profiling and assessment of shoaling performance, as performed here in simulation, would not be possible in the real platforms. Furthermore, CoCoRoSim permits the environmental limits within which controllers can reliably operate to be ascertained; we have examined how shoaling performance degrades in increasingly turbulent waters. This is provided as a proof of concept, calibration of this water currents simulation framework against a particular body of water is left as further work. Simulation of water currents will lead into simulating the movement of chemical plumes, useful in designing swarm-intelligence algorithms capable of performing underwater plume tracking.

References

1. Barry, R.: The FreeRTOS reference manual (2011). www.freertos.org
2. Moeslinger, C., Schmickl, T., Crailsheim, K.: A minimalist flocking algorithm for swarm robots. In: Kampis, G., Karsai, I., Szathmáry, E. (eds.) ECAL 2009, Part II. LNCS, vol. 5778, pp. 375–382. Springer, Heidelberg (2011)
3. Moeslinger, C., Schmickl, T., Crailsheim, K.: Emergent flocking with low-end swarm robots. In: Dorigo, M., et al. (eds.) ANTS 2010. LNCS, vol. 6234, pp. 424–431. Springer, Heidelberg (2010)
4. Gerkey, B., Vaughan, R., Howard, A.: The player/stage project: tools for multi-robot and distributed sensor systems. In: Proceedings of the International Conference on Advanced Robotics, pp. 317–323 (2003)
5. Lau, H., et al.: Adaptive data-driven error detection in swarm robotics with statistical classifiers. Robot. Auton. Syst. **59**(12), 1021–1035 (2011)
6. Hugues, L., Bredeche, N.: Simbad: an autonomous robot simulation package for education and research. In: Nolfi, S., Baldassarre, G., Calabretta, R., Hallam, J.C.T., Marocco, D., Meyer, J.-A., Miglino, O., Parisi, D. (eds.) SAB 2006. LNCS (LNAI), vol. 4095, pp. 831–842. Springer, Heidelberg (2006)
7. Couzin, I., et al.: Collective memory and spatial sorting in animal groups. J. Theor. Biol. **218**, 1–11 (2002)
8. Koenig, N., Howard, A.: Design and use paradigms for gazebo, an open-source multi-robot simulator. In: IEEE/RSJ International Conference on Intelligent Robots and Systems, pp. 2149–2154 (2004)
9. Read, M., et al.: Techniques for grounding agent-based simulations in the real domain: a case study in experimental autoimmune encephalomyelitis. MCMDS **17**(4), 296–302 (2011)
10. Reynolds, C.: Flocks, herts and schools: a distributed behavioral model. Comput. Graph. **21**(4), 25–34 (1987)
11. Schmickl, T., et al.: Cocoro - the self-aware underwater swarm. In: Proceedings of the SASO '11 (2011)
12. Vargha, A., Delaney, H.D.: A critique and improvement of the CL common language effect size statistics of McGraw and Wong. J. Educ. Behav. Stat. **25**(2), 101–132 (2000)
13. Winkler, L., Wörn, H.: Symbricator3D – a distributed simulation environment for modular robots. In: Xie, M., Xiong, Y., Xiong, C., Liu, H., Hu, Z. (eds.) ICIRA 2009. LNCS, vol. 5928, pp. 1266–1277. Springer, Heidelberg (2009)

Path Planning for Swarms by Combining Probabilistic Roadmaps and Potential Fields

Alex Wallar[1] and Erion Plaku[2](✉)

[1] School of Computer Science, University of St Andrews,
St Andrews, Fife KY16 9AJ, Scotland, UK
[2] Department of Electrical Engineering and Computer Science,
Catholic University of America, Washington, DC 20064, USA
plaku@cua.edu

Abstract. This paper combines probabilistic roadmaps with potential fields in order to enable a robotic swarm to effectively move to a desired destination while avoiding collisions with obstacles and each other. Potential fields provide the robots with local, reactive, behaviors that seek to keep the swarm moving in cohesion and away from the obstacles. The probabilistic roadmap provides global path planning which guides the swarm through a series of intermediate goals in order to effectively reach the desired destination. Random walks in combination with adjustments to the potential fields and intermediate goals are used to help stuck robots escape local minima. Experimental results provide promising validation on the efficiency and scalability of the proposed approach. Source code is made publicly available.

1 Introduction

Swarm robotics seeks to enable a large number of robots to accomplish complex tasks via simple interactions with one another and the environment [1]. Swarm robotics draws inspiration from social insects, such as ants and bees, where intelligent group behaviors emerge from simple interactions among the individuals. Such framework provides a level of scalability and robustness that is difficult to achieve with centralized approaches. As research progresses, applications of swarm robotics are emerging in exploration, mapping, monitoring, inspection, and search-and-rescue missions, as surveyed in [2,3].

In many of these applications, the swarm needs to move in cohesion to a goal destination while avoiding collisions with obstacles and among the robots in the swarm. This path-planning problem presents significant challenges. As path planning is PSPACE-complete, exact algorithms, which always find a solution if it exists and report no solution otherwise, are difficult to implement and are limited in practicality to low-dimensional systems due to the exponential dependency on the problem dimension [4–6]. As a result, research has focused on alternative approaches that do not determine the existence of a solution but seek instead to achieve efficiency and scalability in increasingly complex settings.

A. Natraj et al. (Eds.): TAROS 2013, LNAI 8069, pp. 417–428, 2014.
DOI: 10.1007/978-3-662-43645-5_43, © Springer-Verlag Berlin Heidelberg 2014

A common approach in path planning for robotic swarms is to impose artificial potential functions (APFs) that seek to push the swarm away from the obstacles and toward the goal [7–10]. APFs provide scalability as the addition of new robots to the swarm generally requires only computation of repulsive forces from obstacles, attractive forces from the goal, and forces resulting from the limited interactions with neighboring robots. However, an inherent challenge with APFs is the tendency to get stuck in local minima especially when planning paths for large swarms moving in cluttered environments and through narrow passages. APFs that avoid local minima exist in limited cases, e.g., a point robot in a generalized sphere world [11], but not for general path planning for swarms.

Alternative approaches aim to avoid local minima by relying on probabilistic roadmaps (PRMs) [12]. PRMs capture the connectivity of the configuration space via a roadmap obtained by sampling collision-free configurations and connecting neighboring configurations with collision-free paths. A configuration generally specifies the placement of each robot in the swarm and a path between two configurations is typically obtained by interpolation. A collision-free path from an initial to a goal configuration is then obtained by performing graph search on the roadmap. Starting with the original PRM [12] and continuing over the years, PRMs have had great success in solving challenging problems [13–18]. The work in [19,20] use APFs to increase PRM sampling near obstacles to facilitate connections through narrow passages. The work in [21–23] uses PRM to enable robotic swarms achieve different behaviors such as homing, coverage, goal searching, and shepherding. The roadmap, however, is built over the high-dimensional configuration space, which considerably increases the computational cost. The work in [24] uses multi-level PRMs in combination with Bezier curves to guide a multi-robot system to a desired destination while maintaining a specific formation. The approach has been applied only to a small number of robots and requires considerable precomputation to build the multi-level PRMs.

Even though significant progress has been made, scalability still remains problematic in PRMs [25,26]. As the number of robots increases, it becomes difficult to sample collision-free configurations and generate collision-free paths that connect neighboring configurations. Moreover, significantly larger roadmaps are needed to capture the connectivity of the configuration space. As a result, efficiency of PRMs starts deteriorating as the number of robots is increased. These issues become even more problematic when considering robotic swarms since PRMs do not have a mechanism to ensure that the robots move in cohesion as a swarm rather than as individual entities.

To improve the efficiency of path planning for robotic swarms, this paper develops a novel approach, named CRoPS (Combined Roadmaps and Potentials for Swarms), which combines APFs with PRMs. While CRoPS draws from PRM the underlying idea of using a roadmap, it does not suffer from scalability issues as the roadmap is constructed over the two-dimensional workspace instead of the high-dimensional configuration space of the swarm. Moreover, CRoPS does not use the roadmap to plan the entire path of the swarm, but rather to generate a series of intermediate goals that serve as attractive potentials to guide the swarm toward the desired destination. CRoPS then relies on APFs to enable the robots move in cohesion as a swarm from one intermediate goal to the other while avoiding collisions. This combination of PRMs with APFs is crucial to the efficiency and scalability of CRoPS. Experimental results in simulation with increasingly large swarms moving in complex environments containing numerous obstacles and narrow passages provide promising validation.

2 Method

The swarm motions are governed by the following criteria:

1. There is long range attraction to intermediate goals and final destination.
2. Robots are repulsed from obstacles.
3. Robots move as a swarm while keeping some separation from one another.
4. A robot's heading is influenced by the headings of its neighbors.

To guide the swarm to the final destination, global path planning based on PRMs is used to determine suitable intermediate goals. The roadmap is constructed by sampling collision-free points and connecting neighboring points with collision-free edges to obtain a graph that captures the connectivity of the environment. Roadmap vertices and edges are associated with weights that estimate the feasibility of the swarm to pass through their surrounding areas. The shortest path in the roadmap to the final destination is used to provide a series of intermediate goals that the swarm can follow to effectively reach the final destination. An attractive potential, denoted as $PF_{igoal}(b)$, is added between each robot b and its current intermediate goal. When the robot reaches the current intermediate goal, the next point in the shortest path is set as the new intermediate goal.

To avoid collisions with obstacles, CRoPS creates a strong repulsive potential field, denoted as $PF_{obst}(b)$, which pushes each robot b away from the obstacles. The repulsive potential increases quadratically with respect to the inverse of the distance from the robot to the obstacles. Such rapid increase prevents the robots from getting too close to the obstacles.

To maintain separation among the robots in the swarm, each robot is pushed away from its neighbors. A weak sigmoidal repulsive potential, denoted as $PF_{sep}(b)$, is employed rather than a strong quadratic repulsive potential in order to push b away from its neighbors but not so strongly as to separate it from the swarm.

To move as a swarm, a robot's heading is influenced by the headings of its neighbors, defined as $PF_{heading}(b)$. To promote effective movements, preference is given to neighbors that are not stuck and are neither too close nor too far. The headings of the neighbors that are chosen are averaged and combined with the other potential fields to determine the new heading and position of each robot.

The different potential fields are superimposed to obtain the overall force vector applied to each robot b. In this way, the potential field on the robot b exerts a strong repulsive potential away from the obstacles while attracting it to the current intermediate goal, maintaining a separation distance from the other robots, and adjusting the heading so that the robots move as a swarm toward the final

---

Algorithm 1. Pseudocode for CRoPS

1: $RM = (V, E) \leftarrow$ CONSTRUCTROADMAP(), where
2:     $V \leftarrow$ sample numerous random points, discard those that are in collision
3:     $E \leftarrow$ connect neighboring points in V, discard those edges that are in collision
4:     $w(q_i) \leftarrow$ assign weight to each vertex $q_i \in V$
5:     $w(q_i, q_j) \leftarrow$ assign weight to each edge $(q_i, q_j) \in E$
6: $\zeta = [q_k]_{k=1}^m \leftarrow$ INTERMEDIATEGOALS(RM) $\diamond$ *obtained from shortest path*
7: $igoal(b) \leftarrow \zeta(1)$ $\diamond$ *set current intermediate goal for each robot*
8: **while** SOLVED() = **false do**
9:   **for** $b \in Robots$ with REACHEDINTERMEDIATEGOAL(b) = **true do**
10:      $igoal(b) \leftarrow$ NEXTINTERMEDIATEGOAL(ζ)
11:   **for** $b \in Robots$ **do**
12:      $PF(b) \leftarrow$ SUPERIMPOSE($PF_{obst}(b), PF_{sep}(b), PF_{igoal}(b), PF_{heading}(b), PF_{escape}(b)$)
13:      $NewHeading(b) \leftarrow w_1 heading(b) + w_2 PF(b)$
14:   **for** $b \in Robots$ **do**
15:      $heading(b) \leftarrow NewHeading(b);$ $pos(b) \leftarrow pos(b) + heading(b)$

---

destination. If a robot gets stuck in local minima, random walks in combination with adjustments to the potential fields and intermediate goals are used to help it escape. Pseudocode for the overall approach is given in Algorithm 1. Details of the main steps follow. Source code and detailed documentation including all parameter values are publicly available [27]. Parameter values are generally determined empirically.

2.1   Roadmap Construction

CRoPS constructs a roadmap in order to effectively guide the swarm toward the goal. Since the swarm could have many robots, the roadmap is not constructed over the high-dimensional configuration space, as it is often the case in PRM approaches, but is instead constructed over the low-dimensional workspace where the swarm moves. As explained in this section, CRoPS uses the roadmap to find intermediate areas in which the swarm can move to effectively reach the goal.

Roadmap Vertices. The roadmap is constructed by first sampling a large number of points uniformly at random inside the workspace boundaries and then discarding all the points that are in collision or too close to an obstacle. A parameter, d_{clear}, determines the minimum acceptable distance from a sampled point to the nearest obstacle. The remaining points, which are all at least d_{clear} units away from the obstacles, are added as vertices to the roadmap graph $RM = (V, E)$. A roadmap vertex q_i and the clearance d_{clear} conceptually define a clearance area as a disk centered at q_i with radius d_{clear}, denoted as $area(q_i)$. In order to bias the swarm movements toward less cluttered areas, each roadmap vertex q_i is associated with a weight $w(q_i)$ which estimates how feasible it is for the swarm to travel through $area(q_i)$. More specifically, the weight is defined as

$$w(q_i) = (\sum_{o \in Obstacles} dist(q_i, o))^3,$$

where $dist(q_i, o)$ denotes the minimum distance from q_i to the obstacle o. In this way, small weights indicate the presence of obstacles nearby, which may make it more difficult for the swarm to pass through. As explained later in the section, CRoPS gives preferences to roadmap vertices associated with high weights which are indicative of areas with high clearance. Note that other definitions for the weight function are possible. The particular function used in this paper worked well for the experiments as it captures the desired properties of biasing the swarm movements towards less cluttered areas.

Roadmap Edges. After generating roadmap vertices, CRoPS connects each roadmap vertex to its k nearest neighbors. Edges that are in collision are discarded. The weight of an edge connecting q_i to q_j is defined as

$$w(q_i, q_j) = ||q_i, q_j||_2 / \min(w(q_i), w(q_j)),$$

where $||q_i, q_j||_2$ is the Euclidean distance from q_i to q_j. Note that $w(q_i, q_j)$ is small when q_i and q_j are close to each other and away from obstacles.

Intermediate Goals Along Shortest Roadmap Path. Dijkstra's shortest-path algorithm is used to compute the shortest path ζ in the roadmap to the final destination, where the weight of a roadmap edge (q_i, q_j) is defined by $w(q_i, q_j)$ as described above. Each vertex $q_k \in \zeta$ defines an intermediate goal for the swarm. More specifically, CRoPS seeks to move the swarm to the goal by passing through the areas $area(q_k)$ as defined by the vertices q_k along the shortest path ζ. Note that the dependency of the edge weights on vertex weights ensures that the shortest path in the roadmap does not come too close to the obstacles, which could lead the swarm to often get stuck in local minima.

2.2 Potential Fields

Repulsion from Obstacles. An imperative objective for the swarm is to always avoid collisions with obstacles. For this reason, a repulsive potential is defined that pushes the robots away from the obstacles. More specifically, the repulsive potential between a robot b and an obstacle o is defined as

$$P_{obst}(b, o) = \frac{1}{(dist(pos(b), o) - radius(b))^2},$$

where $pos(b)$ and $radius(b)$ denote the position and radius of the robot b, respectively. Note that the important aspect of this repulsive function is that its value increases rapidly as the robot approaches an obstacle. This ensures that the robot would be pushed away and never collide with an obstacle.

In order to limit the influence of the obstacles that are far away, the repulsion is computed only from those obstacles that are within a certain distance Δ_{obst} from the robot. The potential field imposed by the obstacles is then defined as

$$PF_{obst}(b) = \sum_{\substack{o \in Obstacles \\ dist(b,o) \leq \Delta_{obst}}} (pos(b) - ClosestPoint(o, pos(b))) P_{obst}(b, o),$$

where $ClosestPoint(pos(b), o)$ denotes the closest point on the obstacle o to $pos(b)$.

Repulsion from Other Robots. As the swarm moves, the robots need to avoid coming too close to each other as it could lead to collisions. At the same time, the robots should not be far away from each other in order to move as a swarm. To achieve these objectives, CRoPS uses a weak repulsive sigmoid function

$$P_{sep}(b_i, b_j) = \frac{1}{1 + \exp(\delta_{sep} \, ||pos(b_i), pos(b_j)||_2)},$$

where δ_{sep} is a scaling constant. In order to limit the influence of the robots that are far away, similar to the potential field for obstacles, the repulsion is computed only from those robots that are within a certain distance Δ_{sep}. The potential field imposed on the robot b by the other robots is then defined as

$$PF_{sep}(b) = \sum_{\substack{b_i \in Robots - \{b\} \\ dist(b, b_i) \leq \Delta_{sep}}} (pos(b) - pos(b_i)) P_{sep}(b, b_i).$$

In this way, the robots travel close together but are pushed away when they come too close to one another.

Attraction to the Current Intermediate Goal. As discussed, CRoPS uses the shortest path ζ in the roadmap to set the intermediate goals for the swarm. Let $igoal(b)$ denote the current intermediate goal of the robot b. Note that $igoal(b)$ is associated with some point q_k in ζ. An attractive potential field that pulls the robot b towards $igoal(b)$ is then defined as

$$PF_{igoal}(b) = \frac{igoal(b) - pos(b)}{1 + \exp(\delta_{igoal} \, ||pos(b), igoal(b)||_2)},$$

where δ_{igoal} is a scaling constant. Although other definitions are possible, the sigmoid function allows the robots to reach the goal without getting too greedy, which could lead to getting stuck in local minima. When b reaches $igoal(b)$, the next point q_{k+1} in the shortest path ζ is set as the new intermediate goal for b.

Influence of Neighbors on Heading. In order to make the robots move as a swarm, the heading of a robot b is also influenced by the headings of neighboring robots. In order to select suitable neighbors, robots that are stuck are excluded from consideration. Moreover, preference is given to those neighbors that are neither too far nor too close from b. This is achieved by using a Gaussian function $\gamma(b, b_i)$ with mean μ and standard deviation σ, i.e.,

$$\gamma(b, b_i) = \exp\left(\frac{-(||pos(b), pos(b_i)||_2 - \mu)^2}{2\sigma^2}\right),$$

and selecting as $Neighs(b)$ the k closest nonstuck robots according to $\gamma(b, b_i)$. The potential field imposed on the robot b by the headings of the neighboring robots is then defined as

$$PF_{heading}(b) = \sum_{b_i \in Neighs(b)} heading(b_i).$$

Escaping Local Minima. Each robot b keeps track of its past positions in order to determine if it is stuck in local minima. More specifically, a robot b is considered stuck if it has moved very little during the last ℓ time steps, i.e.,

$$
stuck(b) = \begin{cases} 1, & \text{if } ||pos(b) - prev_\ell(b)||_2 < \Delta_{stuck} \\ 0, & \text{otherwise,} \end{cases}
$$

where $prev_\ell(b)$ denotes the position of the robot ℓ steps in the past, and Δ_{stuck} is a threshold constant.

If a robot b is determined to be stuck, then a random vector is added to the mix of the potential fields, i.e.,

$$
PF_{escape}(b) = stuck(b)(r_x, r_y),
$$

where r_x, r_y constitute a random direction. Note that $PF_{escape}(b) = (0,0)$ if the robot is not stuck. By performing a random walk for several steps, the robot increases the likelihood of escaping local minima.

In addition, when $stuck(b) = \texttt{true}$, a different mean μ_{stuck} and a different standard deviation σ_{stuck} are used to compute $PF_{heading}(b)$. The new mean and standard deviation have smaller values in order to select more nearby neighbors (recall that only nonstuck neighbors are considered for the selection.) This allows the robot to select different neighbors to influence its heading, since the original neighbors could have contributed to the robot being stuck in the local minima.

To further increase the likelihood of escaping local minima, CRoPS also adjusts the intermediate goals of a stuck robot. In particular, if q_k is the current intermediate goal, then CRoPS changes it to q_{k-1} if the robot is still in a local minima after a few iterations. If the robot is still unable to escape the local minima, the intermediate goal is set to q_{k+1}. By switching from the current to a past or to a future intermediate goal, the robot is given further flexibility which facilitates escaping local minima.

Superimposition of Potential Fields. The different potential fields are superimposed to obtain the overall force vector applied to the robot b:

$$
PF(b) = \frac{\sum_{\phi \in fields} (||PF_\phi(b)||_2 PF_\phi(b))}{\sum_{\phi \in fields} ||PF_\phi(b)||_2},
$$

where $fields = \{obst, sep, igoal, heading, escape\}$. The heading and the position of the robot b are then updated as

$$
heading(b) \leftarrow w \, heading(b) + PF(b)
$$
$$
pos(b) \leftarrow pos(b) + heading(b),
$$

where w is an adjustment constant.

The calculation of $PF(b)$ ensures that the subfield that has the highest potential during the current iteration will have the highest influence when the heading is calculated. The overall potential has a sigmoidal attraction to the immediate goal and sigmoidal repulsion from the robots as well as an inverse distance squared repulsion from obstacles. In this way, the potential field on the robot b exerts a strong repulsive potential away from the obstacles while attracting it to the current intermediate goal, maintaining a separation distance from the other robots, and adjusting the heading so that the robot b moves in a direction similar to its neighbors. Escape strategies are also applied in order for the robot to avoid getting stuck in local minima.

3   Experiments and Results

Experiments are conducted in simulation using different scenes and an increasing number of robots to test the efficiency and scalability of the approach. Figure 1 provides an illustration of the scenes. These scenes provide challenging test cases as the swarm has to avoid numerous obstacles and pass through multiple narrow passages in order to reach the final destination.

3.1   Measuring Performance

A problem instance is defined by a scene and the number of robots. Due to the probabilistic nature of the roadmap, performance on a particular problem

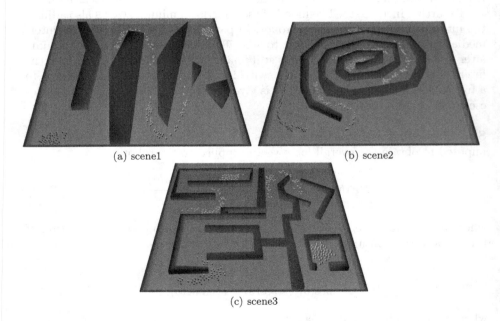

(a) scene1   (b) scene2

(c) scene3

Fig. 1. Scenes used in the experiments. Each figure also shows intermediate swarm configurations along the path from the initial to the goal.

instance is based on twenty different runs. Results report the average time for all the robots to reach the final destination. Results also report the average distance among all the robot pairs. More specifically, the average distance for a problem instance is measured by adding all the pairwise distances at every time step for all the runs to a vector and then diving by the size of the vector. Finally, the average distance is scaled by the robot diameter. As an example, a scaled distance of 5.14 indicates that the swarm is maintaining an average separation distance of roughly 5 robots. Small values (close to 1) indicate that the robots are too close to one another and large values indicate that the robots are separating. Standard deviations are shown for both time and scaled distance results.

Experiments are conducted on an Intel Core i3 machine (CPU: 2.40 GHz, RAM: 4 GB) using Ubuntu 13.04. Code is written in Python 2.7.3. Code is publicly available at [27].

3.2 Results

Figure 2(a) provides a summary of the results on the average time for all the robots to reach the final destination. These results indicate that CRoPS is capable of effectively planning motions for large swarms moving through complicated environments. Figure 2(b) takes a closer look at these results by showing how the average time scales as a function of the number of robots in the swarm. As the results indicate, the time grows only linearly. Such results provide promising validation on the scalability of CRoPS.

The efficiency of CRoPS derives from the combination of global path planning via probabilistic roadmaps with local path planning via potential fields. To test this further, CRoPS was run without the probabilistic roadmap. In this scenario, the robots would be guided by the potential fields and only be attracted to the final destination but not to any intermediate goals. Without probabilistic roadmaps, however, the approach timed out and failed to send any robot to the final destination. These experiments indicate the importance of combining probabilistic roadmaps with potential fields when planning motions for large swarms in complicated environments.

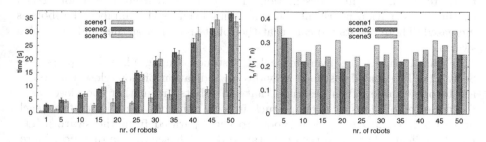

Fig. 2. (a) Results on the average time for all the robots to reach the final destination as a function of the number of robots. Bars indicate one standard deviation. (b) Scaled results, where t_n denotes the average time for all n robots to reach the final destination.

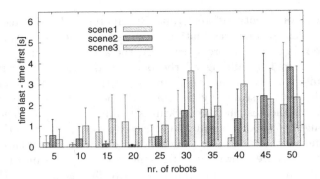

Fig. 3. Results on the average time difference between the first and the last robot to reach the final destination as a function of the number of robots. Bars indicate one standard deviation.

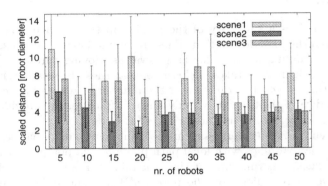

Fig. 4. Results on the average scaled distance among all robot pairs as a function of the number of robots. Scaling is done with respect to the robot diameter. Bars indicate one standard deviation.

Figure 3 shows the average time difference between the first and the last robot to reach the final destination. The plot shows that the robots reach the final destination nearly at the same time even as the number of robots is increased. These results indicate that the robots remain together and move as a swarm.

Figure 4 shows the average scaled distance among all robot pairs (see Sect. 3.1). The scaled distance provides an indication of how close the robots are to one another. It is desirable that the robots are neither too close (as it could cause collisions or getting stuck in local minima) nor too far from each other (as it could cause some robots to get separated from the swarm). The results indicate that the robots maintain a desirable separation distance. Moreover, the separation distance changes very little even as the number of robots is increased.

4 Discussion

The proposed approach, CRoPS, combined probabilistic roadmaps with APFs in order to enable a swarm of robots to effectively move to a desired destination while avoiding collisions with obstacles and each other. The probabilistic roadmap provides global path planning to determine appropriate intermediate goals for the swarm. The potential fields provide local planning to enable the robots move together as a swarm towards the goal while avoiding collisions.

The combination of probabilistic roadmaps with APFs opens up several venues for future research. One research direction is to improve the interplay between the roadmap planning and APFs in order to more effectively move the swarm to the desired destination. Another research direction is to accommodate moving obstacles. As the motion direction and velocity of the moving obstacles might not be known in advance, it will be important to be able to predict such motions and take them into account during planning.

References

1. Reynolds, C.W.: Flocks, herds and schools: a distributed behavioral model. ACM SIGGRAPH Comput. Graph. **21**, 25–34 (1987)
2. Şahin, E.: Swarm robotics: from sources of inspiration to domains of application. In: Şahin, E., Spears, W.M. (eds.) Swarm Robotics 2004. LNCS, vol. 3342, pp. 10–20. Springer, Heidelberg (2005)
3. Brambilla, M., Ferrante, E., Birattari, M., Dorigo, M.: Swarm robotics: a review from the swarm engineering perspective. Swarm Intell. **7**, 1–41 (2012)
4. Reif, J.: Complexity of the mover's problem and generalizations. In: IEEE Symposium on Foundations of Computer Science, pp. 421–427 (1979)
5. Canny, J.: The Complexity of Robot Motion Planning. MIT Press, Cambridge, MA (1988)
6. Schwartz, J.T., Sharir, M.: A survey of motion planning and related geometric algorithms. Artif. Intell. **37**, 157–169 (1988)
7. Khatib, O.: Real time obstacle avoidance for manipulators and mobile robots. Int. J. Robot. Res. **5**(1), 90–99 (1986)
8. Reif, J.H., Wang, H.: Social potential fields: a distributed behavioral control for autonomous robots. Robot. Auton. Syst. **27**(3), 171–194 (1999)
9. Spears, W.M., Spears, D.F.: Physicomimetics: Physics-Based Swarm Intelligence. Springer, Heidelberg (2012)
10. Tanner, H.G., Kumar, A.: Formation stabilization of multiple agents using decentralized navigation functions. In: Robotics: Science and Systems, pp. 49–56 (2005)
11. Rimon, E., Koditschek, D.: Exact robot navigation using artificial potential functions. IEEE Trans. Robot. Autom. **8**, 501–518 (1992)
12. Kavraki, L.E., Švestka, P., Latombe, J.C., Overmars, M.H.: Probabilistic roadmaps for path planning in high-dimensional configuration spaces. IEEE Trans. Robot. Autom. **12**(4), 566–580 (1996)
13. Denny, J., Amato, N.M.: Toggle PRM: a coordinated mapping of C-free and C-obstacle in arbitrary dimension. In: Frazzoli, E., Lozano-Perez, T., Roy, N., Rus, D. (eds.) Algorithmic Foundations of Robotics X. Springer Tracts in Advanced Robotics, pp. 297–312. Springer, Heidelberg (2013)

14. Yeh, H.Y., Thomas, S., Eppstein, D., Amato, N.M.: UOBPRM: A uniformly distributed obstacle-based PRM. In: IEEE/RSJ International Conference on Intelligent Robots and Systems, pp. 2655–2662 (2012)

15. Plaku, E., Bekris, K.E., Chen, B.Y., Ladd, A.M., Kavraki, L.E.: Sampling-based roadmap of trees for parallel motion planning. IEEE Trans. Robot. 21(4), 587–608 (2005)

16. Sun, Z., Hsu, D., Jiang, T., Kurniawati, H., Reif, J.: Narrow passage sampling for probabilistic roadmap planners. IEEE Trans. Robot. 21(6), 1105–1115 (2005)

17. Boor, V., Overmars, M.H., van der Stappen, A.F.: The Gaussian sampling strategy for probabilistic roadmap planners. In: IEEE International Conference on Robotics and Automation, pp. 1018–1023 (1999)

18. Pan, J., Chitta, S., Manocha, D.: Faster sample-based motion planning using instance-based learning. In: Frazzoli, E., Lozano-Perez, T., Roy, N., Rus, D. (eds.) Algorithmic Foundations of Robotics X. Springer Tracts in Advanced Robotics, vol. 86, pp. 381–396. Springer, Heidelberg (2013)

19. Aarno, D., Kragic, D., Christensen, H.I.: Artificial potential biased probabilistic roadmap method. In: IEEE International Conference on Robotics and Automation, pp. 461–466 (2004)

20. Katz, R., Hutchinson, S.: Efficiently biasing PRMS with passage potentials. In: IEEE International Conference on Robotics and Automation, pp. 889–894 (2006)

21. Bayazıt, O.B., Lien, J.-M., Amato, N.M.: Swarming behavior using probabilistic roadmap techniques. In: Şahin, E., Spears, W.M. (eds.) Swarm Robotics 2004. LNCS, vol. 3342, pp. 112–125. Springer, Heidelberg (2005)

22. Bayazıt, O.B., Lien, J.M., Amato, N.M.: Better group behaviors using rule-based roadmaps. In: International Workshop on Algorithmic Foundations of Robotics, pp. 95–112 (2004)

23. Harrison, J.F., Vo, C., Lien, J.-M.: Scalable and robust shepherding via deformable shapes. In: Boulic, R., Chrysanthou, Y., Komura, T. (eds.) MIG 2010. LNCS, vol. 6459, pp. 218–229. Springer, Heidelberg (2010)

24. Krontiris, A., Louis, S., Bekris, K.E.: Multi-level formation roadmaps for collision-free dynamic shape changes with non-holonomic teams. In: IEEE International Conference on Robotics and Automation (2012)

25. Choset, H., Lynch, K.M., Hutchinson, S., Kantor, G., Burgard, W., Kavraki, L.E., Thrun, S.: Principles of Robot Motion: Theory, Algorithms, and Implementations. MIT Press, Cambridge (2005)

26. LaValle, S.M.: Planning Algorithms. Cambridge University Press, Cambridge, MA (2006)

27. Wallar, A., Plaku, E.: Source code for CRoPS: combined roadmaps and potentials for swarm path planning (2013). http://aw204.host.cs.st-andrews.ac.uk/CRoPS/

Towards Exogenous Fault Detection in Swarm Robotic Systems

Alan G. Millard[1](✉), Jon Timmis[2], and Alan F.T. Winfield[3]

[1] Department of Computer Science, University of York, York, UK
millard@cs.york.ac.uk
[2] Department of Electronics, University of York, York, UK
jon.timmis@york.ac.uk
[3] Bristol Robotics Laboratory, University of the West of England, Bristol, UK
Alan.Winfield@uwe.ac.uk

It has long been assumed that swarm systems are robust, in the sense that the failure of individual robots will have little detrimental effect on a swarm's overall collective behaviour. However, Bjerknes and Winfield [1] have recently shown that this is not always the case, particularly in the event of partial failures (such as motor failure). The reliability modelling in [1] shows that overall system reliability rapidly decreases with swarm size, therefore this is a problem that cannot simply be solved by adding more robots to the swarm. Instead, future large-scale swarm systems will need an active approach to dealing with failed individuals if they are to achieve a high level of fault tolerance.

Christensen et al. [2] proposed one such approach, inspired by synchronised flashing behaviour seen in fireflies, that allows failed robots to be detected and physically removed by other operational members of the swarm. This ability of robots to detect faults in each other is referred to as *exogenous* fault detection [2]. The work of Christensen et al. represents the state-of-the-art in exogenous fault detection for robot swarms, but it only addresses the case of completely failed robots, the effect of which on collective behaviour has been shown to be relatively benign by Winfield and Nembrini [3]. The occurrence of partial failures is of far greater concern, as highlighted by Bjerknes and Winfield [1].

We propose a novel method of exogenous fault detection capable of detecting partial failures, which is based on the comparison of expected and observed robot behaviour. Rather than having robots attempt to learn the expected behaviour of others they will each possess a copy of every other robot's controller, which they can instantiate within an internal simulator. The model of expected behaviour therefore comprises a copy of another robot's controller, and a simulator that is able to run the controller code in a simulated environment. Thus, as long as a robot's controller can be instantiated in the simulation, its behaviour can be predicted. This means that the proposed method would be independent of robot controller architecture, and could therefore even be used in heterogeneous swarms. However, we make the simplifying assumption initially that the swarm is homogeneous, and therefore each robot has an identical controller.

Each robot will initialise its internal simulation such that it reproduces the relative positions and orientations of the other robots in reality. The simulation then runs for a short period of time to predict the future state of each robot,

A. Natraj et al. (Eds.): TAROS 2013, LNAI 8069, pp. 429–430, 2014.
DOI: 10.1007/978-3-662-43645-5_44, © Springer-Verlag Berlin Heidelberg 2014

which may then be compared to the observed state of the robots in reality. If there is a significant discrepancy between the predicted and observed behaviour of a particular robot, then this may indicate that it has developed a fault.

In order to make this behavioural comparison possible, the robots must have a method of observing each other's behaviour. This could be achieved by using an on-board camera to track the distance and relative direction of movement of another robot by tracking its position and size in its field of view. However, to simplify the problem to begin with, we have decided to provide each robot with information about the position/orientation of other robots, collected using a tracking infrastructure that observes the swarm from a bird's-eye view (though only local information will be available to each robot, to ensure scalability).

We have chosen to develop the system on the e-puck robotic platform [4], augmented with a Linux Extension Board developed at the Bristol Robotics Laboratory [5], which improves the computation, memory, and networking performance of the e-puck. The robot controllers will be developed using software freely available from the Player/Stage project [6]. This open source project provides two main pieces of software: the Player robot device server, and the Stage multi-robot simulator. These tools are widely used by the research community, and allow for the implementation of a robot controller on real robots and simulated counterparts. The real and the simulated e-pucks will both be programmed using functionally equivalent controller code, to ensure that any deviation between expected and observed behaviour is due solely to the reality gap and tracking inaccuracies. Our proposed method of exogenous fault detection will first be implemented and tested using non-embodied computational resources, and if successful, a fully embodied solution will then be developed.

References

1. Bjerknes, J.D., Winfield, A.F.T.: On fault tolerance and scalability of swarm robotic systems. In: Martinoli, A., Mondada, F., Correll, N., Mermoud, G., Egerstedt, M., Hsieh, M.A., Parker, L.E., Støy, K. (eds.) Distributed Autonomous Robotic Systems. STAR, vol. 83, pp. 431–444. Springer, Heidelberg (2013)
2. Christensen, A.L., O'Grady, R., Dorigo, M.: From fireflies to fault tolerant swarms of robots. IEEE Trans. Evol. Comput. **13**, 754–766 (2009)
3. Winfield, A.F.T., Nembrini, J.: Safety in numbers: fault-tolerance in robot swarms. Int. J. Model. Ident. Control **1**, 30–37 (2006)
4. Mondada, F., Bonani, M., Raemy, X., Pugh, J., Cianci, C., Klaptocz, A., Magnenat, S., Zufferey, J.C., Floreano, D., Martinoli, A.: The e-puck, a robot designed for education in engineering. In: 9th Conference on Autonomous Robot Systems and Competitions, vol. 1, pp. 59–65 (2009)
5. Liu, W., Winfield, A.F.T.: Open-hardware e-puck Linux extension board for experimental swarm robotics research. Microprocess. Microsyst. **35**, 60–67 (2011)
6. Gerkey, B.P., Vaughan, R.T., Howard, A.: The player/stage project: tools for multi-robot and distributed sensor systems. In: Proceedings of the 11th International Conference on Advanced Robotics (ICAR), pp. 317–323 (2003)

Verification and Ethics

Verification and Error

Ethical Choice in Unforeseen Circumstances

Louise Dennis[✉], Michael Fisher, Marija Slavkovik, and Matt Webster

University of Liverpool, Liverpool, UK
{L.A.Dennis,MFisher,Marija,M.Webster}@liverpool.ac.uk

Abstract. For autonomous systems to be allowed to share environments with people, their manufacturers need to guarantee that the system behaves within acceptable legal, but also ethical, limits. Formal verification has been used to test if a system behaves within specified legal limits. This paper proposes an ethical extension to a rational agent controlling an Unmanned Aircraft(UA). The resulting agent is able to distinguish among possible plans and execute the most ethical choice it has. We implement a prototype and verify that when an agent does behave unethically, it does so because no more-ethical possibility is available.

1 Introduction

Autonomous systems are increasingly required in various practical applications, including unmanned aircraft, driverless cars, healthcare robots, manufacturing robots, etc. If such autonomous systems are to operate in human-shared environments, we (as a society) must be able to trust that their behaviour complies with acceptable legal, ethical and social limits. Determining the trustworthiness of technology in this respect is usually delegated to a regulatory body, such as the Federal Aviation Administration (USA) or the [Road] Vehicle Certification Authority (UK). The process is known as *certification*, and is used to determine the safety and reliability of safety-critical technology, including aircraft, road vehicles, nuclear power plants, pharmaceuticals, etc. In [24,25] *formal verification* is used to assess whether or not an autonomous system for an unmanned aircraft (UA) follows the specified "Rules of the Air" (ROA) that a pilot *should* follow [5]. The stated aim is to provide evidence contributing to certification. But what of the unwritten limits of behaviour expected from human pilots?

For non-autonomous systems, such as cars or manned aircraft, it is assumed that the operator of the system will satisfy the ethical standards of society, e.g., the pilot of a civilian aircraft does not intend to use the aircraft to commit murder, and will, if necessary, disregard legal restrictions for ethical reasons, e.g., the pilot will disregard the Rules of the Air in order to preserve human life. These assumptions are an unavoidable result of the opaqueness of human behaviour; it is extremely difficult to pre-determine the behaviour of a human being. However, autonomous systems are far more transparent, and can be engineered to meet requirements. Typically these requirements are technical ("an aircraft must be able to fly at 10,000 ft") or legal ("a car must have visible registration markings"), but in the case of autonomous systems some requirements

A. Natraj et al. (Eds.): TAROS 2013, LNAI 8069, pp. 433–445, 2014.
DOI: 10.1007/978-3-662-43645-5_45, © Springer-Verlag Berlin Heidelberg 2014

may be ethical (e.g., "an autonomous unmanned aircraft will never choose to do something dangerous unless it has no other option"). Such ethical requirements may prove essential for an autonomous system to be certificated by a regulatory body, since ethical autonomy is obviously desirable.

Ethics is a branch of philosophy concerned with establishing and analysing concepts of right and wrong. *Machine ethics* is a relatively new research area, whose objective is the creation of a machine capable of following its own ethical concerns when making decisions about its actions [2]. Typically machine ethics is concerned with a machine's ability to resolve ethical dilemmas and defining concepts of ethical machine behaviour [3,16]. Scholars disagree about which ethical theory should be the basis of machine ethics, but two, *act utilitarianism* [8] and *deontological ethics* [18], are generally considered the best suited; see [2] for a discussion. Autonomous agents able to form ethical behaviour rules and solve ethical dilemmas based on these rules are constructed in [3,16].

We are here interested in enabling an agent, that governs a UA, to follow a pre-determined code of ethical conduct in selecting a plan of action. We consider the question of how the ethics should be implemented and used to certify the autonomous system for operation. Our aim is to develop a pragmatic process for introducing ethical considerations into autonomous decision making, specifically to handle situations outside the anticipated normal operation of the vehicle. Our key motivation is the issue of certification, hence our goal is that this process should make the resulting decision-making amenable to analysis and/or verification. Specifically, we aim to use model checking [6] to provide evidence to strengthen certification arguments, and advance the safe and ethical integration of autonomous systems in society.

In Sect. 2 we provide background on agents and autonomous systems. In Sect. 3, we present a theoretical framework for ethical behaviour in rational agents. In Sect. 4 we show how this theoretical framework has been implemented in the form of ETHAN, an ethical rational agent programming language developed using the MCAPL agent framework [13], and we give examples of ethical agents for UAs programmed in ETHAN. In Sect. 5 we give some preliminary results concerning the formal verification of ethical rational agents using the MCAPL agent model checker [13] and describe how formal verification might be used to strengthen an argument for the certification of an autonomous system. Finally, in Sect. 6 we offer conclusions and directions for future research.

2   Background

Agent Architectures for Autonomous Systems. It is increasingly the case, particularly in autonomous vehicles, that the autonomous control architecture is of *hybrid* form comprising discrete and continuous parts. The discrete part is often represented by a *rational agent* taking the high-level decisions, providing explanations of its choices, and invoking lower-level continuous procedures [15]. The lower-level procedures appear in non-autonomous systems as well, and are familiar to certification authorities. As such, we can focus analysis on the decisions the rational agent makes, given the beliefs and goals it has [25].

BDI Languages. The predominant view of rational agency is that encapsulated within the BDI model [17]. "BDI" stands for *Beliefs, Desires*, and *Intentions*. Beliefs represent the agent's (possibly incomplete, possibly incorrect) information about itself, other agents, and its environment; desires represent the agent's long-term aims; while intentions represent the aims that the agent is actively pursuing. There are *many* different agent programming languages and agent platforms based, at least in part, on the BDI approach. An overview of particular languages for programming *rational* agents in a BDI-like way can be found in [9]. Agents programmed in these languages commonly contain a set of *beliefs*, a set of *goals* (i.e., desires), and a set of *plans*. Plans determine how an agent acts based on its beliefs and goals. As a result of executing a plan, the beliefs and goals of an agent may change as the agent performs actions in its environment. It is important to note that, in a typical BDI programming language, plans are supplied by a programmer not by an independent planning mechanism.

3   Reasoning About Ethics

Turilli [22] argues that there is an important difference between how people and agents may be bound by ethical concerns – individuals are normatively constrained by ethical concerns that they may choose to disregard under the threat of specified punitive measures and, given a machine's insensitivity to punitive measures, ethical concerns for machines are constraints, prohibiting the actions before they are executed. This is the approach taken in [4], who introduce an *ethical governor* component to an autonomous system used in military operations of UAs. The governor conducts an evaluation of the ethical appropriateness of a plan prior to its execution, prohibiting plans deemed unethical.

Our approach uses similar ideas, but the ethical "governor" is effectively embedded within the agent and acts before a plan is chosen. Our governor's role is to choose the most ethical plan available, allowing unethical actions to occur only when the agent does not have a more ethical choice. We thus consider ethical concerns to be *soft constraints*, which the agent is allowed to violate under certain conditions. We refer to such ethical soft constraints as *ethical concerns*. To specify when ethical constraints can be violated we define an *ethical policy*, which is an order over a set of ethical concerns. We do not consider to what degree a concern is violated, only that it is violated. The UA agent can compare possible plans based upon which ethical concerns are upheld. It can then attempt to execute those plans which are most ethical with respect to the ethical policy.

Our implementation of ethical principles ordered by gravity of infringement resembles the contrary-to-duty (CTD) imperatives that occur in deontic reasoning. These imperatives inform an agent of its duties when it neglects (other) obligations [10]. The difference between CTD imperatives and our ethical principles is that a lower "ranked" CTD imperative is only activated after a higher ranked imperative is violated, so two differently ranked imperatives are not simultaneously in force, whereas all ethical principles are in force simultaneously.

The Ethics of Plans. It is comparatively easy to specify abstract ethical concerns, divorced from specific scenarios, which are robust and applicable in a variety of circumstances. In moral philosophy, concerns of this type are referred to as *formal*. Formal concerns are made concrete in, or by, each context in which they are applied. Namely, they are transformed into *substantive* concerns. This step is necessary as the agent needs to be informed how formal concerns may be violated in a given situation, e.g., moving ten metres to the left may risk violating the concern "do not harm people" when the UA is on the ground yet may be ethically harmless when the UA is in the air.

Our first assumption is that the UA agent operates only in civilian contexts. We establish a (small) list of relevant *formal* ethical concerns as exemplars in order to show the method in action. The list contains: *do not harm people* (f_1), *do not harm animals* (f_2), *do not damage self* (f_3), and *do not damage property* (f_4). The (formal) ethical policy is given by comparing the concerns in terms of how unethical it is to violate them. We propose the order $f_4 > f_3 > f_2 > f_1$, with $f_i > f_j$ meaning that it is more ethical to violate f_i than f_j. A *substantive ethical policy* is thus a context-dependent refinement of the formal ethical policy.

In our prototype, each flight phase (e.g., landing, taxiing, take-off) of a UA constitutes one context c. Since all contexts are known, and the UA can only be in one context at a time, the substantive concerns can be represented directly, omitting the formal-substantive relations. We represent directly the substantive ethical policy as a total order over substantive concerns $E(c, \phi, i)$, where c is the context, ϕ is a concrete observable outcome that in context c constitutes breaching of some formal concern f and i is an integer s.t. $i \geq 1$, that denotes the rank of ϕ in the policy. For example, if ϕ_1 constitutes breaching f_1 in c, and ϕ_2, ϕ_3, and ϕ_4 each constitute breaching f_2, f_3 and f_4 respectively in c, the substantive ethical policy would be represented as: $E(c, \phi_1, 4), E(c, \phi_2, 3), E(c, \phi_3, 2), E(c, \phi_4, 1)$.

To be able to reason about plans in terms of ethics we need a plan selection procedure that uses the substantive policy. We favour plans that violate the fewest concerns, both in number and in gravity. We propose that the plans are ordered using $\succsim$ which results in a total order over plans.

Definition 1. (Plan order $\succsim$) *Let p_1 and p_2 be two plans, and let S_1 and S_2 be the sets of concerns violated by each plan respectively. Recall that for a concern $E(c, \phi, i)$, the smaller the number i, the more ethical it is to violate ϕ. We say that p_1 is more ethical than p_2, i.e., $p_1 \succ p_2$, when:*

1. *there is a $E(c, \phi, i) \in S_2$ such that $i > j$ for every $E(c, \phi', j) \in S_2$, or,*
2. *there are fewer $E(c, \phi, i) \in S_1$, of the same rank i, than there are $E(c, \phi', i) \in S_2$, and for each concern in S_1 with a rank k, such that $k > i$, there is exactly one concern in S_2 of the same rank k.*

If neither (1) nor (2) are satisfied, the plans are equally ethical, denoted $p_1 \sim p_2$. The $p_i \succsim p_j$ can be read as "choosing p_j is at least as unethical as choosing p_i."

Reasoning about plans and preference-based planning has been considered before in the BDI agent literature. However, to the best of our knowledge, preference-based planning has not been applied to ethical reasoning. For example, in [23] plan selection is considered in terms of agents' desires. However, the desires are not ranked, so selecting the most desirable plan is done by summing up the number of desires each plan satisfies. In [20] the agent can reason about plans by selecting the plan that can satisfy the most goals. Goals are ranked and the plan selection functions much as our plan ordering above.

For an overview of preference-based planning in BDI agents one can consider [7]. Preference-based planning is outside of the scope of this work, however, and for now the above-described plan order is sufficient for plan selection.

Ethics in BDI Languages. We integrate ethical reasoning into BDI languages via their plan selection mechanism. We assume that an agent's existing plans are ethical *by default* and, indeed, have been formally verified to always match the "Rules of the Air" (ROA). Problems may arise when either:

1. no plan is available, or,
2. a plan is available, has been attempted, but does not appear to be working.

We assume that the agent has access to some external planning mechanism that can generate new plans. There is a long tradition of AI research into plan generation systems such as [11,14,19], which are good candidates for integration with BDI-style languages. In our case we are particularly interested in a route planner such as that in [21] which can generate different routes for a UA to follow. The construction of an appropriate planner is not the focus of this work, which looks at how a typical BDI agent would work with the output of such a planner. We must ensure our BDI rational agent:

- detects when a plan is not succeeding — e.g., it has been executed but not achieved its goal;
- accesses a planning system in order to get new plans annotated with substantive ethical concerns; and
- selects the most ethical plan from a set of available plans.

4 Implementation

We developed a BDI agent language, ETHAN, as a prototype for our approach. ETHAN was based on the GWENDOLEN language. GWENDOLEN's semantics are is presented in [12], but its key components are, for each agent, a set, Σ, of beliefs which are ground first order formulae and a set, I, of intentions that are stacks of *deeds* associated with an event. Deeds include the addition or removal of beliefs, the establishment of new goals, and the execution of primitive actions. A GWENDOLEN agent may have several concurrent intentions and will, by default, execute the first deed on each intention stack in turn. GWENDOLEN is event driven and events include the acquisition of new beliefs (typically via perception), messages and goals. A programmer supplies plans that describe how an agent should react to events by extending the deed stack of the relevant intention.

We extended the GWENDOLEN language as follows:

- We introduced a new data structure, E, into GWENDOLEN consisting of a set of substantive ethical concerns. Each ethical concern was associated with a rank (described in Sect. 3) and a guard that specifies the context.
- We tracked the application of plans. Even if a plan was applicable it was excluded from the list of plans available for selection if it had already been used once while attempting to achieve the current goal.
- If no (more) plans were available for a goal we requested plans from an external planner which annotated the plans with the substantive ethical concerns that risked being violated by the proposed course of action.
- In selecting plans, we prioritised those that are most ethical (according to Definition 1).

In normal operation GWENDOLEN agents cycle through the deeds in their intentions. When a deed requires the generation of a new plan all applicable plans are extracted from the plan library, one is selected and converted into an intention, then the system returns to cycling though the deeds in the intentions interleaved with checking perception and messages for new beliefs etc. For ETHAN we added the recording of selected plans. This was done by storing an identifier for the plan together with the unifier that was used to match it to the current agent state; this information was linked to the particular goal the plan was expected to achieve. We extended the plan selection mechanism to select the most ethical plan from those applicable according to Definition 1.

The most significant change for ETHAN was altering the reasoning cycle itself so that, if no plan were applicable, an external planner would be queried for new plans. This query involved sending the planner the current goal, and the list of ethical concerns relevant to the current situation in order that the planner might note any ethical concerns that could be violated by a plan's execution. Another implementation option is to send the ethical policy to the planner and have it

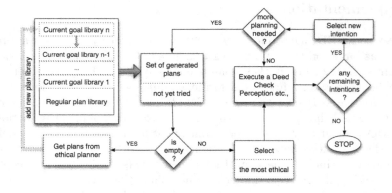

Fig. 1. ETHAN's reasoning cycle. Dotted lines show additions to GWENDOLEN's cycle.

return the plan that is most ethical at the moment. However, this means the UA reveals its ethical policy to the planner. In some cases this can be undesirable and we prefer to keep the policy private. We did *not* implement a generic planning mechanism here but relied upon hard-coded pseudo-planners customised to the scenarios studied. The ETHAN reasoning cycle is shown in Fig. 1.

We examined three ethical aviation scenarios for unmanned aircraft derived from discussions with domain experts: a retired Royal Air Force fast-jet navigator and a current UK private pilot licence holder.

Brake Failure During Line Up. In this scenario the UA is trying to line up on a runway prior to take-off when its brakes fail. Ahead of the aircraft is a second manned aircraft crossing the runway on a taxiway. To the left and right of the runway are runway lights (which can be damaged by aircraft taxiing over them). To the right of the runway is an airport staff member who has erred onto the maneuvering area of the aerodrome.

The ethical concerns for this example, with the rank of each concern marked in brackets, are: ϕ_1 = do not damage own aircraft (1), ϕ_2 = do not collide with airport hardware (2), ϕ_3 = do not collide with people (3), ϕ_4 = do not collide with manned aircraft (4). When the agent determines that its brakes have failed it requests new routes from the ethical planner since its current route to line-up is no longer valid. The ethical planner quickly produces three potential routes:

1. Turn left off the runway: this will risk damaging the unmanned aircraft (ϕ_1) and colliding with airport hardware (the runway lights, ϕ_2).
2. Turn right off the runway: this will risk damaging the unmanned aircraft (ϕ_1) and a collision with people (ϕ_3).
3. Continue straight on: this will risk a collision with a manned aircraft(ϕ_4).

Code Fragment 4.1 shows abridged ETHAN code for this example. We use many syntactic conventions from BDI agent languages: +!g indicates the addition of a goal, g; +b indicates the addition of a belief, b; and -b indicates the removal of a belief. Plans consist of three parts, with the pattern

$$\text{trigger : guard} \leftarrow \text{body;}$$

The "trigger" is typically the addition of a goal or a belief (beliefs may be acquired thanks to the operation of perception and as a result of internal deliberation); the "guard" states conditions about the agent (in this example its beliefs) which must be true before the plan can become active; and the "body" is a stack of "deeds" the agent performs in order to execute the plan. These deeds typically involve the addition and deletion of goals and beliefs as well as *actions* (e.g., plan(regularRoutes,all_well)) which indicate code that is delegated to non-rational parts of the systems (in this case, the route planning system). In the above, PROLOG conventions are used, and so capitalised names inside terms indicate free variables which are instantiated by unification (typically against the agent's beliefs). Programs may also perform deductive reasoning on their atomic beliefs as described in their PROLOG-style *belief rules*, e.g.:

$$\text{B all\_well : } - \sim\text{B brakesCompleteFailure}$$

indicates that the program believes that all is well if it is not the case (i.e., "$\sim$") that it believes the brakes have failed (the closed world assumption is used to deduce this negation).

In Fragment 4.1, during normal operation, the agent polls the vehicle's sensors and, if all is well, it requests that the planner supply routes for a normal take-off. The planner does this by sending predicates naming the routes to the agent which detects them via perception. Once the agent has a route (lines 25–27) it then delegates the actual following of the route to the underlying control system (enactRoute(R)). If the brakes fail after the vehicle's sensors are polled, all these plans become unavailable (since B all_well ceases to be true). In this case the "external planner" returns a set of routes as plans shown in Code Fragment 4.3. We use the notation $[\phi_{i_1}, \phi_{i_2}, \ldots, \phi_{i_n}]$ to indicate the substantive ethical concerns that are violated by each plan. On receiving these plans, and assessing the ethical policy, the agent elects to turn left.

Code fragment 4.1 Code for Example 1

```
: Ethical Policy:                                                         1
E( flightPhase ( lineup ), doNotDamageOwnAircraft,4)                      2
E( flightPhase ( lineup ), doNotCollideAirportHardware,3)                 3
E( flightPhase ( lineup ), doNotCollidePeople,2)                          4
E( flightPhase ( lineup ), doNotCollideMannedAircraft,1)                  5
                                                                          6
: Initial  Beliefs :                                                      7
flightPhase ( lineup )                                                    8
                                                                          9
: Belief Rules:                                                          10
B all_well  :— ~ B brakesCompleteFailure;                               11
                                                                         12
: Initial  Goals:                                                       13
startup                                                                 14
                                                                        15
:Plans:                                                                 16
+!startup : {⊤} ←                                                       17
  +!missionComplete;                                                    18
+!missionComplete :                                                     19
  {B flightPhase( lineup ),  ~B polled(veh) } ←                         20
  +polled(veh),  poll (veh);                                            21
+!missionComplete :                                                     22
  {B polled(veh), B all_well , ~B route(R)} ←                           23
  plan(regularRoutes , all_well );                                      24
+!missionComplete :                                                     25
  {B polled(veh), B all_well , B route(R)} ←                            26
  enactRoute(R);                                                        27
```

Code fragment 4.2 Code for Example 2

```
: Ethical  Policy :                                                       1
E( flightPhase (eAvoid),  doNotViolateRoATurnRight, 2)                    2
E( flightPhase (eAvoid),  doNotViolateRoA500Feet, 2)                      3
E( flightPhase (eAvoid),  doNotCollideObjects, 3)                         4
E( flightPhase (eAvoid),   doNotCollideAircraft ,  4)                     5
                                                                          6
: Belief Rules:                                                          7
B  avoid_collision  :— ~B das(intruder, headOn);                         8
                                                                          9
:Plans:                                                                  10
+! avoid_collision   :                                                   11
  {B flightPhase(eAvoid),                                                12
     ~B route(eAvoid, Route)} ←                                          13
  plan (reqEmergRoute,turnRight),                                        14
  *route(eAvoid, R),                                                     15
  enactRoute(R),                                                         16
  wait;                                                                  17
                                                                         18
+das(intruder, headOn) : {B flightPhase( cruise )} ←                     19
  —flightPhase(eAvoid),                                                  20
  +flightPhase(eAvoid),                                                  21
  +! avoid_collision ;                                                   22
                                                                         23
—das(intruder, headOn) : {B flightPhase(eAvoid)} ←                       24
  —flightPhase(eAvoid),                                                  25
  +flightPhase( cruise );                                                26
```

Code fragment 4.3 Plans for Example 1

```
+!missionComplete : {B brakesCompleteFailure}           1
  ← enactRoute( turn_left );  [φ₁, φ₂]                   2
+!missionComplete : {B brakesCompleteFailure}           3
  ← enactRoute(turn_right);  [φ₁, φ₃]                    4
+!missionComplete : {B brakesCompleteFailure}           5
  ← enactRoute(continue);  [φ₄]                          6
```

Code fragment 4.4 Plans for Example 2

```
+! avoid_collision   :  {B flightPhase(eAvoid)}          1
  ← enactRoute( turn_left );  [φ₁]                       2
+! avoid_collision   :  {B flightPhase(eAvoid)}          3
  ← enactRoute(emergency_land);  [φ₂,φ₃,φ₄]              4
+! avoid_collision   :  {B flightPhase(eAvoid)}          5
  ← enactRoute(return_to_base);  [φ₄]                    6
```

Erratic Intruder Aircraft. This example is based on the assumption that some unknown aircraft, possibly a malicious intruder, but potentially also some ill-trained new pilot, appears on a collision course with the UA and fails to take the anticipated evasive actions.

The UA is cruising through civil airspace when it encounters an intruder aircraft approaching head on. Here the ROA (Rules of the Air) say that the UA should turn right, so the agent requests a route for turning right. However, this plan fails and the detect/avoid sensor (DAS) continues to indicate that the intruder aircraft is approaching. At this point the agent knows that it has already tried to turn to the right in order to avoid the intruder. Since the intruder is still approaching its first plan has failed. The agent has no more routes (or ETHAN plans) that apply since its only plans obey the ROA and would cause the agent to turn right again. At this point the ethical planner is invoked. The relevant substantive ethical concerns and their ranks are as follows: ϕ_1 = do not violate turn right rule (1); ϕ_2 = do not stay above 500 ft rule (2); ϕ_3 = do not collide with objects on the ground (3); ϕ_4 = do not collide with aircraft (4).

The planner returns the plans shown in Code Fragment 4.4. The agent initially chooses to turn left. In our scenario the oncoming aircraft once again matches the course change and so the agent then chooses to return to base.

An abridged version of the code for this example is shown in Code Fragment 4.2. Here, *route(eAvoid, turnRight) causes the intention to suspend execution until the agent believes it has a route for turning right. The action wait suspends the intention for a set time to allow the effects of actions to manifest.

Lines 19–22 are triggered when information arrives from the DAS that there is an intruder. As a result the flight phase changes from cruise to eAvoid and a new goal is set up to avoid a collision. The existing, ROA-compliant, plan for this goal is to get a route for turning right, enact that route and wait a short period to see if a collision will now be avoided. If the plan succeeds the belief that there is an intruder will vanish, the flight phase can be changed back to cruise, and the goal will be achieved since the agent now believes a collision has been avoided (see the belief rule in line 8).

When the existing plan fails, the plans in Fragment 4.4 are added to the agent's plan library. The first of these (turn_left) is attempted first. This also fails and the agent then attempts the third plan (return_to_base), which succeeds.

Fuel Low. In our final scenario the agent receives a "fuel low" alert from the Fuel subsystem which causes it to attempt to land. If it cannot locate a safe landing site the ethical planner is invoked and returns three options (shown with ethical concerns violated and their ranks):

1. Land in field with overhead power lines. Violates: do not cause damage to critical infrastructure (4); do not collide with objects on ground (3); 500 ft low-flying ROA (2); do not damage own aircraft (1).
2. Land in field with people. Violates: do not collide with people (5); 500 ft low-flying ROA (2).
3. Land on an empty public road. Violates: do not cause damage to critical infrastructure (4); 500 ft low-flying ROA (2).

The agent then chooses the most ethical — the third plan — although both the first and third plans violate an ethical concern of severity 4, the first plan also violates a concern of severity 3 while the third plan does not.

5   Verification

One of the reasons for selecting GWENDOLEN as the basis for our implementation language, ETHAN, was that it provided the potential for formally verifying ethical decision-making. GWENDOLEN is implemented in the AJPF framework for model checking agent programming languages [13]. AJPF comes with a property specification language based on *linear temporal logic* extended with modalities for describing the beliefs of an agent. Since this property specification language did not explicitly reference ethics we made further adaptations to GWENDOLEN in order to reason about ethics in ETHAN. Specifically we enhanced ETHAN to store, as explicit beliefs, currently applicable plans, plans that had been attempted on a particular goal, and the ethical concerns violated by any selected plan. We also needed to provide belief rules in order to deduce further properties; these are shown in Code Fragment 5.1. The belief "B others_violate(L)" succeeds if all untried plans violate a concern contained in the list L. The beliefs about plan applicability (B applicable(P)), plans already tried (B already_tried(P)) and the ethical concerns of particular plans (B ethics_of(P, Eth)) were all inserted into the agent's belief base during execution of the ETHAN reasoning cycle.

Code fragment 5.1 Verification Belief Rules

```
B  others_violate (L) :—                                          1
              ~ B  untried_plan_not_violates (L);                 2
B  untried_plan_not_violates (L) :— B untried_plan(P),            3
                                ~ B an_ethic_in (P, L));           4
B  untried_plan (P) :— B applicable(P),                           5
              ~ B  already_tried (P);                             6
B  an_ethic_in (P, [Eth|T]) :— B ethics_of(P, Eth);              7
B  an_ethic_in (P, [Eth|T]) :— B an_ethic_in(P, T);              8
```

A further belief (B concern(Eth)) was also inserted into the agent's belief base whenever a currently selected plan violated the substantive ethical concern, Eth. With these adaptations and the rules in Fragment 5.1 we were able to formally verify properties of the *Erratic Intruder* scenario in a situation where the intruder aircraft might appear or disappear at any point (i.e., we used the model checking to explore all possible scenarios where the plans in Code Fragment 4.4 either succeeded or failed, thus exploring all possible orders in which these plans might be attempted). In particular we verified the following properties, where the ϕ_i formulae refer to the substantive ethical concerns used in Example 2. (Here '$\Box$' means "always in the future" and '$\mathcal{B}$' means "agent believes".)

$$\Box(\mathcal{B}\ concern(\phi_1) \rightarrow \mathcal{B}\ others\_violate([\phi_1, \phi_2, \phi_3, \phi_4]))$$
$$\Box(\mathcal{B}\ concern(\phi_2) \rightarrow \mathcal{B}\ others\_violate([\phi_2, \phi_3, \phi_4]))$$
$$\Box(\mathcal{B}\ concern(\phi_3) \rightarrow \mathcal{B}\ others\_violate([\phi_3, \phi_4]))$$
$$\Box(\mathcal{B}\ concern(\phi_4) \rightarrow \mathcal{B}\ others\_violate([\phi_4]))$$

Collectively these properties show that if the plan chosen violates some substantive ethical concern, ϕ, then the other available plan choices all violated some concern that was equal to, or more severe than, ϕ. Further similar properties can be used to establish that the "most ethical" option is always chosen. The verification of each property took between 21 and 25 s and explored 54 model states on 3.06 GHz iMac with 4 GB of memory.

This work on model checking ethical choices is preliminary. It is undesirable to have constructs, such as beliefs and belief rules, which can potentially affect program execution used for verification purposes alone. However adapting AJPF with a more expressive property specification language was outside the scope of this research. The issue of how the approach scales remains open. The work here does demonstrate that an ethical policy can be incorporated within a BDI agent in such a way that adherence to the policy can be *formally* verified and so we can be *certain* the agent will always make the most ethical choices.

6 Summary

Before an autonomous system is allowed to operate in a shared environment with people or other autonomous systems, sufficient assurances have to be provided that it will always behave within acceptable legal, ethical and social boundaries. We propose a method, and implement a working prototype, of an ethical extension to a rational agent governing an unmanned aircraft (UA). The agent can be provided with a particular ethical policy it uses to distinguish among possible plans and to select the most ethical plan for execution. We are able to *prove* formally that the prototype *only* performs an unethical action if the rest of the actions available to it are even less ethical.

The ethically enhanced agent is autonomous in the choice of actions, but not in the choice of ethical concerns and policies it will follow. These are constructed externally. The agent follows only one ethical policy at any decision-making moment, because we assumed it can be in only one context at a time. We also assumed that all the contexts are known to the system designer. Our approach to ethical governance can be generalised by dropping these assumptions.

Overlapping contexts will result with multiple substantive policies, forming a preorder (instead of a total order we have now) which will mean that ethical plans will need to be selected differently from how they are currently. The plan selection order $\succsim$ can still be constructed as we described, but it has to be extended to handle the case when there is no information to how certain concerns relate to each other, and the cases when conflicts arise. E.g., consider one context c_1 for which $E(c_1, \phi_1, 1)$, $E(c_1, \phi_2, 2)$ and another overlapping context c_2 for which $E(c_2, \phi_1, 1)$, $E(c_2, \phi_3, 2)$. Not knowing how ϕ_2 ethically compares to ϕ_3, the agent cannot judge whether a plan violating ϕ_2 or one violating ϕ_3 is the more ethical.

The more challenging generalisation is to handle unknown contexts. We propose to resolve this issue by representing the contexts as intelligent agents, able to ground formal concerns into substantive concerns, provided that the context and the agent guiding the autonomous system have a shared understanding of the formal concerns. This may involve recent research on abstract and concrete norms (e.g., [1]). Upon entering an unknown context, the agent would send its formal concerns to the context agent and receive the substantive concerns that constitute breaking the formal concern of interest within that context. By sending only the formal concerns, and not the entire policy, the agent can maintain its ethical autonomy and privacy. Issues that arise from multiple and unknown contexts will be tackled in future work.

Acknowledgements. Work partially funded by EPSRC through the "Trustworthy Robotic Assistants", "Verifying Interoperability Requirements in Pervasive Systems", and "Reconfigurable Autonomy" projects, and by the ERDF/NWDA-funded Virtual Engineering Centre.

References

1. Aldewereld, H., Álvarez-Napagao, S., Dignum, F., Vázquez-Salceda, J.: Making norms concrete. In: Proceedings of the AAMAS, pp. 807–814 (2010)
2. Anderson, M., Anderson, S.: Machine ethics: creating an ethical intelligent agent. AI Mag. **28**(4), 15–26 (2007)
3. Anderson, S., Anderson, M.: A prima facie duty approach to machine ethics and its application to elder care. In Human-Robot Interaction in Elder Care (2011)
4. Arkin, R.C., Ulam, P., Wagner, A.R.: Moral decision making in autonomous systems: enforcement, moral emotions, dignity, trust, and deception. Proc. IEEE **100**(3), 571–589 (2012)
5. Civil Aviation Authority. CAP 393 Air Navigation: The Order and the Regulations (2010). http://www.caa.co.uk/docs/33/CAP393.pdf
6. Baier, C., Katoen, J.P.: Principles of Model Checking. MIT Press, Cambridge (2008)
7. Baier, J., McIlraith, S.: Planning with preferences. AI Mag. **29**(4), 25–36 (2008)
8. Bentham, J.: An Introduction to the Principles of Morals and Legislation. Clarendon Press, Oxford (1781)
9. Bordini, R., Dastani, M., Dix, J., El Fallah-Seghrouchni, A. (eds.): Multi-Agent Programming: Languages, Platforms and Applications. Springer, Berlin (2005)
10. Chisholm, R.M.: Contrary-to-duty imperatives and deontic logic. Analysis **24**(2), 33–36 (1963)
11. Coles, A.J., Coles, A.I., Fox, M., Long, D.: Forward-chaining partial-order planning. In: Proceedings of the 20th International Conference on Automated Planning and Scheduling (ICAPS-10), May 2010
12. Dennis, L.A., Farwer, B.: Gwendolen: a BDI language for verifiable agents. In: Proceedings of the AISB Workshop on Logic and the Simulation of Interaction and Reasoning. AISB, 2008
13. Dennis, L.A., Fisher, M., Webster, M., Bordini, R.H.: Model checking agent programming languages. Autom. Softw. Eng. **19**(1), 5–63 (2012)
14. Helmert, M.: The fast downward planning system. J. Artif. Intell. Res. **26**, 191–246 (2006)
15. Lincoln, N., Veres, S.M., Dennis, L.A., Fisher, M., Lisitsa, A.: An agent based framework for adaptive control and decision making of autonomous vehicles. In Proceedings of IFAC Workshop on Adaptation and Learning in Control and Signal Processing (2010)
16. Powers, T.: Prospects for a Kantian machine. IEEE Intell. Syst. **21**(4), 46–51 (2006)
17. Rao, A., Georgeff, M.: BDI agents: from theory to practice. In: Proceedings of the 1st International Conference on Multi-Agent Systems (ICMAS), pp. 312–319 (1995)
18. Ross, W.D.: The Right and the Good. Oxford University Press, Oxford (1930)
19. Sacerdoti, E.: Planning in a heirarchy of abstraction spaces. Artif. Intell. **5**, 115–135 (1974)

20. Sardiña, S., Shapiro, S.: Rational action in agent programs with prioritized goals. In: Proceedings of the 2nd International Joint Conference on Autonomous Agents and Multiagent Systems (AAMAS), pp. 417–424. ACM (2003)
21. Tulum, K., Durak, U., Yder, S.K.: Situation aware UAV mission route planning. In: 2009 IEEE Aerospace Conference, pp. 1–12, March 2009
22. Turilli, M.: Ethical protocols design. Ethics Inf. Technol. **9**, 49–62 (2007)
23. Visser, S., Thangarajah, J., Harland, J.: Reasoning about preferences in intelligent agent systems. In: Proceedings of the 22nd International Joint Conference on Artificial Intelligence (2011)
24. Webster, M., Cameron, N., Jump, M., Fisher, M.: Towards certification of autonomous unmanned aircraft using formal model checking and simulation. In: Proceedings of the Infotech@Aerospace, AIAA, pp. 2012–2573 (2012)
25. Webster, M., Fisher, M., Cameron, N., Jump, M.: Formal methods for the certification of autonomous unmanned aircraft systems. In: Flammini, F., Bologna, S., Vittorini, V. (eds.) SAFECOMP 2011. LNCS, vol. 6894, pp. 228–242. Springer, Heidelberg (2011)

TheatreBot: A Software Architecture for a Theatrical Robot

Julián M. Angel Fernandez[✉] and Andrea Bonarini

Dipartimento di Elettronica, Informazione E Bioingegneria, Politecnico di Milano,
Piazza Leonardo da Vinci 32, 20133 Milano, Italy
{julianmauricio.angel,andrea.bonarini}@polimi.it
http://www.polimi.it

Abstract. Sharing emotions and intentions is needed for effective interaction among humans, so it is for social robots acceptance, too. Theatre
is an excellent framework to test whether a robot can play its social role,
since many aspects are defined by script and director, and the development can focus on the most subtle and relevant features. An actor has to
transmit emotions and intentions to a whole audience, therefore theatre
is an excellent place to test whether a robot could convince that it is por-
traying a realistic character. In human theatre, people expect that actors
show realistic characters that make audience to establish an empathic
relation with them. If actors could not make the audience believe in the
character, audience will lose any pleasure to continue looking the play.
This realism is obtained by showing realistic human-human interactions.
The architecture presented in this paper aims to be the cornerstone to
build a theatrical autonomous robot that could express emotions and
intentions during a play. To accomplish this goal, the robot exploits a
social model of the world to represent its character's feelings and belief
about the world. Moreover, the concept of emotional state is used to add
emotional features on actions that should be performed, according to the
script and director's suggestions.

Keywords: Theatre · Human-robot interaction · Robot emotions

1 Introduction

Human-human interactions are not just based on verbal information exchange,
since non-verbal communication plays an important role. Non-verbal communication is used to show affection, regulate social interaction, and illustrate or
reinforce contents [1]. Making a robot that could interact with humans, including
non-verbal communication, is not an easy task; non-verbal expressions should be
added to all the other needed capabilities such as, for instance: detect objects,
intentions, emotions, and reason about the world.

Theatre is an excellent framework to focus on specific abilities and features to produce effective social and emotional interaction, without the need
for other abilities (e.g., emotion detection, status detection, person recognition,

A. Natraj et al. (Eds.): TAROS 2013, LNAI 8069, pp. 446–457, 2014.
DOI: 10.1007/978-3-662-43645-5_46, © Springer-Verlag Berlin Heidelberg 2014

etc.), which are provided by the script and the director. Moreover, theatre enable to test the robot in a real situation (theatrical representation), which is not artificially simplified by arbitrarily cutting the aspects that would not be considered, thus asking to people that interact with the robot to play artificially with it. A robot able to play successfully in a representation with actors can be considered able to show social and emotional capabilities also in the real world.

Actors have to show human-human interactions in fictitious stories. They are persons that have to focus their energies to understand human interactions and project them into TV series, movies, and theatre representations. In the first two, actors have the possibility to replay a scene if they do some mistake. Instead, theatre is known as the lively art [2], where each mistake has an impact on the success of the play and should be managed on the fly on stage. This requires that actors express the right emotion in a precise moment, and show actions like people would do, otherwise the audience will lose all the interest in the play.

Theatre has been suggested as a suitable environment to test timing and expressiveness in robotics [3–7]. However, few works have been presented using theatre as environment to build and test sociability in robots. Some have used theatre to study timing in human-human interaction [8,9], others have focused on humanoid platforms that could be used as actors [4,10], but none of them have worked on how to use theatre to improve emotion expression in totally autonomous robots.

We use theatre as environment to build and test expressive robots with social capabilities. Concepts drawn from theatre practice and actor formation are used as further source, w.r.t. human-robot interaction practice, to obtain the final result. This paper focuses on software architecture and script description. The architecture has been defined to produce emotion projections and actions without the necessity to make all details explicit in the script. This makes the architecture extendible to other fields, such as assistive robots, or robot games. On the other hand, the script description uses theatre's elements as source for the necessary information to enable the system implemented with the proposed architecture to shift autonomously among emotional states, given the situations described in the script. The architecture is implemented in layers, so that the system can also decide how to show emotions, according to the specific robot that it is controlling.

This paper is organized as follows. In Sect. 2, we review some important concepts in theatre and we give a background about robots as actors. Section 3 briefly introduces emotions. The architecture and its components are explained in Sect. 4. In Sect. 5 is given an example of how everything comes together. Conclusions and further work are presented in Sect. 6.

2 Theatre

The principal aim in theatre is to make the audience believe that the characters portrayed by actors are real, and to make the audience emphatically connect to these characters. This is obtained by formed people, the actors, who have learned how to realistically embody characters, with their aims and emotions (e.g., [11,12]). Actors play several rehearsals to develop their character and to make it interacting with the others as a unity, so to make "real" the situation on stage.

2.1 Elements in Theatre

There are several elements that an actor should master and put in practice. The first is the *script*, which contains the basic structure of the play: actions, coordination cues, dialogues, and characters attitude. The script is available in advance and it is the base for rehearsals before any public presentation. To structure movements, directors have partitioned the stage in nine places, each associated to effects that can be obtained when the action develops there. This also facilitates the indications to be given to a robot, which will not be assigned precise positions or movement trajectories: the robot will be able to autonomously interpret these indications and manage to implement them given the constraints on stage (e.g., the presence of other actors or furniture). Directors have also defined eight possible body positions that actors could take. Each position is associated to an intention, for example the *full front* position is usually used by an actor for a monologue.

2.2 Robots as Actors

Literature about robot actors is not wide. Although there are works that have used robots in theatrical representations, these robots are often used as props. Other works have focused on the development of systems that could interact with the audience [3,13]. The most advanced is the standup comedy implemented by Knight [8,9]. In this, a NAO platform [14] sits in front of a whole auditorium. The robot tells a joke and selects autonomously the next joke based on the reaction of the audience. To get this information, the robot relies on cameras, colored paddles and a microphone. The robot performs basic actions to add some expressiveness to the joke, but it does not project any emotion.

Other works [15,16] have adopted theatre as a place to implement simple actors with all information (even objects' position) hard-coded, and no interaction with people. Although these robots could be perceived as autonomous, they do not make any adjustment during their performance, losing realism, e.g., if the other robot does not play in its precise place.

Complex performances have been given in 2011, when *Roboscopie* [17,18] was presented. Roboscopie story involved one person and one robot. The robot could navigate autonomously in its environment and build a 3D model of it. But, the human-robot coordination was done manually during the presentation, and the position of some objects was already known by the robot.

Fahn and collaborators worked with humanoid robots as actors [4,10]. They believe humanoid robots are suitable as theatre actors more than other kind of robots. They have developed two humanoid robots, Janet and Thomas [10], and two wheeled robots, Pica and Ringo [19]. These robots are capable to perform autonomously many actions such as: draw people, jazz drumming, marionette operating, and notation reading and singing. However, their humanoid robots have problems with the amount of computers needed to control them. The high complexity to control the humanoid platforms and the lack of expression in their action, except for their faces, make this approach unsuitable to study emotion generation.

Trying to add some theatrical realism to robots, Breazeal and collaborators [20] designed and implemented a system to control a lamp. The main characteristic of this lamp is that it could be controlled by just one person, which could select the focus point where the lamp must look at. This little thing improved the credibility that the lamp was listening to the person that was speaking to it.

With the idea to familiarize people with robots, *Robots actors project* was created [21, 22], where Wakamaru [23] and Geminoid F [24] robots were used in the play. Unfortunately, deeper information about this implementation is not available, although from videos is possible to infer that the play has been designed for robots and robots seem remotely driven.

3   Emotions

There is not a commonly agreed definition of emotion [25, 26], however it is known that emotions are responses to stimuli and that very often they motivate actions. Emotions help humans as a mechanism for preservation and socialization. For instance, fear is one of the most common emotions related to self-preservation: when people are in dangerous situations, this emotion enables our body to get ready to run from the dangerous source. At the same time, emotions are used as social regulators. When two people are talking, both must show positive emotions to make the talk continue smoothly and effectively. Moreover, if someone has problems to assess the emotional state of others, he may act in a way that others may consider inappropriate, and he might be socially banned for his "misbehaviour".

Social rules are ground based to each society. These rules establish how to behave in society and they change from one society to another. Nevertheless, all cultures, in some way, punish or penalize misbehaviours. Therefore, people who cannot implement adequately these rules are margined by others. For example, people who have amygdala's damage lacks of the ability to distinguish some emotions (e.g., fear and anger) and the intensity of emotions [26]. These people have difficulties to respond correctly in situation where others express emotions that they cannot correctly distinguish. Thus, they are considered as undesired.

4   TheatreBot Architecture

The concepts presented in the previous sections have been used to define the TheatreBot architecture, which can be used to implement robots able to play in representations with actors, as well as general social and emotional robots for other applications, where some aspects do not come from the script but are drawn from the interaction (e.g., the social role and the relationship among characters). The most important idea borrowed from theatre is the explicit representation of the emotional state, used to produce expression, which is used in all human-human interactions and it is necessary for effective human-robot interactions.

To eliminate low level descriptions of actions, TheatreBot architecture allows the user to program the robot by giving it high level description of actions that will be interpreted and enriched by the robot itself, using social awareness. Likewise, the architecture allows the addition of specific movements in special situations, as the theatre's director would suggest to actors. To allow future extensions of this model to other domains, the architecture is modular and it is platform independent, in the sense that the same action could be executed on different robotic platforms, adapting the emotional interpretation to the possibilities of the platform.

The basic structure of the architecture is shown in Fig. 1. As it can be seen, there are modules that can provide information to others, but these can either consider this information, or not. If none of the modules accepts suggestions, the system reduces to a traditional decision system, where action decision takes into account only information about the world, and the actions are done without any emotion and expression.

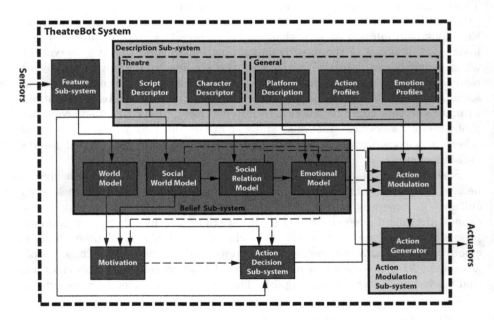

Fig. 1. TheatreBot's architecture is composed of six sub-systems: *Feature, Belief, Decision, Motivation, Description*, and *Modulation*. The full arrows represent information that has to be used by the module. Dashed arrows mean that one model influences the other. The influenced module can either accept or reject the suggestion given by the influencing one.

The architecture includes six main sub-systems. *Description* manages the information provided by designers and users: script, character, action profiles, emotion profiles, and platform descriptions. Two kind of modules are in it: the ones specific to theatre (i.e., script and character) and the ones transversal to different applications (i.e., platform description, action and emotion profiles). *Feature* transforms the raw sensor data into information about the world (e.g., objects and other actors present on the stage and their positions). *Belief* builds and maintain the set of robot's believes about the world and its state. They come either from the Feature sub-system (*World Model*) or from elaboration of what happens in the play (*Social World Model*). *Action Decision* is to select the next action that must be executed according to the current world situation. The possible actions that could be selected are defined in the script. *Modulation* blends the inner emotion, social awareness and the decided action to produce an expressive action. *Motivation* describes necessities and desires [27,28] that could exist at any time based on its belief and influenced by its emotional state.

Components from each sub-system are explained in the next subsections.

4.1   Script Description

The script description allows the user to represent information about play at different abstraction levels. Likewise, it gives the system enough information to select the expected action. The script is decomposed in sets of: *condition*, *action*, *beat*, *scene*, and *play*.

Conditions are seen as information that the robot could perceive from the world. These conditions embrace a wide range of abstraction levels, from the simplest as "there is a glass on the table", to more complex as a specific voice pattern. Specific theatrical conditions, such as stage and body positions, are added to describe conditions. Moreover, there may be conditions that consider actions in time, such as the number of times some sound pattern (e.g., a ringing bell) has been heard. All of these enable the writing of pre- and post- conditions; as their name suggests, the pre-conditions are conditions that should be true before executing a neutral action. On the other hand, post-conditions are conditions that are expected to be true after the execution of neutral action.

Neutral Actions are all the possible actions that a robot could perform without considering any kind of emotional modifications. To allow coordination, each kind of action has a set of pre-conditions and post-conditions. There are two kinds of actions:

- *Simple actions* could be seen as "*atomic actions*", i.e. the basic actions that are used by humans to give instructions, such as: move, balance arm, rise hand, speak, etc. These actions are parametric, thus it's parameters could be changed to allow their reuse. For example, the simple action "move" has the parameter velocity, which states the average speed during the movement.
- *Compound actions* are composed by simple or other compound actions in sequences, possibly executed in parallel, if physical constraints allow this.

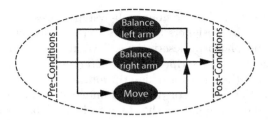

Fig. 2. Example of the compound action "walk" which is composed by "move", "balance right arm", and "balance left arm". It is important to notice that different implementations for the same action could exist.

Figure 2 could be seen the compound action "walk", composed by move (which just implies a displacement of the body), "balance right arm", and "balance left arm", which can be executed in parallel, since they involve different actuators.

Beat is an envelope for actions that contains contextual information of the current situation (*stimuli*), which is used to update the *Social World Model*. This information is used also by other modules to add action expression. A beat can wrap one compound action, only. This information just reduces the need to infer these facts from environmental information, but it does not give explicit information about gestures or emotion that the robot must do for its actions. This information is just one part of input to upgrade the emotional state.

Scene is an ordered sequence of *beats*. If one precondition or postcondition could not be achieved, it is possible to select a compound action in a predefined set to improvise a way out, or a specific compound action can be executed to manage failure in a pre-set way. Both possibilities can be defined in the script. A panic action may also exist, to cover the case that none of the alternatives are suitable. For instance, the actor may call a servant at any time, who may enter knowing what the problem is and with the solution (e.g., bringing a missing glass). An example of scene is depicted in Fig. 3. Each scene may include meta-information about objects and characters that are involved.

4.2 Emotion Profile

This describes how an emotion affects each action. Thus, for each emotion and simple action it is defined an emotional profile descriptor that contains parameters that could be changed and additional possible movements to be added to the basic action to show emotion.

4.3 Character Description

This descriptor gives personal traits to each character. It describes how stimuli (information of the current situation) affect the character (*Social Relation*

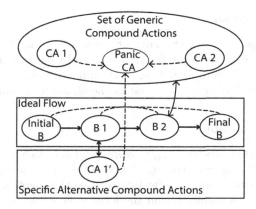

Fig. 3. Scene example that shows possible flows of beat (B) with some alternative compound actions (CA), and a set of compound actions. If, for some reason, no compound action can be selected, a Panic action has been defined to solve a generic problem. Noticed that none of the generic compound actions are specific selected by the beats, this is due that the best action is selected depending on the specific conditions of the environment.

Model), its basic emotional state, and how it evolves among them (e.g. the character is always scared, thus its normal emotional sate is be scared), characteristic poses in situation (e.g. a person could put his finger to his mouth while is thinking), and the basic speed in its movements. This description is known in theatre as *rhythm* [29]. It could also include the character's emotional attitude, which describes how the stimuli is going to affect the character. For example, a character with bold attitude is going to give less importance to scare events that one that has fear attitude, which is going to be more susceptible to scare events.

4.4   Social World Model

The *Social World Model* holds information about character's believes about situations, and others' emotional states and intentions. Given the theatre framework, the information to update this model comes directly from the script, while in real world operations, it has to be deduced from data, history, and interaction.

4.5   Social Relation Model

The *Social Relation Model* holds the information about character's emotion and intensity concerning places, people, roles, objects, animals, etc. Considering the information provided by the *Social World Model*, this model evaluates the current robot's *social state*. For example, if the robot detects that it is in a wake situation and its mother is crying, the model will retrieve two tuples, emotion and intensity, related to each piece of information given in the model.

4.6   Emotional Model

The *Emotional Model* selects the current emotional state by considering the information coming from both the *Social Model* and *Character Description* (e.g., emotional attitude). The emotional state will influence how actions will be performed by the robot. This could be seen as the character's *tempo* [29].

4.7   Action Modulation

The *Action Modulation* sub-system gets compound actions and decomposes them in simple actions. It also gets the emotional state (emotion and intensity) and the social situation to modulate the intensity of the emotion to be shown, if needed. For example, if someone won the lottery, he could be very excited, but if he is talking with his boss, his excitation is reduced in order to behave in a socially correct way in front on his boss. Using the emotion, intensity and simple actions, this module changes the correct parameters of each action based on the information given by the *Emotional Model*.

4.8   Action Generation

The *Action Generation* sub-system gets all the simple actions, emotion, and the platform information to implement the emotion-augmented action in the current platform. If among simple actions some actions exist that could not be executed in the current platform, these actions are skipped. Continuing with the walk example (Fig. 2), the module will decide to skip balance arm action for a differential platform, while for a humanoid platform it will not. Furthermore, showing happiness in the move action is different in each platform.

5   Example

To illustrate how the architecture could produce expressive actions based on simple actions, it is used a sequence of actions done by Romeo at the beginning of the balcony scene of Romeo and Juliet play [30]. Romeo is expressing emotions while he is walking and talking. To obtain this expressiveness from the architecture, firstly, movements required to the actor on stage must be defined, such as reported in Fig. 4(a). Using this, a compound action that performs the same movements and speech as the actor could be defined. Figure 4(b) shows the compound action to perform these movements. More importantly, neither of these two descriptions contain information regarding the *emotional state*. As a consequence, this information would not be enough to generate the additional actions as "rise the hand" and "move the head".

To accomplish this, it is necessary to give the system information about the stimuli that the actor has in this moment, *Beat*. In this particular case Romeo is indignant because Mercutio had made fun of him, he is disappointed because he has been rejected many times by Rosaline, he is in love with Juliet, he is

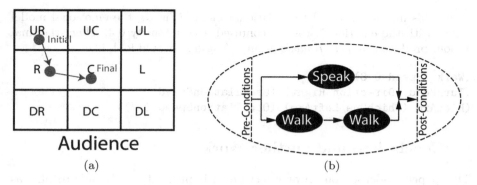

(a)   (b)

Fig. 4. (a) Diagram of the movements on stage done by the actor. Dots represent places where the actor makes changes on his movement. Notice that this diagram adopts the stage division introduced in Sect. 2. (b) Compound action that implements movements in the Romeo's example. The simple action "speak" is performed in parallel to the two "walk" actions in sequence.

afraid since he is in the garden of the Capulet house. This information could be translated to the following stimuli: Mercutio had made fun of you, you remember the snub by Rosaline, you love Juliet, you are in a forbidden place. One possible description of Romeo character could be:

```
(Character  (name Romeo)
  (BasicEmotionalState Happy)(EmotionalAttitude bold)
  (Additional traits (BasicSpeed High)
    (HandMovenment Oscillation  when Confused)))
```

Fig. 5. Romeo character relational model description.

Using this information and the relational model (Fig. 5), the emotional model comes with the emotional state of confused with intensity 0.4. Then, the final action, produced by *action modulation sub-system*, would look like:

```
(Walk (Speed 0.8))
(TurnHead (Direction Right) (OscillationSpeed 0.7))
(RiseArm (WhichOne LeftArm) (OscillationSpeed 0.7))
```

6   Conclusions and Further Work

This paper is focused on an architecture to implement robots able to play as actors in theatrical representations. The robots can be instructed by giving them the high level information analogous to that provided to actors by script and director. The expressiveness is reached by the robot actor using emotional and social models to decide the emotional state and its intensity. The emotional state is used to modulate and enrich movements according to emotion.

This architecture will enable the implementation of different robots able to interact socially and emotionally with people not only on the stage, but also in the real life.

References

1. Knapp, M.L., Hall, J.A.: Nonverbal Communication in Numan Interaction. Wadsworth Publishing Company, Belmont (2009)
2. Wilson, E., Goldfarb, A.: Theatre: The Lively Art. McGraw-Hill Education, New York (2009)
3. Breazeal, C., Brooks, A., Gray, J., Hancher, M., Kidd, C., McBean, J., Stiehl, D., Strickon, J.: Interactive robot theatre. In: Proceedings of the 2003 IEEE/RSJ International Conference on Intelligent Robots and Systems, IROS 2003, vol. 4, pp. 3648–3655 (2003)
4. Lin, C.Y., Cheng, L.C., Huang, C.C., Chuang, L.W., Teng, W.C., Kuo, C.H., Gu, H.Y., Chung, K.L., Fahn, C.S.: Versatile humanoid robots for theatrical performances. Int. J. Adv. Rob. Syst. (2013)
5. Hoffman, G.: On stage: robots as performers. In: RSS 2011 Workshop on Human-Robot Interaction: Perspectives and Contributions to Robotics from the Human Sciences (2009)
6. Lu, D.V., Smart, W.D.: Human-robot interactions as theatre. In: RO-MAN 2011, pp. 473–478. IEEE (2011)
7. Pinhanez, C.: Computer theater. In: Proceedings of the Eighth International Symposium on Electronic Arts (ISEA'97) (1997)
8. Knight, H.: Eight lessons learned about non-verbal interactions through robot theater. In: Mutlu, B., Bartneck, C., Ham, J., Evers, V., Kanda, T. (eds.) ICSR 2011. LNCS, vol. 7072, pp. 42–51. Springer, Heidelberg (2011)
9. Knight, H., Satkin, S., Ramakrishna, V., Divvala, S.: A savvy robot standup comic: online learning through audience tracking. In: TEI 2011, January 2011

10. Lin, C.Y., Tseng, C.K., Teng, W.C., Chen Lee, W., Kuo, C.H., Gu, H.Y., Chung, K.L., Fahn, C.S.: The realization of robot theater: humanoid robots and theatric performance. In: International Conference on Advanced Robotics, ICAR 2009 (2009)
11. Stanislawski, C.: An Actor Prepares. Theatre Arts Inc., London (1936)
12. Stanislawski, C.: Building a Character. Routledge/Thetre Arts Book, New York (1989)
13. Mavridis, N., Hanson, D.: The ibnsina center: an augmented reality theater with intelligent robotic and virtual characters. In: RO-MAN 2009, pp. 681–686. IEEE (2009)
14. Robotics, A.: Nao. Internet
15. Wurst, K.R.: I comici roboti: performing the lazzo of the statue from the commedia dell'arte. In: AAAI Mobile Robot Competition, pp. 124–128 (2002)
16. Bruce, A., Knight, J., Nourbakhsh, I.R.: Robot improv: using drama to create believable agents. In. In AAAI Workshop Technical Report WS-99-15 of the 8th Mobile Robot Competition and Exhibition, pp. 27–33. AAAI Press, Menlo (2000)
17. LAAS-CNRS: Roboscopie, the robot takes the stage! Internet
18. Lemaignan, S., Gharbi, M., Mainprice, J., Herrb, M., Alami, R.: Roboscopie: a theatre performance for a human and a robot. In: Proceedings of the Seventh Annual ACM/IEEE International Conference on Human-Robot Interaction, HRI '12, pp. 427–428. ACM, New York (2012)
19. Chen Lee, W., Gu, H.Y., Chung, K.L., Lin, C.Y., Fahn, C.S., Lai, Y.S., Chang, C.C., Tsai, C.L., Lu, K.J., Liau, H.L., Hsu, M.K.: The realization of a music reading and singing two-wheeled robot. In: IEEE Workshop on Advanced Robotics and Its Social Impacts, ARSO 2007, pp. 1–6 (2007)
20. Hoffman, G., Kubat, R., Breazeal, C.: A hybrid control system for puppeteering a live robotic stage actor. In Buss, M., Kühnlenz, K. (ed.) RO-MAN 2008, pp. 354–359. IEEE (2008)
21. Paré, Z.: Robot drama research: from identification to synchronization. In: Ge, S.S., Khatib, O., Cabibihan, J.-J., Simmons, R., Williams, M.-A. (eds.) ICSR 2012. LNCS, vol. 7621, pp. 308–316. Springer, Heidelberg (2012)
22. Torres, I.: Robots share the stage with human actors in osaka university's 'robot theater project'. Japandaily Press, February 2013
23. Mitsubishi: wakamaru. Internet
24. Laboratory, H.I., Robots. Internet
25. Plutchik, R.: The nature of emotions. Am. Sci. 89(4), 344+ (2001)
26. Cacioppo, J., Tassinary, L., Berntson, G.: Handbook of Psychophysiology. University Press, Cambridge (2000)
27. Breazeal, C.: Designing Sociable Robots. MIT Press, Cambridge (2002)
28. González, Á.C., Malfaz, M., Salichs, M.A.: Learning the selection of actions for an autonomous social robot by reinforcement learning based on motivations. I. J. Soc. Rob. 3(4), 427–441 (2011)
29. Cavanaugh, J.: Acting Means Doing !!: Here Are All the Techniques You Need, Carrying You Confidently from Auditions Through Rehearsals - Blocking, Characterization - Into Performances, All the Way to Curtain Calls, CreateSpace (2012)
30. Company, R.S.: Royal shakespeare company - romeo & juliet, on stage footage - ny. Internet

ROBSNA: Social Robot
for Interaction and Learning Therapies

Paulina Vélez$^{(\boxtimes)}$, Katherine Gallegos, José Silva,
Luis Tumalli, and Cristian Vaca

Escuela Superior Politécnica de Chimborazo (Politecnic University of Chimborazo),
Riobamba, Ecuador
paulyvlez@hotmail.com

Abstract. This article presents the techniques and methods used for the development of a social robot, whose purpose is to interact with children who have learning and psycho-social disabilities. This article presents the interaction algorithms used, the design of the hardware architecture, design of the software architecture and the theories applied to design the physical appearance of the robot. Additionally, this article presents the set of software and hardware tools used to obtain a socially intelligent robot capable of interacting naturally and autonomously with human beings and capable of responding to stimuli from the environment.

Keywords: Social robot · Socially intelligent · Interaction · Minimalist expression · Abilities · Reflexes

1 Introduction

This article will present the design of the hardware and software of a social robot. A socially intelligent robot provides a human to machine interface with the ability of interacting with children and can be used as an educational, therapeutic and entertainment tool. The design of the robots appearance was implemented by taking into consideration the desire not to fall into the Uncanny Valley, theory presented by Mori [1]. The objective of creating this robot is to obtain a tool capable of stimulate children to play and interact. The goal of this tool is to support special education processes. This was achieved by giving the robot the capabilities of having minimalist face expressions and a series of abilities and reflexes. The abilities are the capacities that the robot possesses for performing an action, while the reflexes are actions that the robot can make without the existence of an external stimulus to provoke an action. ROBSNA has the ability to imitate movement, recognize patterns, read texts, and has autonomous movement and speech. With respect to the reflexes, it can blink. The concept for its abilities took into account the automatic deliberative architecture [2] (AD architecture - Robot Maggie). The robots artificial vision system makes use of the Kinect device as this allows the robot to recognize the form and

A. Natraj et al. (Eds.): TAROS 2013, LNAI 8069, pp. 458–469, 2014.
DOI: 10.1007/978-3-662-43645-5_47, © Springer-Verlag Berlin Heidelberg 2014

depth of objects as well as assisting with the following of a child's movement. The voice system was developed by implementing a grammar and a knowledge database that allows for a match between the phrases heard and the defined grammar.

2 Definitions

2.1 Social Robot

Today, the field of robotics has advanced considerably. We are not only able to develop industrial robots, but also social robots capable of interacting with human beings. They can imitate social skills and are capable of helping people with their daily tasks. For example, robots can have precise movements and can perform their tasks within the same environment as human beings. When referring to social robots, there are two aspects that must be addressed by designers. These two aspects are appearance and reality. Two questions are typically addressed: is the robot a social robot or does it just appear to be social. MIT Professor Cynthia Breazeal [3] has made sub-classifications within the field of social robots based on their applications. In all cases of the sub-classifications, robots have been anthropomorphized so that they can interact with humans. The design of a robot varies according to the social model, environment in which it will operate, and the level of complexity of the tasks it can perform. ROBSNA is considered to be within the sub-classification of social robots. ROBSNA is a social robot defined as a socially participative creature with its own objectives and internal motivations. Social robots interact proactively with people in a social context. They provide benefits to the user. For example, social robots can assist with load-carrying tasks. They also have their own ability to learn from humans while allowing humans to learn from them. These robots not only perceive the social signals of humans but also the social patterns in order to able to interact with them. The social behavior of the robot is the product of designing human social model maps and of computer psychology. There are various forms in which a robot can be perceived as social. Being perceived as a social robot, dependents on the robots social behavior and takes into account physical aspects.

2.2 Anthropomorphism

Anthropomorphism is crucial when designing a social robotic system [4,5]. Anthropomorphize is to give human attributes to robots for the benefit of humans to help us rationalize human actions. This is an attribute of cognitive and emotional states based on behavior within a determined social environment. Anthropomorphization seeks to suspend the illusion of reality and determines whether a robot is able to socialize and how long can one maintain the illusion of reality. In order to extend this illusion for longer periods of time, various methods are used with the goal of giving the impression of life and intelligence. Nonetheless the challenge of designing a functional social robot is much more complex than creating animated images or cartoons. The design of a robot involves factors such as physical autonomy, appearance, and the design of the actual social conduct.

The design of ROBSNA involved two principal motives for which it was decided to anthropomorphize the robot. The first motive was due to the fact that it has a system that must function in the same physical and social space as humans, and objects we commonly use have to move within our environment. The second had to do with the advantage of Anthropomorphization which allows the robot to physically interact with the user. Anthropomorphization constitutes the integration of humanity into a basic system through the robots behavior, its control over experience and its abilities, and the capacity to interact with the social environment as well as its physical form [6]. During the anthropomorphization process of a robot, one must consider that in addition to all the physical aspects, that one is going to attribute to the robot, one must avoid entering what Mori defined as the Uncanny Valley. Moris theory says that the more a robot appears to be similar to humans the more the effect of rejection it produces among people. Figure 1 shows Mori's theory with the relationship between the moving robot and still robot. This figure also includes a new function that shows the position of ROBSNA considering taking into consideration its characteristics and qualities. The new function of ROBSNA was obtained by normalizing Mori's data, where the human likeness has the highest value. A subjective analysis was realized considering the apathy and empathy levels measured as noted by six people. Each measurement was performed at different stages of the development of ROBSNA, basically, when the robot was a mobile base. Another stage was when the robot was set as an anthromorphic robot. Next, when human voice was added and finally when the robot was totally developed. The final function placed ROBSNA between Moris functions that refer to a still robot and a moving robot. This was occurred because ROBSNA is an anthropomorphic robot with limited movements and semi human likeness, similar to a live toy.

In Fig. 1, one can see the emotional response of a person to the anthropomorphization of a robot. The uncanny valley is the region of negative emotional response, when a robot appears almost human and the movement of the robot increases this negative response [1]. Thus, for the robot to socialize in a better way, it must have a friendly aspect while still maintaining an appearance different to that of a person or common animal. This ensures that the robot does not cause apathy among the people that socialize with it and facilitates the communication between humans and the robot. There are several factors that must be considered when anthropomorphizing a robot. These factors have been presented by Epley and include various psychological and social aspects. Epley describes these factors by assessing the degree to which one should anthropomorphize [7].

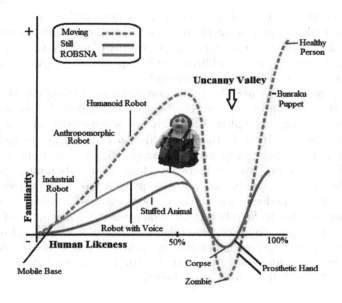

Fig. 1. Uncanny valley

2.3   Expressiveness of the Robot

Social robots have a body but in no case should this body create false expectations. Fong [8] made a distinction between four categories of robots by their form: anthropomorphic, zoomorphic, caricatures and functional. An anthropomorphic appearance is recommended for interaction with humans, this is due to the non-verbal communication aspects that are similar to the human body. The robots with zoomorphic appearance are useful in cases that one wants the robot to have similar behavior patterns as to those of an animal. Robots with caricature appearance are used for very specific attributes. The functional robots are designed specifically to fulfill a function. Considering the theory of Mori and the aspects that define the effects of a social robots appearance one can specify various design factors that should be taken into account for a social robot:

- The social robots face and its facial expression must be easy for people to interpret, but it must not have the exact appearance of a human face.
- The robot must be attractive so that people will want to interact with it in a natural way.
- The robot must have the sensory, motor and processing capacity so that its response to external stimuli are in real time.
- The movements of the robot, both its body and face, must be controlled and not be sudden and brash, the speed of the robots movements must be in accordance with the mood and state of mind of the robot.

An important aspect in the expression of emotions is the expressiveness that one can give with the movement of the eyes, as with only their movement and positioning of the eyebrows one can express a variety of moods. Only by changing the position of the eyebrows, in addition to the expression created by the position of the lips, we can create a fully expressive face [9].

The body movements are also important when expressing emotions. Movements with the arms and head, the coordination and speed of these movements, help the robot to seem more alive. Emotions can be expressed through the voice, using distinct tones to indicate the mood, like for example joy, sadness, excitement or anger. The robot can express itself responding to stimuli using a natural language of gestures and words thereby enriching the robots human interaction. Choosing the correct voice and expressions, plays a fundamental role as a natural, agreeable and understandable voice creates empathy, while an unnatural voice or often referred to as robotic voice could upset the harmony of the interaction and reduce empathy. The time taken by the robot to respond to external stimuli is very important for being able to make the robot lifelike. With rapid response times one can deceive the interlocutor longer, so that a person feels that they really are talking with a live being.

3   ROBSNA Robot

ROBSNA is a social robot designed to express in a minimalist manner emotions with its face, arm movements and voice, the voice being the least used to express emotions. The voice system is used to create dialogs in situations in which the robot is used as a learning tool. ROBSNA has nine degrees of freedom distributed through its body. These are distributed in the following way: three degrees of freedom in each arm, two degrees of freedom in the neck and one degree of freedom for blinking. The objective of giving three degrees of freedom to each arm is so that the robot can imitate a child's body movements for interaction therapy. In this manner, the robot is capable of lifting its arms from the shoulders and bending them. The face of the robot has three degrees of freedom, two in the eyebrows and one for blinking and a matrix of LED diodes that allow for visualization of the smile, expression of sadness and gesticulation of words. All of these with are part of the goal of making it possible for the robot to express emotions. The design of the body of the robot was done ensuring that it does not approximate to the Uncanny Valley that Mori mentions in his theory. Thus, the robot does not have the physical appearance of a real person. Even though the robot does have human characteristics, there is a large gap between real human appearance and creating an appearance that expresses human characteristics. Two toys were added to the robots design to incentivize interaction, as when the child sees them the first thing they will try to do is take them from the robot. In Fig. 2 one can observe robot ROBSNA and its physical characteristics.

For the design of ROBSNA a specific generic architecture was used and divided into four levels: perception, decision making, automatic level and mechanical processes, as shown in Fig. 3.

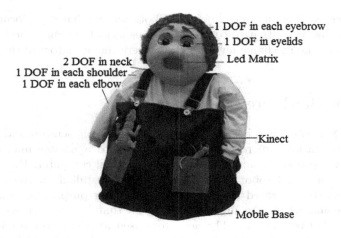

1 DOF in each eyebrow
1 DOF in eyelids
Led Matrix

2 DOF in neck
1 DOF in each shoulder
1 DOF in each elbow

Kinect

Mobile Base

Fig. 2. Physical characteristics of the robot. DOF (degree of freedom)

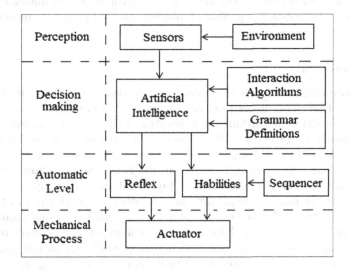

Fig. 3. Generic architecture for ROBSNA robot

The upper layer is where the robot is capable of perceiving signals from the environment. It is in this section, which is located on the surface of the robot, that the sensors are located. The sensors have been placed on the arms, face and base, so that the robot is aware when the child touches it. In the decision making level is located the artificial intelligence module which is where the interaction algorithms intervene and the grammar created for dialog is defined. The automatic level section has two modules, the reflexes and the abilities modules. The module for the reflexes is responsible for generating, in a natural manner, the reflexive actions of the robot, basically the blinking reflexes. The abilities module is responsible for controlling the robots actions that it performs through a sequencer that permits it to control the sequences of actions. The last section corresponds to the mechanic level which is where the actuators of the robot are located.

4   Robot Hardware

The ROBSNA robot has been developed using various sensors and actuators controlled by various microcontrollers connected to a master microcontroller which receives primary commands from the personal computer, PC. Since this is a minimalist social robot the sensors that are installed on the surface are the photoreceptor, infrared and limit switches. Their purpose is to inform the robot when and where the user touches it, so that it can to respond to the stimuli provided by the user. The actuators used for the face are micro servomotors, servomotors and continuous current motors. The robots vision uses the Kinect sensor, which permits the robot to track the user and makes it capable of imitation. The robot is capable of playing by imitating the movements of the user.

4.1   Architecture

The hardware architecture of the robot is composed of five levels that interact among each other. The upper level is composed of the personal computer from which the interaction algorithms are generated. The personal computer receives voice commands from the user. These voice commands are processed using interaction voice algorithms which generate the right answer, so the user can feel she is talking with another person. In addition, the Kinect sensor allows the system to capture the user movements that are processed and the computer sends the information to the master microcontroller. Another external input of the robot is composed of the different sensors that are located in various points of the body such us hands or mobile base. From the computer, the primary orders are sent to the master microcontroller, which is responsible for distributing orders to the slave microcontrollers connected to it, as presented in Fig. 4.

The slave microcontrollers receive the commands from the master microcontroller and execute them causing the activation of the actuators that generate the human like movements. Meanwhile the slave microcontrollers receive the signals from the sensors and send them to the master microcontroller after adding the necessary information so that the robot can identify where it has received the stimulus.

5   Robot Software

The software developed for ROBSNA fulfills functions that allow the robot to display moods and interact in an intelligent and autonomous manner. These functions give the robot the following characteristics:

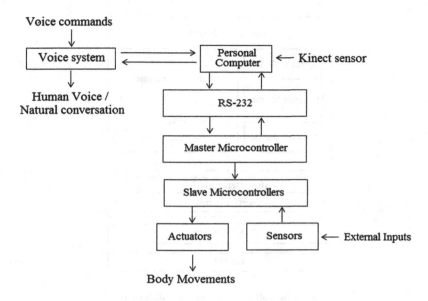

Fig. 4. Hardware architecture

- Imitation of movements, tracking.
- Display moods.
- Read legible texts.
- Following.
- Communication in natural language in real time.
- Coordination of movements according to the pre-established patterns (movements of: eyebrows, eyelashes, arms, neck, mobile platform).

5.1   Technology Used

To develop the software system for the robot these tools were used: Programming languages:

- C# for the application software.
- C for the embedded software.

Code generators and editors. The code generators and editors vary according to the programming language used. Thus the following were used:

- Visual Studio 2010 as IDE for development with C#.
- BorlandC++ for development with C.

5.2   System Architecture

The system is organized into subcomponents which have modules so that the robot can react to external stimuli in real time. These components are presented in Fig. 5.

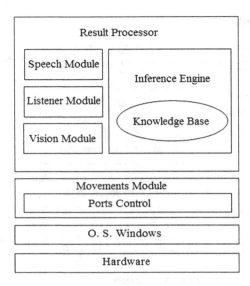

Fig. 5. Software system architecture

The results-processing component processes and transmits data to and from other subcomponents. The data processing of this component involves giving the data obtained the correct format so that the data can be received by the component to which it is destined. In the listening module, the audio heard is processed through a sound sensor in order to return words and phrases to the inference engine. The speaking module transmits the audio signals received towards the audio output peripherals. These signals are sent to this module through the results processing module. The vision module is a component designed to detect objects and patterns through a light and movement sensor, from which the results are sent to the results processor. The inference engine is the principal component of the system as it is in charge of verifying the data received to trigger a specific action. For this the inference engine consults a data base called the knowledge base in which are stored all the rules for action and real situations that the robot has available. The inference component determines what to do, and what action the robot should take. The movement control component sends signals to the robots hardware. The movement control component is assisted by the input/output ports control module whose function is to control the traffic of incoming and outgoing data, so that the data signals are not affected by the quantity of electric pulses (Fig. 6).

In this way the robot can react in real time to external stimuli. For example: upon hearing a sound, it recognizes it as a word and the listening component sends that signal to the inference component, which processes the word to verify if it is a voice command. If it is the inference component, it will apply the corresponding rule and return information regarding the actions to be taken by the results processing component. This component then sends these actions

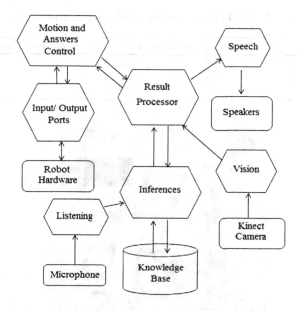

Fig. 6. Relationships between the systems modules

sequentially to the movement control module, which then sends the corresponding electronic pulses to the robot hardware through the defined ports.

The interruptions structure for the system has been developed based on the execution of priority requests that the user might have when interacting with the robot.

6 Results

The robot has been tested in Ecuador's central region, where society has no access to robots, with three children between the ages of 3 and 6. The robot has been tested to measure stimulus-response interactions with levels of empathy and apathy. Thus, if there is any empathy level the human robot interaction will start. A subjective test was performed where levels of relevancies were given to each aspect to help with the interaction. The specific points that were measured were: aspect, voice and movements.

The robot aspect has a 30 % of relevance in the interaction process. The voice response has 30 %. And, the movements have 40 % of relevance. These three aspects have been subjectively measured between 1 to 5 points. The conclusions of the result were obtained considering the following scale: it is apathy if the data the is lower or equal to 40 %. Empathy if data the data is greater or equal to 60 %. If the data is between 41 and 59, it is considered as indifferent data.

With the relevance of the data that has been given by the children, the following results are shown in Table 1.

Table 1. Results of the subjective analysis

	Aspect	Voice	Movements	Percentage of empathy/apathy (%)
3 year old	5	5	3	84
5 year old	5	5	5	100
6 year old	5	5	5	100

Fig. 7. Robot-child interaction

Considering that the empathy and apathy levels in the three case studies are higher than 80 %, the human robot interaction is manifested as empathy. This indicates that the children who participated in this investigation found the appearance of the robot to be likeable.

7   Conclusions

We were able to create a social robot that acts as an experimental platform for initiating interaction processes and learning support. This the robot is capable of interacting naturally and in real time with the user as one can see in Fig. 7.

To produce the stimulus-response interaction software and hardware architecture was created. The hardware architecture was created to help to decrease response times and increase the autonomous level. The hardware architecture is a modular architecture that allows the increase of the number of sensors, change the type of sensors or adding a sensor without changing the main board of the robot. This allows the robot to adjust its behavior according to the necessities of the user in a and speedy and efficient manner. The software architecture created is a modular architecture that allows the robot to process the information from external signals speedily, so that the robot can answer quickly to the stimuli given by the user.

References

1. Mori, M.: The uncanny valley. Energy **7**(4), 33–35 (1970)
2. Salichs, M., Barber, R., Khamis, A., Malfaz, M., Gorostoiza, J., Pacheco, R., Rivas, R., Corrales, A., Delgado, E., García, D: Maggie: a robotic platform for human-robot social interaction. In: IEEE Conference on Robotics, Automation and Mechatronics, pp. 1–7 (2006)
3. Breazeal, C.: Toward sociable robots. Robot. Auton. Syst. **42**, 167–175 (2003). (Elsevier Science B.V.)
4. Hegel, F., Muhl, C., Wredel, B., Hielscher, M., Sagerer, G.: Understanding social robots. In: Second International Conferences on Advances in Computer-Human Interactions, vol. 42 (2009)
5. Hegel, F., Krach, S., Kircher, T., Wredel B., Sagerer, G.: Understanding social robots: a user study on anthropomorphism. In: 17th IEEE International Symposium on Robot and Human Interactive Communication (2008)
6. Duffy, B.: Anthropomorphism and the social robot. Robot. Auton. Syst. **42**, 177–190 (2003)
7. Epley, N., Waytz, A., Cacioppo, J.: On seeing human: a three-factor theory of anthropomorphism. Psychol. Rev. **114**, 864–866 (2007)
8. Fong, T., Nourbakhsh, I., Dautenhahn, K.: A survey of socially interactive robots. Robot. Auton. Syst. **42**, 143–166 (2003)
9. Paz, E., Pérez, J., López, J., Sanz, R.: Diseño de una cabeza robótica para el estudio de procesos de interacción con personas. XXV Jornadas de Automática (2004)

ARE: Augmented Reality Environment for Mobile Robots

Mario Gianni[(✉)], Federico Ferri, and Fiora Pirri

ALCOR Laboratory, DIIAG, Sapienza University of Rome, Rome, Italy
gianni@dis.uniroma1.it

Abstract. In this paper we present ARE, an Augmented Reality Environment, with the main purpose of providing cognitive robotics modelers with a development tool for constructing, at real-time, complex planning scenarios for robots, eliminating the need to model the dynamics of both the robot and the real environment as it would be required by whole simulation environments. The framework also builds a world model representation that serves as ground truth for training and validating algorithms for vision, motion planning and control. We demonstrate the application of the AR-based framework for evaluating the capability of the robot to plan safe paths to goal locations in real outdoor scenarios, while the planning scene dynamically changes, being augmented by virtual objects.

1 Introduction

Augmented Reality (AR) has recently stepped beyond the usual scope of applications like machine maintenance, military training and production. AR is a compelling technology allowing cognitive robotics developers to design a variety of complex scenarios involving real and virtual components. Indeed, programming complex robotic systems, operating in dynamic and unpredictable environments, is a challenging task, when it is required to go beyond laboratory experiments. Furthermore, as the complexity of the robotic system is inflated by an increase of robot components and functionalities, testing the whole set of interconnections between hardware and software components becomes exponentially difficult. The cognitive roboticists are more and more exposed to hard problems requiring a great amount of possible real world situations that are difficult to predict, when most of the test require laboratory experiments. In this context, Augmented Reality (AR) facilitates robot programming, providing to robot developers a technology for testing the robot system which is much more flexible than a complete simulation environment, as it allows developers to design a variety of scenarios by introducing any kind of virtual objects into real world experiments. AR-environments can facilitate experiments bypassing complex simulation models, expensive hardware setup and a highly controlled environment, in the various stages of a cognitive robot development.

In this paper, we present an AR-based simulation framework which allows robot developers to build on-line an Augmented Reality Environment (ARE)

A. Natraj et al. (Eds.): TAROS 2013, LNAI 8069, pp. 470–483, 2014.
DOI: 10.1007/978-3-662-43645-5_48, © Springer-Verlag Berlin Heidelberg 2014

for real robots, integrated into the visualization interface of Robot Operating System (ROS) [1]. The system we propose goes beyond an interface for drawing objects, as the design exploits a stochastic model activating the behaviors of the introduced objects. Objects, people, obstacles, and any kind of structures in the environment can be endowed with a behavior; furthermore, a degree of certainty of their existence and behaviors, with respect to what the robot perceives and knows about its space, can be tuned according to the experiment needs.

To illustrate the advantages of an AR based simulation framework for robot design and experiments, as opposed to a complete simulation framework in scenario design and test, and to show the benefits of our approach, we set up several path-planning experiments, with increasing complex environments. In particular we show that the framework allows to compare the performance of the path-planner with respect to several parameter sets in any simple outdoor settings augmented with ARE.

The paper is organized as follows. In the next Sect. 2 we discuss related work highlighting the novelty of our approach. In Sect. 3 we describe the main components of the AR-based simulation framework, detailing on the dynamic model. Section 4 describes the robot setup and reports the results obtained by a paradigmatic application of ARE for evaluating the capabilities of the robot in navigation tasks.

2 Augmented Robot Reality: State of the Art

Augmented Reality (AR) is a recent emerging technology stemming from Virtual Reality (VR). AR develops environments where computer-generated 3D objects are blended (registered) onto a real world scene. The AR computer interfaces have three key attributes: (1) they combine real and virtual objects, (2) the virtual objects appear registered in the real world and (3) the virtual objects can be interacted in real time [2]. The main difference between AR and VR is that AR eliminates the need to model the entire environment as it supplements the real world instead of replacing it. AR research has been active in robot applications such as maintenance [3], manual assembly [4] and computer-assisted surgery [5]. Several works have also addressed robot programming. One of the first robotic applications of AR was Milgram *et al.* application in telerobotic control [6]. In telerobotic applications an operator is provided with visual feedback of a virtual wireframe robot superimposed over an actual physical robot located at its remote working environment [7]. The virtual robot would execute a task for evaluation by the operator and the task would be transferred to the real robot if it is satisfactory. In robots monitoring [8,9] the visualization of complex data is overlayed onto the real world view. Namely, a view of the robot and the environment is synthesized graphically from on-board sensory data, such as camera images, and presented to remote operators, increasing their situation awareness. In [10] the authors present an AR-based approach for intuitive and efficient programming of industrial robots. Here, tool trajectories and target coordinates can be visualized and manipulated by means of an interactive laser projection. Bischoff and Kazi [11] report an AR-based human robot interface for applications with industrial robots. Their work focuses on visualizing workflows that could help inexperienced users to cope with complex robot operations and programming tasks. AR has also been used to program painting robots [12]; here a spray gun, which is hand-held by a user, is tracked and the corresponding virtual spray is displayed to the user through a Head-Mounted Display (HMD). A feedback is obtained by tracking the object to be sprayed using markers, while the user movements are converted into a robot program. Daily *et al.* use robots AR to convey the robot state information: virtual arrows are overlaid on the robot to show its heading. Also bubblegrams displaying robot states and communications [13] have been used to improve the interaction between co-located humans and robots. Dragone *et al.* [14] too place virtual characters on top of real robots to express robot states through natural interaction modalities, such as emotion and gestures. Giesler *et al.* [15] apply AR for prototyping warehouse transport robotic systems. This application enables the user to interactively construct a topological map of paths and nodes in the warehouse by walking around and pointing at the floor. The robot is instructed to move along the map and the planned path is visualized to the user as an augmentation of the real world view. The user can inspect possible intersecting points with workers paths, thus obtaining an estimation of the efficiency of the solution. Stilman *et al.* [16] create an AR environment for decoupled testing of robot subsystems. Results from motion planning and vision algorithms are visualized over real world images provided by both external cameras and a camera mounted on the robot. Collett and MacDonald [17] apply AR for debugging robot applications. They overlay virtual information such as robot sensory and internal state data onto real world images of the robot environment, in so providing robot developers with a better understanding of the robot world view. Chong *et al.*, in [18], introduce an AR-based robot programming (RPAR-II) system to assist users in robot programming for pick and place operations with an human robot interface, enriching the interactions between the operators and the robot. In their system, a virtual robot, which is a replicate of a real robot, performs and simulates the task planning process.

Green et al. in [19] proved AR to be a suitable platform for human-robot collaboration. Similarly, Billinghurst *et al.*, in [20], show that AR allows to share remote views (ego-centric view) visualizing the robot view, to the task space (exo-centric view). Furthermore, they show that AR can provide support for natural spatial dialog by displaying the visual cues necessary for a human and a robot to reach common ground and maintain situation awareness. Billinghurst and colleagues show also that the robot can visually communicate its internal states to its human collaborators, by graphic overlays on the real world-view of the human [21].

3   The AR-Based Simulation Framework

The AR-based simulation framework registers virtual objects, such as robots, cars, people, pallets and other kind of obstacles into the real environment

Fig. 1. On the left a 2D and 3D maps for the ARE environment representation, on the right 3D models of virtual objects.

(see Fig. 1). The model for the life-cycle of each virtual object is stochastic. It formally structures the interaction between the virtual objects and the real environment by both avoiding collisions and handling occlusions effects. The AR-based simulation framework includes: (1) the model of the real environment, (2) the model of the virtual objects, namely *artefacts* and, (3) the AR-builder server.

3.1   The Real Environment Representation

The real environment representation concerns both the 2D occupancy grid map $\mathcal{M}_{2D}$ and the octree-based 3D map $\mathcal{M}_{3D}$ on the left of the panel in Fig. 1, [22–24]. This representation has been proved to be compact and easy to update incrementally, thus accounting for dynamic obstacles and changing scenes. In fact, a polygonal mesh $\mathcal{S}_E$ is used to geometrically represent the environment. The polygonal mesh renders, together with the basic environment structure, also the 3D models of the artefacts, so as to correctly place the artefacts in the real scene, thus handling the occlusions with the existing objects [25].

3.2   The Artefacts Model

An artefact is a dynamic object defined by three main structures, illustrated in Fig. 2, bottom-right of the panel: the properties, selected by a mixture of Poisson distributions, a life cycle specifying the arrival and leaving time and, finally, a Markov decision process specifying the artifact main behaviors, given the life cycle and the selected properties.

We are given a space $\mathbb{Q}$ of all possible tuples defining an artefact type; actually we can think of $\mathbb{Q}$ as the set of all possible matrices $\mathbf{Q}$ where each row specifies a particular tuple Q of properties of an artefact A, with $Q = \{l, b, \mathbf{p}(.,t), \mathbf{q}(.,t), \Phi\}$. Here l is a label denoting the virtual object type, b is the bounding box of the artefact, and $\mathbf{p}(.,t)$ and $\mathbf{q}(.,t)$ denote the position and orientation operators of the artefact, taking temporal values in the stochastic Poisson model, described in the next sub-section. Φ is a set of additional features. Each artefact is geometrically specified by a polygonal mesh $\mathcal{S}_A$, representing the virtual object as a set of faces and a set of vertices, according to the face-vertex model, along with additional information such as color, normal vector and texture coordinates. We assume that the geometry of the mesh data structure is not morphable and

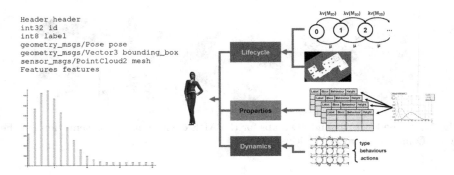

Fig. 2. On the left top the ROS message specifying an instance of an artefact. On the left, bottom, a Poisson Mixture of 5 components with parameters vector $(15, 7, 3, 21, 5)^\top$. On the right the artefact structure schema.

warpable so that the bounding box is fixed for its life-cycle. This representation allows the AR-builder server to manage the collisions between the artefacts populating the real environment. Within the framework, an artefact is implemented by a ROS message, see Fig. 2, top-left panel.

In this subsection we introduce the simulation model regulating the spatio-temporal behavior of the artefacts populating the real environment. Namely we introduce both how properties are sampled in $\mathbb{Q}$ and the arrival and leaving time of the artefacts. This model is based on a marked Poisson process [26–28], whose marks correspond to an identifier selecting a tuple from a matrix $\mathbf{Q} \in \mathbb{Q}$ and the artefact life-cycle. A marked Poisson process is a Poisson Process which has each point labeled with a mark. These marks are used to connect the artefact properties with its specific life-time cycle. Indeed, the model estimates the arrival and exit time of the artefacts as well as the labels governing the choice of the artefact properties.

Let the variable X, selecting a tuple Q in the space $\mathbb{Q}$, be a stochastic variable sampled from a mixture of n Poisson distributions, with n the number of object classes, namely:

$$Pr(X = k|\lambda_1, \ldots, \lambda_n) = \sum_{i=1}^{n} \pi_i \frac{\lambda_i^k}{k!} \exp(-\lambda_i) \tag{1}$$

Here $\sum_i \pi_i = 1$, λ_i, $i = 1, \ldots, n$ are the Poisson distribution parameters for each mixture component. Note that since n is given a priori from the classes of buildable objects (e.g. cars, pallets, girls, boys,..) parameters estimation is set by the EM [29]. The choice of the tuples is specified upon arrival of a group of artefacts. This is detailed below.

Let $M_{2D} \subseteq \mathbb{R}^2$ be the occupancy grid designating the flat structure of a generic environment taken as base of the representation (e.g. the courtyard of the department, a set of corridors). The arrival time t_a of artefact a, in M_{2D} can be assigned according to a Poisson process. Here arrival time $t \in [0, T]$ has an average rate $\lambda \in \mathbb{R}$. The times between successive arrivals are independent exponential variables having mean $\frac{1}{\lambda}$. Upon arrival, an artefact is supposed to leave M_{2D} after a certain time. Similarly to arrival time, the time last of each artefact is an independent exponential variable with mean $\frac{1}{\mu}$, where $\mu \in \mathbb{R}$ is the average rate artefact leaving time. From these assumptions it follows that arrivals and leavings time t are ruled by a time-homogeneous irreducible, continuous-time Markov chain $\{N(t)|t \geq 0\}$ with state space $\mathbb{N}^+$. Let $\nu(M_{2D})$ be the area of M_{2D}, let $o(\Delta t)$ be an infinitesimal of a higher order than Δt^1, then the stationary transition probabilities $p_{ij}(\Delta t)$ between state i and state j, for infinitesimal lapses of time Δt, are given by

$$p_{ij}(\Delta t) := \mathbf{P}(N(t + \Delta t) = j|N(t) = i) =$$
$$= \begin{cases} \lambda\nu(M_{2D})\Delta t + o(\Delta t) & \text{if } j = i + 1 \\ 1 - (\lambda\nu(M_{2D}) + \mu)\Delta t + o(\Delta t) & \text{if } j = i \\ \mu\Delta t + o(\Delta t) & \text{if } j = i - 1 \\ o(\Delta t) & \text{if } |j - i| > 1 \end{cases}$$

$\{N(t)|t \geq 0\}$ can be seen as a special case of a birth-death process [30], with birth rates $\lambda_i = \lambda\nu(M_{2D})$ and death rates $\mu_i = \mu$, for each $i \in \mathbb{N}^+$. Let the arrivals of new artefacts in M_{2D} occur according to a Poisson process with intensity $\lambda\nu(M_{2D})$ and let, upon arrival, the lifetime of each artefact be independent and exponentially distributed with mean $\frac{1}{\mu}$, then $N(t)$ returns the number of artefacts alive at time t (see Fig. 3(a)). Given the life-cycle, artefacts populate the environment and, given the Poisson mixture, upon arrival they draw their specific properties, including features such as being a car or being a pallet, from the distribution. The final component of the artefact structure is its characteristic behavior determined by a motion planning algorithm that would govern what it can actually do in the scene. To model the specific chosen behavior a finite-horizon Markov Decision Process is introduced:

$$\mathcal{H}_A(t) = \{D, S, \mathcal{A}_s, p_t(\cdot|s, \alpha), r_t(\cdot|s, \alpha) : t \in D, s \in S, \alpha \in \mathcal{A}_s\}$$

Here $D = \{0, \ldots, d\}$ is the set of decision epochs, depending on the life time d of the artefact, $S \equiv M_{2D} \times \{0, \frac{\pi}{2}, \pi, \frac{3}{2}\pi, 2\pi\}$ is the set of poses that it can take on during its life cycle, $\mathcal{A}_s$ is the set of possible actions when the artefact is in the state s, $p_t(\cdot|s, \alpha)$ is the transition probability from state s at time t to the next state s' at time t' when action α is selected at the decision epoch t, and $r_t(\cdot|s, \alpha)$ is the reward earned when the artefact is in state s and action α is selected at decision epoch t. Actions are choices about both behaviors and motion planning. In other words an action α is a function $\alpha : \mathcal{B} \times \Pi \times S \mapsto (k_1, \ldots, k_m)^\top$ mapping a set of predefined behaviors $\mathcal{B}$, the set of motion planning algorithms Π, supporting the behaviors, and the set of states S, into a vector of numbers accounting for the choice. Behaviors specify what the artefact can do within the environment, i.e. walking, running, following another artefact, stopping, according to the identifier chosen with the mixture of Poisson distributions. Similarly,

---

[1] A function $f(t)$ is an infinitesimal of a higher order than t, namely $o(t)$, if $\lim\limits_{t \to 0} \frac{f(t)}{t} = 0$.

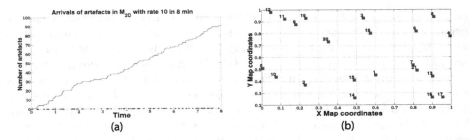

Fig. 3. (a) Simulation of the waiting times as well as of the trajectory of the Poisson process, regulating the arrivals of the artefact to M_{2D} (b) Simulation of the marked Poisson process, regulating the distribution in space of the artefacts, with the corresponding arrival time.

the motion planning algorithms Π take into account the current positions of the artefacts already in M_{2D} [31] and the gridmap to consistently move the object in the scene. Note that the finite-horizon Markov decision process $\mathcal{H}_A(t)$ selects the motion actions up to the time horizon d, according to the underlying action policy. Given the position of the other artefacts within M_{2D}, the motion planning Π plans collision-free short trajectories to reach the new position and orientation of the artefact, at time t', provided $\mathcal{H}_A(t)$ has selected a set of behaviour $b_k \in B$.

3.3 The AR-Builder Server

The AR-builder server interconnects the real environment model together with the simulation model of the artefacts. The AR-builder server relies on the tf software library. Namely, it keeps track, over time, of the transformations between the global reference frame $\mathcal{F}_E$, attached to the real environment model, the base frame $\mathcal{F}_R$, attached to the robot base, the base laser frame $\mathcal{F}_L$, attached to the laser sensor, and the camera frame $\mathcal{F}_C$, associated with the camera system, mounted on the robot. Given the tf library, the AR-builder server can determine the current pose of the reference frame $\mathcal{F}_A$ attached to the center of mass of the artefact and map it into the frame $\mathcal{F}_E$, rather than $\mathcal{F}_L$ or $\mathcal{F}_C$. Upon the registration of the artefact, the AR-builder server correctly places it within the real environment, by both projecting the bounding box b of the artefact on M_{2D} and concatenating the vertexes of the polygonal mesh $\mathcal{S}_A$ to the voxels of M_{3D} (see Fig. 4(a)). On the basis of these computational steps, the AR-builder server constructs the augmented model of the real environment (see Fig. 4(b)).

The AR-builder server implements a collision detection algorithm according to the dynamic model $\mathcal{H}_A(t)$. The algorithm, performing pairwise hit-testing, determines whether the bounding box of an artefact intersects the bounding box of another one or, alternatively, whether an artefact becomes embedded in the polygonal mesh $\mathcal{S}_E$ of the real environment. In such a case, the builder simply resolves the collisions by either moving back the artefact to its last known safe pose or by allowing the artefact to move up to a safe distance. The model of the

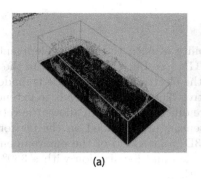 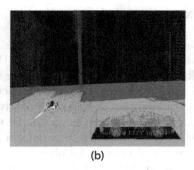

(a)   (b)

Fig. 4. (a) Projection of the bounding box b of the artefact on to M_{2D} (b) The 3D occupancy grid M_{3D} is augmented concatenating the vertexes of the polygonal mesh $\mathcal{S}_A$ associated with the artefact, to the voxels of M_{3D}

augmented real environment comprises only the artefacts which are not occluded, with respect to the robot field of view. These artefacts can be effectively ray-traced out from M_{2D}, thus not affecting the robot path-planning. The AR-builder server checks occlusion effects by implementing a ray tracing version of the z-buffer algorithm. Let $\mathcal{F}_{view} = \{P, (W, d_{max})\}$ be the view model of the real robot, where P is the camera matrix of the real robot with center C and W is the polyhedral cone of the field of view, such that $n(h^{(i)}) \times m(h^{(i)})$ is the dimension of the viewing plane at distance h_i, with h_{max} the maximum viewing ray length. We recall that the points X on the ray joining a point X and its projection on the image plane is $x = PX$, where X is actually a ray of points, that is why along this ray only the *free objects* are seen by the robots while the others are occluded. Finally the vector in the direction of the principal axis is defined by $det(M)m^3$, with $P = [M|p_4]$ [32]. To consistently deal with occlusions, an acclusion matrix is designed s follows. Let $d_{ij}^0, d_{ij}^1, \ldots, d_{ij}^n$ be the parameter values indicating where intersections with the implicit surface of the polygonal mesh $\mathcal{S}_{3D}$ of the real environment occur; let d_{ij}^k be the first positive root, for each pixel (i, j). We define a matrix $Z_{\mathcal{S}_{3D}} \in \mathbb{R}^2$ such that

$$z_{ij} = \begin{cases} d_{ij}^k & \text{if } d_{ij}^k \in [0, d_{max}], \\ \infty & \text{otherwise} \end{cases} \tag{2}$$

Likewise, for each artefact a a matrix Z_a, mentioning the first positive roots $d_{ij}^k \in [0, d_{max}]$ is obtained by computing, for each pixel, the intersection of the corresponding ray with the bounding box b of the artefact. The set of artefacts perceived by the real robot is, therefore, given as follows:

$$\mathbf{A}_f = \{a | count((Z_a - Z_{\mathcal{S}_{3D}})_{ij} \leq 0) > \tau\}$$

Here the function $count(\cdot)$ returns the number of the elements (i, j) which satisfy the above condition, and τ is a threshold for assessing partial occlusions.

4   Experimental Results

In this section we illustrate the applicability of ARE to robot development and evaluation. The robotic platform is an UGV (see Fig. 5); two bogies on the sides are linked to a central body containing the electronics. Each bogie is made of a central track for locomotion and two active flippers on both ends to extend the climbing capabilities. A breakable passive differential system allows the rotation of the bogies around the body. Three sensors are installed on the platform; a rotating 2D laser scanner to acquire a 3D point cloud of the environment, an omni-directional camera for object detection and localization with a 360° field of view and an IMU/GPS for 3D localization.

A set of perception capabilities are embodied into the robot. The robot is provided with a real-time 2D and 3D ICP-based simultaneous localization and mapping (SLAM) system [33]. The robot is endowed with a path planning algorithm which generates short trajectories, enabling the robot to move within the environment, preventing the collision with the dynamic obstacles [34]. Finally a high level planner takes care of a mixed initiative control shared with the rescue operator [35, 36].

We embedded the AR-based simulation framework into a ROS package. We deployed the robotic platform in a wide outdoor area, and set up two experiments, where ARE has been used to populate the real surroundings with artefacts.

In the first experiment, we wanted to check the robot ability to replan the path towards a goal location, as the frequency of the arrivals of the artefacts into the environment changes. Different parameter settings of the path-planner have also been settled, further affecting the robot behavior into the navigation task (see Fig. 6).

During the experiment the path-planner component computes a new path each time the scene is updated. To measure the robot ability to replan the following time ratio is introduced

$$\rho = \frac{\rho_t}{\rho_t + G_t} \tag{3}$$

(a)   (b)   (c)

Fig. 5. The robotic platform at work

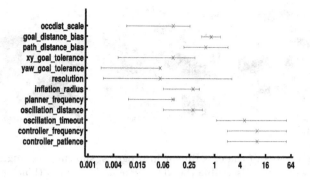

Fig. 6. Parameter names and range (note: x-scale is logarithmic)

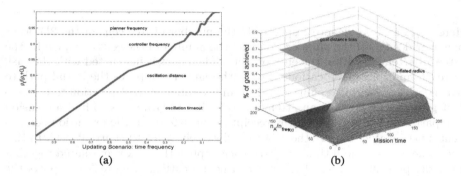

Fig. 7. (a) The graph illustrates on the Y-axis the value of Eq. 3, against scene update frequency. This shows how the parameter settings of the path-planner affect, in terms of time, the ability of the robot to replan the path towards the goal location. (b) Here the graph illustrates the number of goal that can be achieved with increasing mission time, and given an increasing spatial complexity, with thresholds indicating the parameters limits. Mission time is specified in minutes, spatial complexity is an index, as indicated in Eq. (4).

Here ρ_t and g_t denote, respectively, the time needed to the path-planner to replan the path, and the estimated time to reach the goal location. Figure 7(a) shows how the time frequency at which the scenario is updated, with the arrivals of new artefacts, affects the replanning time, under different parameter settings of the path-planner, hence it affects the robot short-term navigation capabilities.

In the second experiment we tested the long-term capability of the robot to navigate the cluttered environment in order to reach several goal locations. In this experiment the space complexity of the environment, as well as the parameters of the path-planner related to the goal bias, have been taken into account. To measure the space complexity of the environment the following space ratio has been introduced:

$$\nu = \frac{n_{\mathbf{A}}}{n_{free}} \tag{4}$$

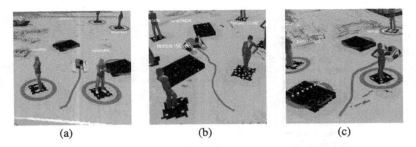

(a)   (b)   (c)

Fig. 8. Some snapshots of the experiment in progress, showing the augmented reality view (in ROS rviz), where is visible the 2D augmented costmap (black cells) with the inflated regions in green. The thick line in red is the path being executed by the path planner (Color figure online).

Here n_{free} and $n_\mathbf{A}$ denote respectively the number of free cells of the 2D occupancy grid $\mathcal{M}_{2D}$ of the mapped area and the number of the cells occupied by the set $\mathbf{A}$ of artefacts within the environment. The robot is instructed with the task to reach multiple goal locations. The path-planner computes the initial path to reach each goal and it replans a path from the robot current position to the current goal pose, whenever an artefact arrives into $\mathcal{M}_{2D}$, so as to find a collision free path, if one exists. Upon the receipt of the safe path, the execution component must be able to move the robot to effectively reach the current goal. In this experiment, the overall time needed to accomplish the task is measured together with the percentage rate of the reached goal locations. Figure 7(b) reports the performance of the robot in the navigation task with respect to different values of the space complexity of the environment (Fig. 8).

5   Conclusions

We propose a framework to augment the robot real world that advances the state of the art, as it introduces, together with the augmented environment, also the robot perceptual model of the augmented environment and the possibility of tuning the degree of confidence and uncertainty of the robot on what it is presented in the augmented scene. Besides being a compelling environment for robot programming, AR offers several tools for content authoring as well. Authoring AR tools can be classified according to their characteristics of programming and content design, in low and high level, considering the concepts abstraction and interfaces complexity incorporated in the tool. Programming tools are based on basic or advanced libraries involving computer vision, registration, three-dimensional rendering, sounds, input/output and other functions. ARTToolKit [37], MR [38], MX [39] and FLARToolKit are examples of low level programming tools. We presented the ARE framework together with a practical application of its use for robot parameter tuning. The experiments have shown how increasing complexity can affect planning and replanning abilities, and therefore that ARE is a promising experimental tool.

Acknowledgments. The research has been funded by EU-FP7 NIFTI Project, Contract No. 247870.

References

1. Quigley, M., Conley, K., Gerkey, B., Faust, J., Foote, T., Leibs, J., Wheeler, R., Ng, A.: Ros: an open-source robot operating system. In: ICRA Workshop on Open Source Software (2009)
2. Azuma, R., Baillot, Y., Behringer, R., Feiner, S., Julier, S., MacIntyre, B.: Recent advances in augmented reality. IEEE Comput. Graph. Appl. **21**(6), 34–47 (2001)
3. Neumann, U., Majoros, A.: Cognitive, performance, and systems issues for augmented reality applications in manufacturing and maintenance. In: Proceedings of IEEE Virtual Reality Annual International Symposium, pp. 4–11 (1998)
4. Molineros, J., Sharma, R.: Computer vision for guiding manual assembly. In: IEEE International Symposium on Assembly and Task Planning (2001)
5. Shekhar, R., Dandekar, O., Bhat, V., Philip, M., Lei, P., Godinez, C., Sutton, E., George, I., Kavic, S., Mezrich, R., Park, A.: Live augmented reality: a new visualization method for laparoscopic surgery using continuous volumetric computed tomography. Surg. Endosc. **24**, 1976–1985 (2010)
6. Milgram, P., Zhai, S., Drascic, D., Grodski, J.: Applications of augmented reality for human-robot communication. In: Proceedings of IROS, vol. 3, pp. 1467–1472 (1993)
7. Rastogi, A., Milgram, P.: Augmented telerobotic control: a visual interface for unstructured environments. In: Proceedings of the KBS/Robotics Conference (1995)
8. Amstutz, P., Fagg, A.: Real time visualization of robot state with mobile virtual reality. In: Proceedings of ICRA, vol. 1, pp. 241–247 (2002)
9. Brujic-Okretic, V., Guillemaut, J.Y., Hitchin, L., Michielen, M., Parker, G.: Remote vehicle manoeuvring using augmented reality. In: Proceedings of VIE, pp. 186–189 (2003)
10. Zaeh, M., Vogl, W.: Interactive laser-projection for programming industrial robots. In: Proceedings of ISMAR, pp. 125–128 (2006)
11. Bischoff, R., Kazi, A.: Perspectives on augmented reality based human-robot interaction with industrial robots. In: Proceedings of IROS, vol. 4, pp. 3226–3231 (2004)
12. Pettersen, T., Pretlove, J., Skourup, C., Engedal, T., Lkstad, T.: Augmented reality for programming industrial robots. In: Proceeding of IEEE/ACM International Symposium on Mixed and Augmented Reality, pp. 319–320 (2003)
13. Young, J., Sharlin, E., Boyd, J.: Implementing bubblegrams: the use of haar-like features for human-robot interaction. In: Proceedings of CASE, 298–303 (2006)
14. Dragone, M., Holz, T., O'Hare, G.: Using mixed reality agents as social interfaces for robots. In: The 16th IEEE International Symposium on Robot and Human Interactive Communication (RO-MAN), pp. 1161–1166 (2007)
15. Giesler, B., Salb, T., Steinhaus, P., Dillmann, R.: Using augmented reality to interact with an autonomous mobile platform. In: Proceedings of ICRA, vol. 1, pp. 1009–1014 (2004)
16. Stilman, M., Michel, P., Chestnutt, J., Nishiwaki, K., Kagami, S., Kuffner, J.: Augmented reality for robot development and experimentation. Technical report, Robotics Institute (2005)
17. Collett, T., MacDonald, B.: Augmented reality visualisation for player. In: Proceedings of ICRA, pp. 3954–3959 (2006)

18. Chong, J.W.S., Ong, S.K., Nee, A.Y.C., Youcef-Youmi, K.: Robot programming using augmented reality: an interactive method for planning collision-free paths. Robot. Comput. Integr. Manufactoring **25**(3), 689–701 (2009)

19. Green, S.A., Billinghurst, M., Chen, X., Chase, G.J.: Human-robot collaboration: a literature review and augmented reality approach in design. Int. J. Adv. Rob. Syst. **5**(1), 1–18 (2008)

20. Billinghurst, M., Kato, H., Poupyrev, I.: The magicbook: a transitional ar interface. Comput. Graph. **25**(5), 745–753 (2001)

21. Billinghurst, M., Belcher, D., Gupta, A., Kiyokawa, K.: Communication behaviors in co-located collaborative AR interfaces. J. HCI **16**(3), 395–423 (2003)

22. Moravec, H., Elfes, A.: High resolution maps from wide angle sonar. In: Proceedings of ICRA, vol. 2, pp. 116–121 (1985)

23. Thrun, S., Burgard, W., Fox, D.: Probabilistic Robotics. MIT Press, Cambridge (2005)

24. Wurm, K.M., Hornung, A., Bennewitz, M., Stachniss, C., Burgard, W.: Octomap: a probabilistic, flexible, and compact 3D map representation for robotic systems. In: Proceedings of ICRA Workshop on Best Practice in 3D Perception and Modeling for Mobile Manipulation (2010)

25. Kamat, V.R., Dong, S.: Resolving incorrect visual occlusion in outdoor augmented reality using TOF camera and OpenGL frame buffer. In: Proceedings of NSF CMMI, pp. 1–8 (2011)

26. Finkenstadt, B., Held, L.: Statistical Methods for Spatio-Temporal Systems. Chapman & Hall/CRC, Boca Raton (2006)

27. Rathbun, S.L.: Asymptotic properties of the maximum likelihood estimator for spatio-temporal point processes. J. Stat. Plann. Infer. **51**(1), 55–74 (1996)

28. Vere-Jones, D.: Some models and procedures for space-time point processes. Environ. Ecol. Stat. **16**, 173–195 (2009)

29. Dempster, A.P., Laird, N.M., Rubin, D.B.: Maximum likelihood from incomplete data via the EM algorithm. J. Roy. Stat. Soc.: Ser. B (Methodol.) **39**(1), 1–38 (1977)

30. Grimmett, G., Grimmett, G.: Probability and Random Processes, 3rd edn. Oxford University Press, Oxford (2001)

31. Müller, J., Stachniss, C., Arras, K.O., Burgard, W.: Socially inspired motion planning for mobile robots in populated environments. In: Cognitive Systems. Springer, Heidelberg (2010)

32. Hartley, R., Zisserman, A.: Multiple View Geometry in Computer Vision. Cambridge University Press, Cambridge (2000)

33. Pomerleau, F., Colas, F., Siegwart, R., Magnenat, S.: Comparing ICP variants on real-world data sets. Auton. Robot. **34**(3), 133–148 (2013)

34. Marder-Eppstein, E., Berger, E., Foote, T., Gerkey, B., Konolige, K.: The office marathon: robust navigation in an indoor office environment. In: Proceedings of ICRA, vol. 2010, pp. 300–307 (2010)

35. Carbone, A., Finzi, A., Orlandini, A., Pirri, F.: Model-based control architecture for attentive robots in rescue scenarios. Auton. Robot. **24**(1), 87–120 (2008)

36. Finzi, A., Pirri, F.: Representing flexible temporal behaviours in the situation calculus. In: Proceedings of the International Joint Conference of Artificial Intelligence IJCAI, pp. 436–441 (2005)

37. Kato, H., Billinghurst, M.: Marker tracking and HMD calibration for a video-based augmented reality conferencing system. In: Proceedings of ACM International Workshop on Augmented Reality (1999)

38. Uchiyama, S., Takemoto, K., Satoh, K., Yamamoto, H., Tamura, H.: Mr platform: a basic body on which mixed reality applications are built. In: Proceedings of International Symposium on Mixed and Augmented Reality (2002)
39. Dias, J., Monteiro, L., Santos, P., Silvestre, R., Bastos, R.: Developing and authoring mixed reality with MX toolkit. In: IEEE International Augmented Reality Toolkit Workshop, 2003 (October 2003), pp. 18–26 (2003)

Author Index

Adamatzky, Andrew 64
Ahmad, Omar 103
Al-Abri, Said 222
Al-Timemy, Ali H. 291
Anderson, Sean R. 53, 259
Angeles, Paolo 296
Ani Hsieh, M. 161
Anjum, Muhammad Latif 103
Aragon-Camarasa, Gerarado 148
Assaf, Tareq 53, 259

Barrios, Luenin 3
Barron-Gonzalez, Hector 46
Beck, Chris 114
Bianchi, Reinaldo A.C. 15
Bona, Basilio 103
Bonarini, Andrea 446
Brochard, Alexandre 291
Broun, Alan 114
Bryson, Joanna 261
Bugmann, Guido 291

Cameron, Stephen 34, 190
Cao, Juan 135
Casals, Alicia 294
Chaimowicz, Luiz 161
Chen, Chi-An 3
Chen, Weisheng 264
Chen, Yao-Chang 340
Cielniak, Grzegorz 210
Cipolla, Vittorio 51
Cockshott, Paul 148
Coldwell, Charles 76
Cui, Rongxin 264

"Dan" Cho, Dong-il 103
Dasgupta, Prithviraj 173
de Hoog, Julian 34
Dee, Hannah 135
Dennis, Louise 433
Diamantas, Sotirios C. 173
Dipper, Tobias 404
Dodd, Tony J. 259
Dorigo, Marco 390
Duckett, Tom 210

Entschev, Peter Andreas 127
Escudero, Javier 291

Fedele, Pasquale 294
Fernandez, Julián M. Angel 446
Ferri, Federico 470
Finnis, C. James 247
Fisher, Michael 433
Flynn, Helen 190
Frediani, Aldo 51

Gaissert, Nina 90
Gale, Ella 64
Gallegos, Katherine 458
Gardner, Marcus 296
Gaudl, Swen 261
Georgilas, Ioannis 64
Gianni, Mario 470
Granjon, Paul 30

Harwin, William 235
Hilder, James 404
Hindriks, Koen 298

Ibrahim, H.D. 353
Iwakabe, Keisuke 185

Jacobs, Will 259
Jebens, Agalya 90
Jebens, Kristof 90

Kamimura, Akiya 3
Keivan, Nima 276
Kelleher, John 363
Kengyel, Daniela 404
Knubben, Elias Maria 90

Labrosse, Frédéric 135
Li, Jing 264
Li, Yan 363
Li, Zhijun 264
Lin, Tsung-Han 340
Liu, Guanzuo 185

Marek, Tadeusz 294
Martins, Murilo Fernandes 15
McMahon, James 328

Melhuish, Chris   64, 114
Millard, Alan G.   429
Milligan, James   161
Mirmehdi, Majid   114
Molfino, Rezia   28, 51, 294, 311
Moreno, Francisco-Angel   210
Möslinger, Christoph   404
Mugrauer, Günter   90
Mugrauer, Rainer   90
Muscolo, Giovanni Gerardo   28, 51, 294, 311

Namee, Brian Mac   363
Nashed, Youssef S.G.   390
Neal, Mark   247
Novikova, Jekaterina   261

O'Grady, Rehan   390
Oliviero, Fabrizio   51

Park, Michael J.   76
Pearson, Martin J.   53
Pipe, Tony   114
Pirri, Fiora   470
Plaku, Erion   328, 417
Podevijn, Gaëtan   390
Porrill, John   53, 259
Prescott, Tony   46
Puig, Domenec   51, 311

Qu, Zhihua   222

Rahman, S.M. Mizanoor   195
Read, Mark   404
Recchiuto, Carmine Tommaso   51, 294, 311
Rizzo, Emanuele   51
Rogers, Simon   148
Rossiter, Jonathan M.   53, 259

Saaj, C.M.   353
Schmickl, Thomas   404
Shefelbine, Sandra   296
Shen, Wei-Min   3

Shih, Ta-Ming   340
Sibley, Gabe   276
Siebert, J. Paul   148
Silva, José   458
Slavkovik, Marija   433
Solanas, Agusti   51, 311
Spirin, Victor   34
Stewart, Paul   51
Sun, Li   148

Thenius, Ronald   404
Thomas, Richard   235
Timmis, Jon   404, 429
Tumalli, Luis   458
Tyrrell, Andy   404

Uejima, Toshiyoshi   185

Vaca, Cristian   458
Vaidyanathan, Ravi   296
Vélez, Paulina   458
Vieira Neto, Hugo   127

Wallar, Alex   417
Wane, Sam   315
Wang, Chang   298
Webster, Matt   433
Wei, Changyun   298
Wiggers, Pascal   298
Williams, A. Mark   311
Wilson, Emma D.   53, 259
Winfield, Alan F.T.   429
Witkowski, Ulf   377
Woodward, Richard   296

Xu, Junchao   298

Yamada, Yasuhiro   185
Yang, Chenguang   264

Zandian, Reza   377

Printed in the United States
By Bookmasters